Geographic Information Science

Mastering GIS: Technology, Applications and Management series

Location-based Services and Geomatic Engineering, Allan Brimicombe and Chao Li (forthcoming)

GIS and Crime, Spencer Chainey and Jerry Ratcliffe (forthcoming)

Landscape Visualisation: GIS Techniques for Planning and Environmental Management, Andrew Lovett, Katy Appleton and Simon Jude (forthcoming)

Integration of GIS and Remote Sensing, Victor Mesev (forthcoming)

GIS for Public Sector Spatial Planning, Scott Orford, Andrea Frank and Sean White (forthcoming)

GIS Techniques for Habitat Management, Nigel Waters and Shelley Alexander (forthcoming)

Geographic Information Science

Mastering the Legal Issues

George Cho
Division of Health, Design and Science
University of Canberra
Australia

John Wiley & Sons, Ltd

Other Wiley Editorial Offices

John Wiley & Sons Inc., 111 River Street, Hoboken, NJ 07030, USA

Jossey-Bass, 989 Market Street, San Francisco, CA 94103-1741, USA

Wiley-VCH Verlag GmbH, Boschstr. 12, D-69469 Weinheim, Germany

John Wiley & Sons Australia Ltd, 33 Park Road, Milton, Queensland 4064, Australia

John Wiley & Sons (Asia) Pte Ltd, 2 Clementi Loop #02-01, Jin Xing Distripark, Singapore 129809

John Wiley & Sons Canada Ltd, 22 Worcester Road, Etobicoke, Ontario, Canada M9W 1L1

Wiley also publishes its books in a variety of electronic formats. Some content that appears in print
may not be available in electronic books.

Library of Congress Cataloging in Publication Data

Cho, George, 1946–
   Geographic information science : mastering the legal issues / George Cho.
     p.   cm. — (Mastering GIS)
   Includes bibliographical references and index.
   ISBN 0-470-85009-4 (cloth : alk. paper) — ISBN 0-470-85010-8 (pbk. : alk. paper)
   1. Geographic information systems—Law and legislation.   2. Geographic information systems.
   I. Title. II. Series.
   K4293.C478 2005
   344′.095—dc22

                          2004028498

British Library Cataloguing in Publication Data

A catalogue record for this book is available from the British Library

ISBN 0-470-85009-4 (HB)
ISBN 0-470-85010-8 (PB)

Typeset in 11/13 pt Times by Integra Software Services Pvt. Ltd, Pondicherry, India

This book is printed on acid-free paper responsibly manufactured from sustainable forestry in which
at least two trees are planted for each one used for paper production.

This book is dedicated to
Marion Cho
and to
Carolyn and David Hardman
for their support and encouragement
over the course of this project

Contents

Table of Cases   xiii

Table of Statutes   xxiii

Acknowledgements   xxxii

Introduction   1

**1   Geographic Information Science: Legal and
     Policy Issues**   **15**
     Learning Objectives   15
     1.1   Introduction   16
     1.2   An Introduction to Law   21
           1.2.1   Common Law, Statutes and Civil Law   21
           1.2.2   Court System   23
           1.2.3   Alternative Dispute Resolution (ADR)   24
           1.2.4   Remedies   25
           1.2.5   International Law   27
     1.3   Key Policy Issues   27
           1.3.1   Factors Influencing Geographic Information
                   Policy   28
           1.3.2   Existing Policy and Policy-making Processes   30
           1.3.3   Data Pricing Policy   30
           1.3.4   Policy on the Legal Protection of Data   31
           1.3.5   Data Preservation   32
           1.3.6   Conclusions   33
           1.3.7   Policy Developments in Australia   33
     1.4   The Geographic Information and Law Nexus   34
           1.4.1   Databases   38
           1.4.2   Data Sharing   40
           1.4.3   Maps   42
           1.4.4   Global Positioning Systems (GPS)   44
           1.4.5   Aerial Photographs and Images   46

Contents

1.5   Geography Really Does Matter   49
Summary   51

2   Sharing Geographic Information and Data   53
Learning Objectives   53
2.1   Introduction   54
2.2   Sharing Geographic Information and Data   55
2.3   Policies on Access to Public Sector Information (PSI)   61
      2.3.1   Australia–New Zealand   64
      2.3.2   United States   72
      2.3.3   European Union   77
      2.3.4   Conclusions   85
2.4   Frameworks for Accessing Geospatial Information   86
      2.4.1   Metadata Content Standards   87
      2.4.2   Clearinghouse and Geolibrary   91
      2.4.3   Access and Exchange Standards   93
2.5   Towards a Global Information Infrastructure (GII)   97
      2.5.1   United States National Spatial Data
              Infrastructure (NSDI): Evolution and Growth   98
      2.5.2   Canadian Geospatial Data Infrastructure
              (CGDI): Private Sector Leadership   100
      2.5.3   European Geographic Information Infrastructure
              (EGII): Balanced Representation   101
      2.5.4   Australian Spatial Data Infrastructure (ASDI)
              Developments   103
      2.5.5   Asia-Pacific and Africa Spatial Data
              Infrastructure (SDI) Efforts   104
      2.5.6   Global Spatial Data Infrastructure (GSDI)
              Strategic Plan   105
Summary   106

3   Geographic Information and Intellectual Property
Rights   109
Learning Objectives   109
3.1   Introduction   110
3.2   The Life of Gigo   110
3.3   Intellectual Property Rights (IPR)   113
3.4   Intellectual Property Rights Protection in Australia   118
3.5   *Quid Pro Quo* and the International Environment for
      Intellectual Property Rights Protection   122
      3.5.1   Intellectual Property Rights Conventions,
              Agreements and Treaties   128

3.5.2   Berne Convention for the Protection of Literary and Artistic Works 1998   128

3.5.3   Rome Convention for the Protection of Performers, Producers of Phonograms and Broadcasting Organizations 1961   129

3.5.4   Agreement on Trade-related Aspects of Intellectual Property Rights (TRIPS) 1995   129

3.5.5   WIPO Copyright Treaty (WCT) and WIPO Performances and Phonograms Treaty (WPPT)—the Internet Treaties   130

3.6   Copyright and Geographic Information   131

    3.6.1   Maps   133

    3.6.2   Electronic Databases   143

    3.6.3   European Union Database Directive   159

    3.6.4   Moral Rights and *Sui Generis* Regimes   166

    3.6.5   Business Methods and Geographic Information Patents   171

    3.6.6   The Digital Agenda   175

3.7   Atypical Developments and Other Legal Issues   178

    3.7.1   The 'Copyleft' Movement and No Rights Reserved   178

    3.7.2   Other Intellectual Property   182

3.8   Infringements, Defences and Remedies   188

3.9   Intellectual Property Rights: Employees and International Research   193

    3.9.1   Intellectual Property Rights in the Work of Employees   193

    3.9.2   Intellectual Property Rights and International Research   196

3.10   Lessons, Litigation and the Fate of Gigo's Code   201

Summary   204

4   Geographic Information and Privacy   **207**

Learning Objectives   207

4.1   Introduction   208

4.2   Philosophical Issues: Nature and Structure of the Problem   210

    4.2.1   Geographic Information Systems are not Personal Data Intensive   210

    4.2.2   Lack of Understanding of Privacy Issues   212

    4.2.3   Ethical Use of Geospatial Technologies   214

Contents

4.3		Privacy: The Legal and Regulatory Framework	217
	4.3.1	The *Privacy Act* 1988 (Cwlth)	219
	4.3.2	The *Privacy Amendment Act* 1990 (Cwlth)	221
	4.3.3	*Data-matching Program (Assistance and Taxation) Act* 1991 (Cwlth)	223
	4.3.4	*Privacy Amendment (Private Sector) Act* 2000 (Cwlth)	223
	4.3.5	Freedom of Information	227
	4.3.6	The Common Law and the Disclosure of Personal Information	228
	4.3.7	Industry Codes of Conduct and Self-regulation	234
	4.3.8	The Regime in the United States	237
	4.3.9	Common Law Privacy in the United States	245
	4.3.10	Evolving Fair Information Privacy Principles	246
4.4		Geospatial Technologies and Privacy Implications	252
	4.4.1	Data Aggregation and Databases	253
	4.4.2	Regulation and Use of Databases	256
	4.4.3	Some Definitions: Location, Tracking and Dataveillance	258
	4.4.4	Geospatial Technology Applications: Home Location	260
	4.4.5	Tracking Movements of Individuals in Space	262
	4.4.6	Tracking Transactions	264
	4.4.7	Tracing Communications	264
	4.4.8	Convergence of Locational and Tracking Technologies	266
	4.4.9	Privacy Risks with Location and Tracking Technologies	268
	4.4.10	Privacy-invasive Technologies (PIT): Privacy-enhancing Technologies (PET) and Privacy-sympathetic Technologies (PST)	270
4.5		Emergent Policy and Practice	272
	4.5.1	European Union Data Protection Directive	277
	4.5.2	European Union–United States Safe Harbour Framework	279
	4.5.3	European Union Data Protection Directive and Implications for Australia, Canada and United Kingdom	283
	Summary		286

5 Geographic Information and Contract Law   **291**
Learning Objectives   291
5.1 Introduction   292
5.2 A Contract is a Meeting of the Minds   296
5.3 Contract *for* Service and Contract *of* Service   307
    5.3.1 Personnel Contracts   308
    5.3.2 Academics and Researchers
        as Employees?   313
5.4 Geographic Information Systems: Product
    or Service?   314
5.5 Licensing   316
    5.5.1 Why is Spatial Data Special?   323
5.6 Liability Implications and the Privity
    of Contract   327
    5.6.1 Exclusion Clauses   328
5.7 Contract Execution: Discharged, Failed Contracts,
    and Remedies   330
5.8 Web-based Contracts   332
    5.8.1 Electronic Transaction Regulations   333
    5.8.2 Electronic Contracts   335
    5.8.3 Summary of Legal Issues   342
    5.8.4 Contract Precedents   347
Summary   349

**6 Geographic Information and Liability
Standards**   **351**
Learning Objectives   351
6.1 Introduction   351
6.2 Legal Standards and Guidelines   353
6.3 Legal Liability Theories   359
    6.3.1 Contract and Strict Products Liability 360
    6.3.2 Tortious Liability   369
    6.3.3 Statutory Liability   377
    6.3.4 Other Liability Theories   378
6.4 Liability Risk Management   381
    6.4.1 Disclaimers   381
    6.4.2 Data Quality Issues   382
    6.4.3 Global Positioning Systems (GPS) and Map
        Quality Standards   383
    6.4.4 Legal Risk Management Strategies   390

Contents

6.4.5   Discussion   391

6.4.6   Minimising Liability and Damage
Claims   393

Summary   395

References   397

Internet URL References   419

Index   423

Table of Cases

Advanced Micro Devices Inc. v Intel Corp. Arbitration Award February 24, 1992, Palo Alto, CA. (CCH ¶60,368).

Aetna Casualty & Security Co. v Jeppesen & Co. 642 F.2d. 339 (9th Cir. 1981).

Albert R Sparaco v Matusky, Skelly Engineers U.S. District Court, S.D.N.Y. 60 F.Supp.2d 247 (1999).

Amazon.com Inc. v Barnesandnoble.com, Inc. 239 F.3d 1343 (Fed. Cir. 2001).

Amazon.com Inc. v Barnesandnoble.com, Inc. 73 F.Supp.2d 1228 (W.D. Wash. 1999).

Ansett Transport Industries (Operations) Pty Ltd v Commonwealth (1977) 139 CLR 54.

Arizona Retail Systems Inc. v Software Link Inc. 831 F.Supp. 759 (D Ariz 1993).

Astley v Austrust (1999) 161 ALR 155.

Atari Games Corp. v Nintendo of America, Inc. 975 F.2d 832 (1992).

Aubrey v . . . ditions Vice-Versa Inc. [1998] 1 SCR 591.

Aubry v Duclos (1996) 141 DLR (4th) 683.

Australian Broadcasting Corporation v Lenah Game Meats Pty Ltd [2001] HCA 63, 15 November 2001.

Australian Chinese Newspapers Pty Ltd v Melbourne Chinese Press (2003) 58 IPR 1.

B P Refinery (Westernport) Pty Ltd v Hastings Shire Council (1977) 180 CLR 266.

baumarkt.de, Oberlandesgericht Düsseldorf 29 June 1999, [1999] Multimedia und Recht 729; [2000] Computer und Recht 184.

Bayley & Co. v Boccacio Pty Ltd and Ors (1986) 8 IPR 297.

Beloff v Pressdram Ltd [1973] 1 All ER 241.

Berlin.Online, Landgericht Berlin 8 October 1998, [1999] Computer und Recht 388.

Beta Computers (Europe) Ltd v Adobe Systems (Europe) Ltd FSR [1996] 367.

Bevan Investments Ltd v Blackhall and Struthers (No. 2) [1973] 2 NZLR 45.

Black, Jackson & Simmons Insurance Brokerage Inc. v International Business Machines Corporation 109 Ill. App.3d 132; 440 N.E.2d 282 (1982).

Blake v Woodford Bank & Trust Co. 555 S.W.2d 589 (Ky. Ct. App. 1977).

Bolam v Friern Hospital Management Committee [1957] 1 WLR 582.

Booth v Electronic Data Systems Corp. (DC Kan. 1992), 4 CCH Computer Cases 46,801.

Bradbury v CBS 287 F.2d. 478 (1961).

British Horseracing Board (BHB) v William Hill Organisation Ltd High Court of Justice Ch. Div. 9 February 2001, Case No. HC 2000 1335 at http://www.courtservice.gov.uk/judgements/judg_home.htm [30 June 2004] and on appeal at http://www.bailii.org/ew/cases/EWCA/Cir/ 2001/1268.html [30 June 2004].

Brocklesby v Jeppesen 767 F.2d. 1288 (9th Cir. 1985) *cert. den.* 474 US 1101 (1986).

Brocklesby v United States 753 F.2d. (9th Cir. 1985).

Brower v Gateway 2000 Inc. 246 AD 2d 246, 37 UCC Rep Serv 2d 54 (NY 1998).

Burrow-Giles Lithographic v Sarony 111 US 53 (1874).

C.Net, Kammergericht (Court of Appeal) Berlin 9 June 2000, [2001] ZUM 70.

California v Ciraolo 106 S.Ct. 1809 (1986).

Caltex Oil (Australia) Pty Ltd v The Dredge 'Willemstad' and *Caltex Oil (Australia) Pty Ltd v Decca Survey Australia Ltd* (1976) 136 CLR 529; 51 ALJR 270.

Campbell v Mirror Group Newspapers [2002] All ER (D) 177 (October).

Campbell v Mirror Group Newspapers [2002] All ER (D) 448 (March); [2003] QB 633. Also at http://www.lawreports.co.uk/qbmarb0.3.htm [10 May 2004].

Caslec Industries Pty Ltd v Windhover Data Systems Pty Ltd Federal Court of Australia No. N G627 of 1990 FED No 580 Trade Practices, 13 August 1992; 43 KB.

Chapelton v Barry Urban District Council [1940] 1 KB 532.

Chin Keow v Government of Malaysia [1967] 1 WLR 813.

Codelfa Construction Pty Ltd v State Rail Authority of NSW (1982) 149 CLR 337.

Commonwealth of Australia v John Fairfax & Sons Ltd [1981] 32 ALR 485.

Commonwealth v John Fairfax and Sons Ltd (1980) 32 ALR 485.

Computer Services Corporation v Ferguson 74 Cal. Rptr. 86 (Cal. 1968).

Craig Carnahan v Alexander Produdfoot Co. World Headquarters & Alexander Proudfoot Co. of Australia (Fla DC App 1991) 3 CCH Comp. Cases.

Cyprotex Discovery Ltd v The University of Sheffield [2004] EWCA Civ. 380 also at http://www.courtservice.gov.uk/judgementsfiles/j2444/cyprotex-v-sheffield.htm [5 August 2004].

Cubby v Compuserve Inc. et al. 776 F.Supp. 135 (S.D.N.Y. 1991).

Cybersell Inc. (AZ) *v Cybersell Inc.* (FL) 130 F.3d 414 (9th Cir. 1997).

*Daniel v Dow Jones & Co.*137 Misc.2d 94; 520 N.Y.S.2d 334 (Cir.Ct. 1987).

De Telegraaf/NOS and HMG, Netherlands Competition Authority (Nederlandse Mededingingsautoriteit), 10 September 1998, Mediaforum 1998, p. 304.

Desktop Marketing Systems Pty Ltd v Telstra Corporation Ltd (Unreported M85/2002, HCA, 20 June 2003).

Desktop Marketing Systems Pty Ltd v Telstra Corporation Ltd [2002] FCAFC 112.

Dictionnaire Permanent des Conventions Collectives, Tribunal de grande instance de Lyon 28 December 1998, [1999] 181 RIDA 325.

Diversified Graphics Ltd v Groves 868 F.2d 293 (8th Cir. 1989).

Douglas v Hello! Ltd [2001] 2 WLR 992; [2001] 2 All ER 289.

Douglas v Hello! Ltd [2003] All ER (D) 209 (April).

Dow Chemical v United States (1986) 106 S.Ct. 1819, 90 Led 2d 226 (1986).

Dow Jones & Co. v Board of Trade 546 F.Supp. 113 (SD NY 1982).

Dow Jones & Co. v Gutnick at http://www.austlii.edu.au/au/cases/cth/high_ct/2002/56.html [10 December 2002].

Dutton v Bognor Regis Urban District Council [1972] 1 QB 373.

E v Australian Red Cross Society (1991) 27 FCR 310.

Editorial Aranzadi, Court of First Instance Elda (Alicante) 2 July 1999 at http://www.onnet.es/aranzadi.pdf [30 June 2004].

Eldred v Ashcroft 537 U.S. 185, 65 USPQ2d. 1225 (2003).

Électre v T.I. Communication et Maxotex, Tribunal de commerce de Paris 7 May 1999 at http://www.juriscom.net/txt/jurisfr/da/tcparis19990507.htm [30 June 2004].

Electronic Data Systems Corp. v Kinder, 360 F.Supp. 1044 (1973).

Entores Ltd v Miles Far East Corporation [1955] 2 QB 327; [1955] 2 All ER 493.

Esso Petroleum Co. Ltd v Harpers Garage (Southport) Ltd [1968] AC 269.

Express Newspapers Inc. Plc. v Liverpool Daily Post and Echo Plc. (1986) 5 IPR 193.

Federal Trade Commission v American Tobacco Co. 264 U.S. 298, 44 S.Ct. 336, 68 L.Ed. 696.

Feist Publications v Rural Telephone Service Company 111 S.Ct. 1282, 113 L.Ed.2d 358, 499 US 340 (1991).

Financial Information Inc. v Moody's Investor Service Inc. 808 F.2d (2nd Cir. 1986).

Financings Ltd v Stimson [1962] 3 All ER 386.

Florida v Riley 488 U.S. 445 (1988).

Fluor Corp. v Jeppesen & Co. 216 Cal. Reptr. 68 (1985).

Forrest v Verizon Communications Inc. D.C. No. 01-CV-1101, 28 August (2002) *Electronic Commerce and Law Report*, v. 7 n. 35, p. 905.

France Télécom v MA Editions, Tribunal de commerce de Paris 18 June 1999 [1999] Droit de I'informatique & des télécoms 57, [1999] Muntimedia und Recht 568.

Furniss v Fitchett [1958] NZLR 396.

Govind v State of Madhya Pradesh (1975) 62 AIR (SC) 1378.

Greaves & Co. (Contractors) Ltd v Baynham, Meikle & Partners [1975] 3 All ER 99.

Greenwood Shopping Plaza Ltd v Beattie (1980) 111 D.L.R. (3d) 257.

Griswold v Connecticut 381 U.S. 479, 14 L.Ed.2d 510, 85 S.Ct. 1678 (1965).

Grosse v Purvis [2003] QDC 151, 16 June 2003.

Groupe Moniteur and others v Observatoire des Marchés Publics, Cour d'appel de Paris 18 June 1999, [2000] 183 RIDA 316.

Gutnick v Down Jones & Co. Ltd [2001] VSC 305 (28 August, 2001).

Hawkes & Son (London) Ltd v Paramount Film Service Ltd [1934] 1 Ch. 593.

Hedley Byrne & Co. Ltd v Heller & Partners Ltd [1964] AC 465.

Helen Remsburg, Administrator of the Estate of Amy Lynn Boyer v Docusearch Inc. U.S. District Court, SC New Hampshire, No. 2002–255 at http://www.courts.state.nh/supreme/opinions/2003/remsb017.htm [15 March 2004].

Henderson v Radio Corporation Pty Ltd (1960) SR (NSW) 576.

Hill v Gateway 2000 Inc. 105 F.3d 1147 (7th Cir. 1997).

Hill v National Collegiate Athletic Association 7 Cal.4th 1 at 30; 865 P.2d 633 (1994).

Hill v Van Erp (1997) 188 CLR 159.

Hospital Computer Systems Inc. v Staten Island Hospital 788 F.Supp. 1351 (D.N.J. 1992).

Hotmail Corp. v van Money Pie Inc. 1988 WL 388389, 47 USPQ 2d 1020 (BNA) (ND Cal. April 16, 1998).

Howell v New York Post Co. 596 N.Y.S.2d 350; 612 N.E.2d 699 (Ct. App. 1993).

Hubbard v Vosper [1972] 1 All ER 1023.

Hughes v McMenamon Civil Action No. 2001-10891-RBC, 28 May 2002 (2002) *Electronic Commerce and Law Report,* v. 7 n. 23, p. 568.

Hutchison Tel. Co. v Frontier Directory Co. 770 F.2d 128, 131 (8th Cir. 1985).

Images Audio Visual Productions Inc. v Perini Building Company Inc. 91 F.Supp.2d 1075 (April 12, 2000, U.S. District Court, E.D. Michigan S.D.)

Independent Broadcasting Authority v EMI (Electronics) Ltd (1980) 14 Build LR 1.

Interstate Parcel Express Co. Ltd v Time-Life International (Nederlands) BV and Anor (1977) 15 ALR 353.

J'Aire Corp. v Gregory 24 Cal.3d 799; 598 P.2d 60 (1979).

Jeweler's Circular Publishing Co. v Keystone Publishing Co. 281 F.83, 89, 95 (2d Cir.) cert. denied 42 SC 464 (1922).

John Richardson Computers Ltd v Flanders [1993] FSR 497.

Junior Books Ltd v Veitchi Co. Ltd [1982] 3 All ER 201.

Katz v United States 389 U.S. 347 (1967).

Kelly v Arriba Soft Corporation 280 F.3d 934 (9th Cir. 2002).

Kelly v Cinema House Ltd (1928–35) MacG. Cop. Cases 362.

Kelly v Morris (1866) L.R. 1 Eq. 697 at 702.

Kelsey-Hayes Company v Ali Malehi 765 F.Supp. 402 (1991).

Kennison v Davie (1986) 162 CLR 126.

Kern River Gas Transmission Co. v Coastal Corp. 899 F.2d 1458 (5th Cir. 1990).

Kidnet/Babynet, Landgericht Köln 25 August 1999, [2000] Computer und Recht 400.

KPN v XSO, President District Court of The Hague 14 January 2000, [2000] Mediaforum 64.

Kraten.com, President District Court of Rotterdam, 22 August 2000, [2000] Mediaforum 344.

Kufos v C. Czarnikow Ltd (The Heron II), [1969] 1 AC 350.

L'Estrange v F Graucob Ltd [1934] 2 KB 294.

Lott v JBW & Friends Pty Ltd [2000] SASC 3.

M A Mortenson Company Inc. v Timberline Software Corporation 998 P 2d 305 (Wash 2000).

Mahon v Osborne [1939] 2 KB 14.

Mars v Teknowledge, High Court, Chancery Division 11 June 1999, [1999] EIPR N-158.

Mason v Montgomery Data Inc. 741 F.Supp. 1282 (SD Tex. 1990).

Mason v Montgomery Data Inc. 765 F.Supp. 353 (SD Tex. 1991).

Mason v Montgomery Data Inc. 976 F.2d. 135 (5th Cir. 1992).

Matthews v Chichory Marketing Board [1938] 60 CLR 263.

McCulloch v Lewis A. May (Produce Distributors) Ltd [1947] 2 All ER 845.

McDonald's System of Australia Pty Ltd v McWilliam's Wines Pty Ltd (No 2) 28 ALR 236.

Medizinsches Lexicon, Landgericht Hamburg 12 July 2000.

Mercantile Bank of Sydney v Taylor (1891) 12 LR (NSW) 252 (affirmed [1893] AC 317.

Micron Computer Systems Ltd v Wang (UK) Ltd unreported 9th May 1990 (QBD).

MIDI-Files, Landgericht Müchen I 30 March 2000, [2000] Computer und Recht 389.

Ministry of Housing and Local Government v Sharp [1970] 2 QB 223.

Moorgate Tobacco Co. Ltd v Philip Morris Ltd (No. 2) (1984) 156 CLR 414; 3 IPR 545.

Natural Business Lists Inc. v Dun & Bradstreet Inc. 552 F.Supp. 89, 92–93 (ND Ill. 1982).

New Jersey v New York 523 U.S. 767 (1998).

New York Times v Sullivan 376 U.S. 254; 11 L.Ed.2d 686; 84 S.Ct. 710 (1964).

Nichols v Universal Pictures (1930) 45 F.2d. 119.

NOS v De Telegraaf, Court of Appeals of The Hague 30 January 2001, [2001] Mediaforum 90.

NVM v De Telegraaf, President District Court of the Hague 12 September 2000, [2000] Mediaforum 395.

NVM v De Telegraaf, President District Court of the Hague 21 December 2000, [2001] Mediaforum 87.

Olley v Marlborough Court Ltd [1949] 1 KB 532.

Olmstead v U.S. 277 U.S. 438 (1928).

P v D [2000] 2 NZLR 591.

Panayiotiou v Sony Music Entertainment (UK) Ltd [1994] ECC 395.

Perre v Apand Pty Ltd (1999) 164 ALR 606.

Peters American Delicacy Co. Ltd v Champion (1928) 41 CLR 316.

Pinch-a-Penny of Pinellas County v Chango [1990–1 Trade Cases 68-961] 557 SO. 2d 940 (Fla 2d. DCA 1990).

ProCD Inc. v Zeidenberg 86 F.3d 1447 (7th Cir. 1996).

R v Broadcasting Standards Commission; ex parte British Broadcasting Corporation [2000] 3 WLR 1327; [2000] 3 All ER 989.

R v Khan [1997] AC 558 at 582.583.

Redrock Holdings Pty Ltd & Ors v Adam Hinkley [2000] VSC 91 (4 April); [2001] VSC 277 (2 August).

Republic Systems and Programming Inc. v Computer Assistance Inc. 322 F. Supp. 619 (D. Conn. 1970).

Robinson v Graves [1935] 1 KB 579.

Rockford Map Publishers Inc. v Directory Service Co. 768 F. 2d. 145 (7th Cir. 1985) cert. denied 106 SC 806 (1986).

Rockford Map Publishers Inc. v Directory Service Company of Colorado Inc., 768 F.2d 145 (7th Cir. 1985).

Ross v Caunters [1980] 1 Ch. 27.

Saloomey v Jeppesen & Co. 707 F.2d 671 (2d Cir. 1983).

San Sebastian Pty Ltd v Minister Administering Environmental Planning and Assessment Act 1979 (NSW) (1986) 162 CLR 340.

Sands & McDougall Pty Ltd v Robinson (1917) 23 CLR 49.

Sandy Koplowitz v Andre Girard, Fla. DCA 4th District, No. 93-1781. August 9, 1995 (CCH ¶47,310).

Saphena Computing Ltd v Allied Collection Agencies Ltd [1995] FSR 616.

Sayre v Moore (1785) 1 East 316, n. 5, 102 ER 139.

Sega Enterprises Ltd v Accolade Inc. (Unrpt. 9th Cir. No. 92-15656, 20 October 1992).

Shaddock and Associates Pty Ltd v Parramatta City Council (1981) 55 ALJR 713.

Shetland Times Ltd v Wills (1996) 37 IPR 71; and [1997] SCLR 160.

Shorter v Retail Credit Co. D.C.S.C., 251 F.Supp. 329 at 330.

Shroeder Music Publishing Co. Ltd v Macaulay [1974] 1 WLR 1308.

Shuey v United States 92 U.S. 73 (1875).

Skinner v Mid-America Pipeline Co. 490 U.S. 212 (1989).

Smith v Maryland 442 U.S. 735 (1979).

Specht v Netscape Communications Corp. 150 F.Supp.2d 585 (S.D.N.Y. 2001); *Electronic Commerce & Law Report*, v. 7 n. 39, p. 999.

Spencer Industries Pty Ltd v Anthony Collins & Anor [2002] APO 4 (18 January 2002); [2003] FCA 542 (4 June 2003).
Sperry Rand Corporation v Kinder 360 F.Supp. 1044 (ND Tex. 1973).
St Albans City and District Council v International Computers Ltd [1995] FSR 686 (QBD), unrept. 26 June 1996 (Ct. of Appeal).
St Albans City and District Council v International Computers Ltd [1995] FSR 686 (QBD), unreported 26 June 1996 (Ct. Appeal).
State Farm Mutual Auto Insurance Co. v Brockhurst 453 F 2d 533 (10th Cir. 1972).
Step-Saver Data Systems Inc. v Wyse Technology and Software Link Inc. 939 F.2d 91 (3rd Cir. 1991).
Stern v Delphi Internet Services Corporation 626 N.Y.S.2d 694 (Sup.Ct. 1995).
Steve Jackson Games v United States Secret Service 816 F.Supp. 432 (W.D. Tex. 1993), aff'd, 36 F.3d 457 (5th Cir. 1994).
Stratton Oakmont, Inc. v Prodigy Services Co. 1995 N.Y. Misc. Lexis 229, 23 *Media L. Rept* 1794 (1995).
Streetwise Maps Inc. v Vandam Inc. 159 F.3d 739 (2d Cir. 1998).
Streisand v Adelman Case No. SC 077 257. Cal. W.D. 31 December 2003 at http://www.californiacoastline.org/streisand/slapp-ruling.pdf [17 May 2004].
Süddeutsche Zeitung, Landgericht Köln 2 December 1998, [1999] Computer und Recht 593.
Sutherland Shire Council v Heyman (1985) 157 CLR 424.
Sutherland Shire v Heyman (1985) 157 CLR 424.
Sutton Vane v Famous Players Film Ltd (1923–28) MacG. Cop. Cases 6.

Telecomputing Services Inc. 1 CLSR 953 (1968).
Tele-Direct Publications v American Business Information (1997) 154 DLR (4th) 328.
Tele-Info-CD, Bundesgerichtshof (Federal Supreme Court) 6 May 1999, [1999] Multimedia und Recht 470.
Telstra Corporation Ltd v Desktop Marketing Systems Pty Ltd (2001) 51 IPR 257; [2001] FCA 612.
Thornton v Shoe Lane Parking Ltd [1971] 2 QB 163; 1 All ER 686.
Ticketmaster Corp v Tickets.com Inc. 54 USPQ 2d (BNA) (1344 US D Cal. 27 March 2000); 2000 WL 525390 (C.D.Cal.2000).
Ticketmaster Corporation v Microsoft Corporation No. 97-3055 DDP (CD Cal. 12 April 1997).

Tiffany Design Inc. v Reno-Tahoe Specialty Inc. 51 U.S.P.Q. 2d 1651 (July 12 1999, U.S. District Court, D. Nevada).

Time Inc. v Hill 385 US 374, 389; 17 L.Ed.2s 456; 87 S.Ct. 534 (1967).

TNT (Melbourne) Pty Ltd v May and Baker (Australia) Pty Ltd (1966) 40 ALJR 189.

Toby Constructions Products Pty Ltd v Computa Bar (Sales) Pty Ltd [1983] 2 NSWLR.

Tournier v National Provincial and Union Bank of England [1924] 1 KB 461.

Trade Mark Cases 100 US 82 (1879).

Triangle Publications Inc. v Sports Eye Inc. 415 F.Supp. 682 (ED Penn. 1976).

Trident General Insurance Co. Ltd v McNiece Bros Pty Ltd (1988) 165 CLR 107; 80 ALR 574.

U.S. v Elcom Trial brief Case No. CR 01-20138 RMW U.S. District Court Northern District of California, San Jose Division, 21 October 2002 at http://www.eff.org/Cases/US_Elcom/ [8 December 2002].

United States Lines Inc. v United States, No. 79 Civ. 4209 (S.D.N.Y. 1983).

United States v Karo 468 U.S. 705 (1984).

United States v Knotts 460 U.S. 276 (1983).

United States v Maxwell 42 M.J. 568 (A.F.C.C.A. 1995).

United States v Penny-Feeny 773 F.Supp. 220 (D. Haw. 1991).

United States v Place 462 U.S. 696 (1983).

United States v Smith No. 91-5077 5th Cir. Nov, 12, 1992.

United States v United States Shoe Corp. 523 U.S. 360 (1998).

Universal City Studios, Inc. v Reimerdes 111 F.Supp.2d 294 (S.D.N.Y. 2000), *aff'd* 273 F.3d 429 (2d Cir. 2001).

Universal City Studios, Inc. v Corey 273 F.3d 429 (2d Cir. 2001).

University of London Press Ltd v University Tutorial Press Ltd [1916] 2 Ch. 601.

UNMS v Belpharma Communication, Court of Brussels, 16 March 1999, [1999] Auteurs & Media 370.

Victoria Park Racing and Recreation Grounds Co. Ltd v Taylor (1937) 58 CLR 479; 43 ALR 597 (HCA).

Warewick Shipping Ltd v Her Majesty the Queen (1982) 2 F.C. 147, affirmed in (1983) 48 NMR 378.

Wilkinson v Downton [1897] 2 QB 57.

Table of Cases

Winter v P.G. Putnam & Sons. 983 F.2d 1033 (9th Cir. 1991).
World Series Cricket Pty Ltd v Parish (1977) 16 ALR 181.

Yahoo! Inc. v La Ligue Contre Le Racisme et L'Antisémitisme, 145 F.Supp.2d 1168 (N.D.Cal.2001); 169 F.Supp.2d 1181 (N.D.Cal.2001), *appeal pending*.

Table of Statutes

Australia

Administrative Appeals Tribunal Act 1975 (Cwlth).
Administrative Decisions (Judicial Review) Act 1977 (Cwlth).
Australia Act 1986 (Cwlth).
Australian Capital Territory Government Service (Consequential Provisions) Act 1994 (ACT).
Australian Wine and Brandy Corporation Act (AWBC) 1980 (Cwlth).
Circuit Layouts Act 1989 (Cwlth).
Commerce in Trade Description Act 1905 (Cwlth).
Commonwealth Authorities and Companies Act 1997 (Cwlth).
Consumer Affairs and Fair Trading Act 1990 (NT).
Contracts Review Act 1980 (NSW).
Copyright Act 1968 (Cwlth).
Copyright Amendment (Digital Agenda) Act 2000 (Cwlth).
Copyright Amendment (Moral Rights) Act 2000 (Cwlth).
Copyright Amendment Act 1984 (Cwlth).
Copyright Amendment Act 1992 (Cwlth).
Customs & Excise Legislation Amendment Act 1988 (Cwlth).
Data-Matching Program (Assistance and Taxation) Act 1991 (Cwlth).
Designs Act 2003 (Cwlth).
Electronic Transactions Act 1999 (Cwlth).
Electronic Transactions Act 2000 (NSW).
Electronic Transactions Act 2000 (NT).
Electronic Transactions Act 2000 (SA).
Electronic Transactions Act 2000 (Vic).
Electronic Transactions Act 2000 (WA).
Electronic Transactions Act 2001 (ACT).
Electronic Transactions Act 2001 (Qld).

Table of Statutes

Electronic Transactions Act 2001 (Tas).
Fair Trading Act 1985 (Vic).
Fair Trading Act 1987 (NSW).
Fair Trading Act 1987 (SA).
Fair Trading Act 1987 (WA).
Fair Trading Act 1989 (Qld).
Fair Trading Act 1990 (Tas).
Fair Trading Act 1992 (ACT).
Financial Management and Accountability Act 1997 (Cwlth).
Freedom of Information Act 1982 (Cwlth).
Freedom of Information Act 1982 (Vic).
Freedom of Information Act 1989 (ACT).
Freedom of Information Act 1989 (NSW).
Freedom of Information Act 1991 (SA).
Freedom of Information Act 1991 (Tas).
Freedom of Information Act 1992 (Qld).
Freedom of Information Act 1992 (WA).
Frustrated Contracts Act 1978 (NSW).
Goods Act 1958 (Vic).
Health Records Act 2001 (Vic).
Information Privacy Act 2000 (Vic).
Invasion of Privacy Act 1971 (Qld).
Occupational, Health and Safety Act 1983 (NSW).
Patents Act 1990 (Cwlth).
Privacy Act 1988 (Cwlth).
Privacy Amendment Act 1990 (Cwlth).
Privacy Amendment (Private Sector) Act 2000 (Cwlth).
Privacy and Personal Information Protection Act 1998 (NSW).
Sale of Goods Act 1895 (SA).
Sale of Goods Act 1895 (WA).
Sale of Goods Act 1896 (Qld).
Sale of Goods Act 1896 (Tas).
Sale of Goods Act 1923 (NSW).
Sale of Goods Act 1954 (ACT).
Sale of Goods Act 1972 (NT).
Surveillance Devices Act 1998 (Vic).
Trade Marks Act 1955 (Cwlth).
Trade Marks Act 1995 (Cwlth).
Trade Practices Act 1974 (Cwlth).
Trade Practices Act 1992 (Cwlth).
Wrongs Act 1958 (Vic).

Canada

Copyright Act 1988 (Can).
Crown Liability Act 1952–53 (Can).
North American Free Trade Agreement Act 1993 (Can).
Personal Information Protection and Electronic Documents Act 2001 (PIPEDA) (Can).

European Union

E.U. 1985 *Directive on Product Liability 85/374/EEC of 25 July 1985 on the approximation of the laws, regulations and administrative provisions of the Member States concerning liability for defective products*, Official Journal L 210, 07/08/1985, p. 0029–0033.

E.U. 1990 Council Convention on Jurisdiction and the Enforcement of Judgement in Civil and Commercial Matters, 1990 O.J. (C 189) 2 (Consolidated).

E.U. 1991 *Directive on the Legal Protection of Computer Programs* 1991 91/250/EEC OJ L 122/42 of 17 May 1991.

E.U. 1995 *Directive on the Protection of Individuals with regard to the Processing of Personal Data and on the Free Movement of such Data*, Brussels: European Commission; Directive 95/46/CE; Official Journal of the European Commission (L 281) and at http://europa.eu.int/comm/internal_market/privacy/docs/95-46-ce/dir1995-46_part1_en.pdf [30 June 2004].

E.U. 1996 *Directive on the Legal Protection of Databases* 96/9/EC O.J. L 7, 27 March 1996.

E.U. 1999 *Directive on Electronic Signatures* approved on the 30th of November 1999.

E.U. 2000 Council Regulation (EC) No. 44/2001 of 22 December 2000 on Jurisdiction and the Recognition and Enforcement of Judgements in Civil and Commercial Matters, 2001 O.J. L 12(1).

E.U. 2000 Data Directive and US Official Journal of European Commission L 215 of 25 August 2000.

E.U. 2002 Data Directive and Canada Official Journal L002, 04/01/2002 pp. 0013–0016 at http://europa.eu.int/comm/external_relations/canada/summit_12_99/e_commerce.htm [30 June 2004] and E.U. 'adequacy' standard agreement at http://europa.eu.int/comm/internal_market/privacy/adequacy_en.htm [30 June 2004].

E.U. 2002 *Directive on Privacy and Electronic Communications* (2002/ 58/EC).

EEC 1981 Council of Europe *Convention for the Protection of Individuals with regard to Automatic Processing of Personal Data*, Brussels: EEC and at http://www.privacy.org/pi/intl_orgs/coe/dp_convention_108. txt [30 June 2004].

Treaties and Conventions

Berne Convention for the Protection of Literary and Artistic Works 1886 as revised in Paris in 1971.

Berne Convention for the Protection of Literary and Artistic Works 1998.

Hague Convention on Jurisdiction and Foreign Judgements in Civil and Commercial Matters Draft at http://www.hcch.net [6 June 2004] and at ftp://hcch.net/doc/gen_pd7e.doc [6 June 2004] under the heading 'Special Commission on General Affairs and Policy', as Preliminary Document No. 7.

ICCPR 1976 *International Covenant on Civil and Political Rights*, New York: United Nations U.N.T.S. No. 14668, v. 999 (1976), p. 171 and at http://www.privacy.org/pi/intl_orgs/un/international_covenant_civil_ political _rights.txt [30 June 2004].

Rome Convention for the Protection of Performers, Producers of Phonograms and Broadcasting Organizations 1961.

UDHR 1948 *Universal Declaration of Human Rights*, 10 December 1948 at http://www.un.org/Overview/rights.html [30 June 2004].

UNCITRAL 1996 Model Law on Electronic Commerce with Guide to Enactment 1996 General Assembly Resolution 51/162 at http:// www.uncitral.org/english/texts/electcom/ml-ecomm.htm [30 June 2004].

UNCITRAL 2002 Working Group on Electronic Commerce Draft Convention on Electronic Contracting—Working Paper 95.

UNCITRAL 2002 Draft Convention on Electronic Contracting at http://www. ag.gov.au [8 January 2004]. Also at http://www.uncitral.org/english/ sessions/unc/unc-35/509e.pdf [8 January 2004] and at http://www. uncitral. org/english/sessions/unc/unc-36/acn9-527-e.pdf [8 January 2004].

UNCITRAL Working Group on Electronic Commerce, Draft Uniform Rules on Electronic Signatures at http://www.uncitral.org/english/sessions/ wg_ec/wp-84.pdf [8 January 2004].

United Nations 1980 *Convention on Uniform Laws on the Formation of Contracts for the International Sale of Goods (Vienna Convention)*.

World Intellectual Property Organization 1996 *WIPO Copyright Treaty (WCT)*, Geneva: WTO and at http://www.wipo.org/clea/docs/en/wo/wo033en.htm [30 June 2004].

World Intellectual Property Organization 1996 *WIPO Performances and Phonograms Treaty (WPPT)*, Geneva: WTO and at http://www.wipo.org/clea/docs/en/wo/wo034en.htm [30 June 2004].

World Trade Organization 1995 *General Agreement on Trade in Services (GATS)*, Geneva: WTO.

New Zealand

Building Act 1991 (NZ).
Contracts (Privity) Act 1982 (NZ).
Copyright Act 1994 (NZ).
Local Government Meetings and Official Information Act 1997 (NZ).
New Zealand Bill of Rights Act 1990 (NZ).
Privacy Act 1993 (NZ).

United Kingdom

Consumer Protection Act 1987 (UK).
Contracts (Rights of Third Parties) Act 1999 (UK).
Copyright (Computer Programs) Regulations 1992 (UK).
Copyright Act 1911 (UK).
Copyright Act 1956 (UK).
Copyright and Rights in Databases Regulations 1997 (UK).
Copyright, Designs and Patent Act 1988 (UK).
Crown Proceedings Act 1947 (UK).
Data Protection Act 1998.
Electronic Communications Act 2000 (UK).
Freedom of Information Act 2000 (UK).
Halsbury's Laws of England (UK) 415–70.
Sale of Goods Act 1979 (UK).
Statute of Anne 1709 (8 Anne c.19).

Supply of Goods and Services Act 1982 (UK).
Unfair Contract Terms Act 1977 (UK).
Unfair Terms in Consumer Contracts Regulations 1994 (UK).

United States

10 U.S.C. § 167; 1791–96 [Armed Forces].
28 U.S.C. §§ 1346(b) and 2671 [Judiciary and Judicial Proceedings].
44 U.S.C. § 1336 [Nautical charts].
49 U.S.C. §§ 741–752 [Transport].
49 U.S.C. App. § 1301–1348(b)(3) [Aeronautical charts].
Alaska Constitution, Art. I, § 22. Privacy Protection.
Arizona Constitution, Art. 2, § 8. Privacy Protection.
Arms Export Control Act (AECA) 1976 22 U.S.C. Chapter 39 § 2778(a)(1).
Berne Convention Implementation Act of 1988 Pub. L. No. 100-568 at 17 USC. §§ 101–810.
Cable Communications Policy Act of 1984, 47 U.S.C. § 551.
California Anti-paparazzi Act, California Civic Code § 1708.8.
California Constitution, Art. I, § 1. Privacy Protection.
California Government Code Annex § 818.8—negligent or intentional misrepresentation.
Children's Online Privacy Protection Act of 1998 (COPPA), 15 U.S.C. § 651-05.
Collections of Information Antipiracy Act (CIAA) 1999 H.R. 354, 106th Cong. (1999).
Communications Act of 1934 47 § U.S.C. 222.
Computer Matching and Privacy Protection Act of 1988, 5 U.S.C. § 552.
Constitution Art. 1, § 8 Cl. 8.
Consumer and Investor Access to Information Act (CIAIA) 1999 H.R. 1858, 106th Cong. (1999).
Copyright Act 1909.
Copyright Act 1976, 17 U.S.C. § 101 (West Supp. 1992).
Digital Millennium Copyright Act 1998 (DMCA).
Drivers Privacy Protection Act of 1974, 18 U.S.C. § 2721.
Electronic Communications Privacy Act of 1986 (ECPA), 18 U.S.C. § 2510-22, 2701-11.

Electronic Freedom of Information Act 1996.

Electronic Signatures in Global and National Commerce Act (e-Sign) 2000 (US).

Fair Credit Reporting Act of 1970, 15 U.S.C. § 1681.

Fair Debt Collection Practices Act of 1977, 15 U.S.C. § 1692-92.

Family Education Rights and Privacy Act of 1974, 20 U.S.C. 1232.

Florida Constitution, Art. I, § 23. Privacy Protection.

Freedom of Information Act (FOIA) 1966, 5 U.S.C. § 552.

General Education Provisions Act of 2002, 20 U.S.C. § 1232.

Gramm-Leach-Bliley Act of 1999, 15 U.S.C. § 6801-10.

Hawaii Constitution, Art. I, § 6. Privacy Protection.

Health Insurance Portability and Accountability Act of 1996 (HIPPA), 42 U.S.C. § 1320.

High Seas Driftnet Fisheries Enforcement Act 1992 Public Law 102-582 I.

Illinois Constitution Art. I, §§ 6, 12. Privacy Protection.

Land Remote Sensing Policy Act 1992.

Location Privacy Protection Act of 2001.

Louisana Constitution Art. I, § 5. Privacy Protection.

Montana Constitution, Art. II, § 10. Privacy Protection.

National Information Infrastructure Act of 1993.

Paperwork Reduction Act 1995.

Privacy Act of 1974, 5 U.S.C. § 552.

Privacy Protection Act of 1980, 42 U.S.C. § 2000.

Privacy Protection for Rape Victims Act of 1978 (US).

Right to Financial Privacy Act of 1978, 12 U.S.C. § 3401-22.

Sony Bono Copyright Term Extension Act (CTEA) 1999.

South Carolina Constitution, Art. 1, § 10. Privacy Protection.

Suits in Admiralty Act 1948, 46 App. U.S.C. 741 *et seq.*

Telecommunications Act of 1996, 47 U.S.C. § 222.

Telephone Consumer Protection Act of 1991 (US).

Uniform Commercial Code (UCC) § 2-204(1), Art. 2.14: Contract for sale of goods.

Uniform Electronic Transactions Act 1999 (US).

Uniform Freedom of Information Act 1986.

Video Privacy Protection Act of 1988, 18 U.S.C. § 2719.

Visual Artists Rights Act 1990 17 U.S.C. § 106A.

Washington Constitution, Art. I, § 7. Privacy Protection.

Wireless Communication and Public Safety Act of 1999 (WCPSA or E911 Act).

Wireless Privacy Protection Act of 2003.

Other Jurisdictions

Austria

Constitutional Law 285 of 1987.

Denmark

Copyright in Literature and Artistic Works, Act No. 158 of 31 May 1961, Art. 49.

Finland

Publicity of Official Documents Act 83 of 1951.
User Charging for Government Service Act 1992.
Copyright in Literature and Artistic Works, Law No. 404 of 8 July 1961, Art. 49.

Hong Kong SAR

Electronic Transactions Ordinance 2000.

Ireland

Freedom of Information Act 1997.

India

Information Technology Act 2000.

Japan

Copyright Act 1970, Law No. 48 of 1970, Art. 12.
Copyright Act 1985.

Personal Data Protection Law 2003.
Law for Partial Amendments to the Copyright Law of 23 May 1986, Art. 2(1) (xter).

Malaysia

Digital Signature Act 1997.

Netherlands

Government Information Act 1980.

Norway

Property Rights in Literary, Scientific or Artistic Works, No. 2 of 12 May 1961.

Singapore

Electronic Signatures Act 1998.

Acknowledgements

This book is written first to help me further understand and clarify the nexus between Geographic information science and systems and the rules of law established by the legal system. A corollary aim is that the book should be useful for GI professionals to help unravel and master the legal issues. This book attempts to answer the 'how to' question when confronted with a problem that straddles both the technology and the law. In taking this view I am grateful for the influence of many writers—both technical and legal—who have helped with trying to define the GI professional's role. In the final analysis it seems to me that this role is first and foremost to explain the facts, garner the evidence, and provide the expertise on technical matters at issue. As a technical expert the GI professional's task is also to unleash the power of the information system to help present the case, whether it be for economic planning of poorer countries, search and rescue, health care facility management, chronicling war crimes, archaeology, navigation, or facility location. Increasingly, a further task that the GI professional is being called upon to perform is that of helping to resolve conflicts either in a court of law or as a mediator between interest groups, for example, environmental interests as against commercial interests. These newly acquired roles are a far cry from earlier preoccupations as GI systems analysts.

In writing this book I wish to acknowledge the patience, thoughtfulness, and tolerance of my wife Marion. She has always remained very encouraging of my efforts and challenges me at every turn to do the best that I can. Her presence has released me from the many domestic and pet-care chores that need to be done. My thanks also to Carolyn and David for their continuing support and their diversions to the snow, the Brumbies and the most welcome arrival of Catherine Isobel Marie.

I wish to also acknowledge the help of a large Australia Research Council grant (1998–2000) that supported the research for this work. Without material assistance of this kind the advance of research and Australia's Action Agenda for the Spatial Industry will be slow in coming to fruition.

Acknowledgements

I gratefully acknowledge the Outside Studies Program Committee of the Division of Health, Design and Science at the University of Canberra, in particular the then Acting Chair, Carmel O'Meara who facilitated time release to complete this book project. The intellectual support of my many friends, colleagues, and students at the University assisted considerably in crystallising my thoughts and ideas for this book. A thank you also to Professor Eugene Clark, Head of the School of Law, who has been instrumental in encouraging an enduring interest in GI and information technology law as well as its applications in e-business through teaching stints and joint research projects and publications. My sincere thanks and appreciation also to Chris Gray and Lorraine Goodwin for their material support in the administration of the GIS survey as well as taking care of all other administrative matters. Their good humour and companionship is much appreciated.

I am also grateful to the authors and publishers who have generously given their permission to quote from their work. I gratefully acknowledge the owners of copyright material who have given their kind permission to use their works cited in this text. Every effort has been taken to trace the owners of copyright material, and in a few cases this has proven to be impossible. I take this opportunity to offer my apologies to any copyright holder whose rights I may have unwittingly infringed.

I would also like to acknowledge the considerable assistance of John Wiley & Sons Publishers and in particular the editorial assistance of Keily Larkins and an anonymous reviewer for their professionalism and dedication in coaxing this project to completion.

George Cho
Mayfair, a house in Kaleen
Australian Capital Territory

Introduction

Geographic information (GI) and the law is a subject worthy of study. GI systems began as a tool for map making and automating cartography, but have since evolved to become an information system in their own right. GI systems have broken new ground, not only in supplying map products, but also augmented services that accompany those products. In today's world it would be rare that GI and systems that drive it remain in the background where geospatial data are used. Everywhere one looks today, from the hand-held mobile phone, to the use of credit cards, the electronic tags on our cars to facilitate paying road tolls or to open electronic gates, and the swipe cards that permit travel on commuter mass transit systems, an information system is all too pervasive. Embedded deep in these electronic gadgets are knowledge-based intelligent tags that can gather information pertaining to time, space, event, and any other transactional type of activity. The 'when' question is as important as the 'where' question because both these elements may mean a revenue stream for the operator. Here geospatial information pertaining to location in space depends to a large degree on positioning systems that give coordinates obtained either from overhead satellites in space or from ground-based stations in combination with satellites.

In their spurt towards maturity, GI systems have also had to deal with legal issues. These legal issues can both be an instigator of change as well as a 'problem' to be considered when applying GI science methods and tools to resolve particular real-world problems. An articulation of the legal principles and their potential use as an ally in the further development of GI science and systems would be of invaluable assistance to all. More importantly, to help both the users and vendors of GI, there is a need for a practical text that provides a simple statement of the legal principles that underpin geospatial information and data.

Geographic Information Science: Mastering the Legal Issues   George Cho
© 2005 John Wiley & Sons, Ltd   ISBNs: 0-470-85009-4 (HB); 0-470-85010-8 (PB)

When GI was first developed and used there were not many publicly known examples of undesirable consequences of a legal nature. However, in time and with a maturing geospatial technology and GI system a re-evaluation is required. This re-evaluation should assess where we have been and what types of legal problems we have had to sort out. A mature GI system will have to promote a more principled approach to the solution of legal problems as well as the evolution of balanced geospatial policies. The case law examples should provide sufficient theory and practice to guide and develop the legal and policy structure in the use of geospatial technologies. These developments will involve all parties in the industry, from governments and public bodies, to corporations and companies, and to users and consumers generally.

This book is written by an academic and is offered as a further attempt in the journey towards a definitive treatise on the subject. There may be parts which some might consider naïve and inadequate and other areas that might need a total re-working. It is here that fellow academics, users of GI and the GI industry participants can provide invaluable input and feedback towards a better mastery of the law and policy on GI.

This edition on GI Science focuses on techniques and tools to help master the legal issues. The text concentrates on detailed legal knowledge obtained from presenting classes and undertaking research in the area of GI and the law, and e-business and the law. The knowledge is also obtained from providing commentary and critique to draft legislation pertaining to the Internet, and in giving opinions to law reform Commissions in Australia. In this edition therefore there are sections that deal with Web-based GI either as static maps with accompanying databases or as dynamic real-time map applications. These methods bring up a real need for an awareness of liability should something go wrong. As these relationships between parties may take place at arms length, the transactions may have to be governed by carefully written agreements and contracts. Protecting property both tangible and intangible, for example copyright and privacy, are further issues that use of distributed GI systems will require.

There is much empirical content in this text that is topical and derived from real-world case studies. Attention is also paid to recent developments in g-commerce (geo-commerce) particularly in relation to location-based services and geodemographic studies and the application of positioning system technology in navigation, tracking, and tracing of persons and things. Together these cases emphasise the fact that GI and geospatial data raise unique legal and policy issues. These issues are not generally discussed in information science, or for that matter in legal studies.

This is a book on law and policy focused on GI. It is about how the law affects the use of GI as much as it is about how GI has influenced the law and policy. An analysis of the legal issues is an elaboration of the policy on the use of GI. However, this book is neither about GI systems as a technology nor about the technical details of its use. The focus here is simply on the legal issues that usage throws up, and how different laws have been fashioned to direct as well as to respond to such eventualities. The general principles guiding the legal framework surrounding GI in various countries are given as examples in order to analyse, explain, and describe both court decisions and the policy behind the law. In particular, Australian law is used as the starting point for a generic discussion of the legal principles, and laws from other jurisdictions are used either to embellish or to give a contrasting viewpoint in similar fact situations.

The use of GI raises issues that can be even more of a legal and policy minefield than traditional information systems. By way of case studies the main focus is on the law and policy issues in employing geospatial information. This book also introduces unique problems confronting the law and policy in GI and is explored in detail in individual chapters. The core subjects of the chapters include the public–private debates of data use in regards to access, sharing, and sale as well as the for-profit and open access to government information; protection of spatial information, databases and property; personal data and informational privacy; liability in the use of information products and services, and contractual relationships.

This book will be invaluable to practitioners, professionals and policy makers because of its emphasis on mastering the legal issues. The topic is also ripe for serious academic research and analytical work in a fast-maturing science within the knowledge economy. Cautionary tales and unseen legal pitfalls are highlighted as well as a description of some unintended consequences in the use of GI. However, the general thrust of this book is as a handbook or short guide to help master the legal topics. It gives an outline of the law, but does not pretend to be a definitive treatise on this rapidly expanding area of information technology law. Where appropriate signposts have been added for further reading and study on specific aspects of the law.

It is my belief that legal questions need to be considered in advance of GIS litigation in general and liability claims in particular. This consideration will enable those involved in the use of private and public sector GI to weight their risks and responsibilities more intelligently. As mentioned previously, this book uses as its foundation Australian legal principles as embodied in the common law and statutes. However, to remain more general, the book focuses on 'fundamental' legal principles and policies that are germane, regardless of jurisdiction. As a case study in applying these legal principles the more general and universal matters are highlighted. Moreover, the common law and statute law traditions of the Australian legal systems are based on English and Scottish law which are also to be found in other jurisdictions elsewhere, especially the countries of the Commonwealth and the U.S. This treatment can also have the advantage of providing the underlying principles from which comparisons with other jurisdictions can be made and also how the law and practice can be further improved. The comparative approach that is adopted throughout this text will show up practical problems that will need to be resolved in order to achieve particular objectives. Cross-jurisdictional comparisons will also force a consideration of very different linguistic and legal cultures.

The material is presented in six interlinked and cross-referenced chapters. The structure adopted is more for organisational reasons than anything else. Each of the legal theories that make up the theme of a chapter, while addressed as a discrete entity, relies on other theories to provide the substance and core principles. No one legal theory can provide all that may be necessary to either defend or prosecute a suit. Thus, some chapters will need to be read in conjunction with others. The organisational structure is designed to assist readers who may be interested in only specific topics. It is therefore recommended that readers go to those chapters of interest in the first instance and then branch out to the other chapters after this initial reading. As a practitioner's text on mastering the legal issues it is best used for pursuing information on a specific topic. This text is not be read from cover to cover. The organisation of the chapters provides a convenient structure for delivering the material in teaching a semester unit in GI systems and the law.

Caveat. This book provides legal information of a general nature only. It is not intended to provide legal advice and any statement in the book should not be acted upon without consultation with appropriate legal counsel. Appropriate legal advice should be sought before acting. In addition, while the legal theory and principles remain unchanged for long periods of time, the law relating to GI systems and geospatial technology is expanding so quickly that constant updates might be required.

There is no summary to this text. However, the following short description of the organisation and content of each chapter may serve as a synopsis for this book. Readers may wish to read a synopsis of each chapter here before proceeding to the chapter itself.

A note about URLs. Where possible I have provided the full universal resource locator (URL) for Internet references. However, some websites use automated techniques that do not display a unique URL for a page, or the URL given may be overly long and complex for referencing purposes. A root URL to the home page is given for complex URL references. As a website may change locations or be restructured, the date on which I last accessed a particular electronic reference is given in square brackets.

Organisation of this Book

Chapter 1 Geographic Information Science: Legal and Policy Issues

The key message in the introduction to this chapter is that geographic information (GI) is intrinsically valuable as a resource that feeds in various ways into a technology, a discipline of study, and a science of knowledge. The application of GI systems as a technology is not simply to produce cartographic products, but to provide a tool for any kind of analysis—physical, socio-economic, political, historical, and legal. This utilitarian technology captures both time and space and is found in various types of studies and applications. In addition, estimates throughout the world demonstrate this technology's economic value and benefit to national economies. The benefit cost ratio of 4:1 in an Australian study on investments in the spatial information industry contrasts with the monetary estimate of £100 million per year in the U.K. for the Ordnance Survey (OS) or to the estimated investment value of public sector information for the entire E.U. of €9.5 billion per year and an economic value estimated at €68 billion per year.

An introduction to law, legal systems, and legal theories that reflect the custom and culture of various jurisdictions follows in the next section. Whether under civil law, common law or statute, the principles of protection and litigation are similar. Here the issue to be resolved is why there is a need for the legal protection of geospatial data and information and whether this need is necessary and justified. To seek answers to these queries, possible legal mechanisms that may be used are discussed and these are considered to have been determined by public policy.

The factors influencing GI policy, extant policies and policy-making processes are discussed next. In particular the discussion focuses on an appraisal and evaluation of different data pricing policy regimes for GI around the world. A concluding part describes geospatial information policy development in Australia with the government's Spatial Industry Action Agenda.

The chapter then discusses the nexus between GI on the one hand and the law on the other. This is an important discussion as it posits the view that there are aspects of the law that determine how GI systems and science have developed and have been implemented. There is also the view that GI systems and science have influenced how a policy should be framed and consequently how the law may be drafted in order to support public policy initiatives. Evidence is provided to show how GI practitioners and scientists have a major role to play in this regard. This section outlines the framework for important legal theories to be heeded. The nexus between information systems and the law is underlined by concerns for privacy, liability, and property. Examples of legal dilemmas as a result of GI applications are evaluated here.

The final section in this chapter reiterates that geography does really matter and that maps can be used either as evidence or as an indispensable tool even in an 'ageographic' borderless electronic environment. Indeed, the ability to graphically portray three-dimensional spaces as well as hyperspaces of several dimensions in magnitude demonstrate the greater utility of GI but also the integrative aspects of the geographer's craft encapsulated in the 'geographic', 'information', 'system' neologism GIS.

Chapter 2 Sharing Geographic Information and Data

In sharing GI and data, public sector managers have already been concerned with, and are now very familiar with, legal matters such as licensing agreements, intellectual property, legal liability, pricing, access and marketing. In a previous period they were addressing technical issues of quality, accuracy, and reliability of data and information. These shifts in focus are a reflection of the evolution and growing maturity of the spatial information industry.

This chapter is about the sharing and commercialisation of GI and data. The adopted model of disseminating GI and data is reflective of underlying data policies extant in that jurisdiction as well as the legal regimes governing all such interactions and transactions. Exchange standards and frameworks would need to be in place to ensure the smooth transfer of GI. Standards to ensure the quality of the data, and metadata to describe data content and assist in discovery of the data are all designed to facilitate the sharing and use of information. For these to take place formal mechanisms such agreements, contracts and licences need to be put in place as well.

GI as a resource, asset, commodity, and infrastructure is explored in the first section. How GI is viewed may provide an insight into the evolution of policy that will guide its use, sharing and sale as a good in the data marketplace. Access to and the commercialisation of spatial information will depend largely on its quality and availability.

Data policies and legal frameworks for accessing data are presented next, with Australian experiences given as examples. The dissemination of public geospatial data is used as a basis for evaluating the general topic of whether to share data or to sell data to recover costs. International comparisons are made, especially with the U.S. open records policy and the public sector information policies of the E.U. An efficient allocation of resources and the need for equity in the use of government resources are some of the prime reasons for policies on access to government information. In the Australasian context there is no overarching access or pricing policy for public sector information. On the other hand, in the U.S. the open records regime is there to ensure that the government is accountable for its actions. In Europe, however, there is in general a cost recovery regime for most member states of the union. Overall there appears to be a growing maturity of the spatial information industry worldwide coupled with converging spatial technology derived from mainstream information management know-how. This maturity has brought about an appreciation of the business benefits in traditional spatial information areas and has sparked a shift in industry dynamics.

An outline for an international framework for developing access policies by way of standards and the use of metadata, clearinghouse, and registries to facilitate exchange, sharing, sale and use of GI, is given next. Such a framework has as its prime objective the promotion of interoperability of GI worldwide.

In the final section possible scenarios for the development of a global information infrastructure (GII) is outlined, together with the role of a global spatial data infrastructure (GSDI). The development of supranational and national spatial data infrastructures (NSDI) is described beginning with that of the U.S., followed by Canada, Australia and New Zealand and then the E.U. Six possible steps in achieving a GSDI are proposed, including marketing the vision, describing the concept and vision, to developing interoperability, community-based developments, and relationship building based on achieving a little at a time and influencing government policy. Success may come from the efforts of champions at the local, regional, national and global levels.

Chapter 3 Geographic Information and Intellectual Property Rights

This chapter uses the storyline of an imaginary GI professional to introduce the topic of intellectual property rights (IPR) and its pervasiveness in an information age generally and in GI in particular. Substantive IPR issues raised in the storyline are addressed in major sections of this chapter. Practice notes offer practical suggestions and timely reminders of the 'do's' and 'don'ts' to avoid litigation and damage awards for infringements. Lessons and litigation and the fate of the GI professional's code are given at the end of this chapter.

Each section of this chapter addresses particular IPR that impinge on GI systems, services, and science. In the first section the question of what constitutes IPR in a general sense is given and whether such rights are any different in an electronic environment are discussed. Intellectual property (IP) refers to the property of the mind or intellect and is a generic name that encompasses a bundle of rights which protect innovation and reputation. In the second section the rights pertaining to copyright, patents, trade marks, designs, confidential information, moral rights and *sui generis* [one of a kind] rights are discussed together with its protective mechanisms. These lay the groundwork for understanding the rationale and purpose of the protection.

In the third section, IPR protection is characterised as a *quid pro quo* for maintaining a proper balance between protecting private rights and property on the one hand and for sharing knowledge, utility, and interests with the public on the other. This section focuses on the international environment of IPR protection by discussing the various conventions, agreements, and treaties designed to harmonise protective measures around the world. These international agreements have influenced domestic laws. Following this section the Australian legislative framework gives the backcloth for facilitating the protection of IPR in a common law jurisdiction. The basic features and characteristics of IPR in Australia is summarised here.

In the fifth section, copyright *per se* is addressed and the main objective to observe its influence on GI—systems, science, and services. In particular this section addresses specific issues relating to maps, to electronic databases, the E.U. Database Directive, moral rights, and *sui generis* schemas that have been proposed. Also discussed are business methods patents and those relating to GI and the implications of the Digital Agenda on GI. Section six addresses other legal issues and atypical developments that provide IPR protection in other ways—the 'copyleft' movement, for example. This section also discusses other IP, such as geographical indications, photographs and fonts. There is also a discussion of infringements, defences to infringements of IPR, and suggests some remedies to these.

The seventh section discusses the issue of ownership of IPR generated by employees and other workers. This discussion is tied to the description of IPR agreements with employees, contractors and academics in universities and research institutions. Given that many GI professionals would be engaged in or involved in some form of consultancy, both domestically and internationally, multi-participant international GI projects and IPR issues are discussed here. This is because there will inevitably arise questions of the ownership of the project data, the economic protection of IPR, the resolution of disputes, questions of jurisdiction and law, and access to the information post-project. In the final section, the fate of 'Gigo's code' is given as a hypothetical legal brief in response to litigation and what lessons may be learnt from the activities undertaken by a GI professional. A conclusion suggests what may transpire in the near term—that contracts and agreements rather than the present legislative model might better serve copyright protection.

Chapter 4 Geographic Information and Privacy

The theme of this chapter is about the relationships between privacy and GI science in terms of the role of regulation, self-regulation, and best practice. This discussion may yield policy guidelines for the protection of information privacy as well as the privacy of individuals. While the claim to privacy may be a relatively recent one, a salient characteristic of past protection has been *ad hoc*. The Australian Constitution has not been vested with powers over privacy protection and the common law protects privacy rights indirectly. Similarly in the U.S. there is a common law right to privacy and as well as amendments to the U.S. Constitution to cover personal privacy matters. Today, however, most jurisdictions have included some statutory protection to privacy.

While the economic and social impacts of advances in geoinformation technology have been overwhelmingly positive, concerns have been raised on the part of individuals about what information is being collected, how it is being used, and who has access to it. These concerns, in turn, have led to calls for policy and regulation. Several key privacy issues are addressed in the four sections in this chapter.

Philosophical and doctrinal issues provide the basis of discussion for grappling with the nature and structure of the problem of privacy in the first section. These issues arise because it is argued that GI systems are not an inherent threat to personal privacy as the systems are not personal data intensive. Another is that there is a lack of understanding of privacy because of the fuzzy thinking about whether we are attempting to protect data or protecting privacy of that information. A final area of contention involves the ethical questions in using geospatial technology and what might be the 'right' thing to do in so far as privacy is concerned.

A second section analyses the legal, regulatory and policy framework that governs the source of the 'right' to privacy. The first half of this section deals with the Australian common law and legislative regime and in particular the development of a regulatory mandate for information privacy principles and national privacy principles. Such principles are important since they govern the practice and processes in information gathering, use, and dissemination. The common law and the disclosure of personal information provide various protections in tort, negligence, and the duty to keep confidences. Industry codes of conduct and self-regulation supplement both the legislation and common law to promote better practice in privacy protection. The second half to this section presents the U.S. regime as a contrast and comparison to Australia's efforts. The presentation serves to demonstrate how rules have developed in tandem and in some cases how the approaches justifying one course of action have differed because of cultural backgrounds, attitudes, and public policy. There is an initial discussion of Federal and State statutes dealing with the protection of privacy, followed by a section on the common law protection of privacy. A concluding part to this section is the assessment of fair information privacy principles evolving in major common law countries. Fair information practices are wholly apposite with GI systems and the manipulation of databases since privacy may be infringed in one of two ways: first, in using data containing personal information; second, privacy may also be infringed when data containing no personal information are aggregated from disparate sources so that information pertaining to individuals may be identified from the data. Five core principles for fair information privacy protection are proposed.

An evaluation is undertaken in the third section of different geospatial technologies that promote intrusiveness, enhance privacy protection or are sympathetic to privacy protection. This evaluation is critical given that GI science is heavily steeped in technology that is based on location and spatial relationships. Geospatial technologies may be used for tracking people, their shopping and travel habits, the places that they go to for recreation, for what duration, and in some instances, making an inference of the purposes of that event. In particular, location-based services (LBS) that rely on the key ingredients of time and space are capable of revealing much. LBS are no different from geodemographics, an information technology that enables marketers to predict behavioural responses of consumers, based on statistical models of identity and residential location. The regulation and use of databases are presented here because of the privacy concerns raised above and in particular the tracking of movements of individuals in space, the tracking of transactions, the tracing of communications, as well as the convergence of locational and tracking technologies.

The final section frames the emergent policy and practice in privacy protection. In particular the E.U. Data Protection Directive is analysed given its wide-ranging impact on many jurisdictions in terms of data protection and data transfer principles. A 'safe harbour' mechanism is given here together with the spectrum of alternatives implemented in Australia, Canada and the U.K. to respond to this data directive. While technology will continue to be both a problem and a solution, technological advances such as LBS, informatics, and GI science, will continue to challenge privacy boundaries.

Chapter 5 Geographic Information and Contract Law

This chapter is about the law of contract and its role in GI science. GI scientists need to have a working knowledge of contract law because this may be the first point of contact between information providers, software consultants and end-users. The law of contract both binds the major players as well as providing the platform for establishing relationships between parties. Knowledge of the law of contract is vital for building sound business practices, and this becomes a foundation for good business relationships. Contract law is about relationship building rather than simply attempting to either drive a hard bargain or as a ploy to get out of conflict situations. It will indeed be too late if the parties were to face each other in a court of law as this may signal a breakdown in the relationship.

This chapter is about those elements of a contract that are both necessary and sufficient conditions for relationship building. The absence of a necessary condition may prevent the contract being formed. Sufficient conditions refer to those elements that determine unequivocally that a contract will be formed. These may include such elements as offer, acceptance, consideration and other criteria for a valid contract. Such elements feature prominently in this chapter.

In the first section the traditional law of contract is promoted in terms of a 'meeting of the minds' where an offer is accepted and sealed with some sort of consideration. Here, well-drafted contracts will help avoid problems in the future as it will be quite expensive for parties to meet in a court of law. The elements of a contract are addressed by way of questions. These questions include: why contract? What are the elements of a valid contract? Must a contract be in writing? Why must there be an intention to create legal relations? Can an offerer prescribe the method of acceptance? Such questions provide a structure for checking off a list of requirements for a valid legal contract. Legal details follow answers to these questions, including methods of acceptance, termination, revocation and lapse of an offer, and the terms of a contract.

A second section addresses the issue of understanding and distinguishing between a contract *for* service and a contract *of* service. The distinction is one between a consultant (contractor) and an employee and the legal implications that flow from this is further elaborated here. Two other types of contracts are discussed here are those pertaining to contractors and those relating to academics and researchers. Closely aligned to this discussion is the third section dealing with whether GI systems offer a product or a service. Again different rules will apply, depending on which side of the ledger the answer lies. Irrespective of the choice, different legal consequences follow since the former may deal with intangibles— IPR, mistake, liability; whilst the latter may relate to personnel services— standards and quality, timeliness and responsibilities.

Licensing is addressed in the fourth section as a means of providing use of information and data without relinquishing ownership. Licensing may also protect the information from misuse, protecting proprietary interests in the information asset, such as intellectual property, protecting personal and informational privacy and confidentiality of the information as well as minimising liability. In this section there is a discussion of the special needs of spatial data to merit the use of agreements and licences rather than that of a formal contract. The very nature of spatial data may require taking this path, either in supplying the data or in purchasing the data.

Legal duties and responsibilities arising from contract may also overlap with consumer protection legislation. In the fifth section the privity of contract is highlighted since only a party to a contract can enforce the contract. The contract confers rights and imposes obligations on parties to the agreement to the exclusion of others. A non-party to the agreement cannot generally be burdened or benefited in law by the formation of a contract. Apart from this, there may be exclusion clauses in a contract that may specifically exempt or exclude liabilities, such as a limit on claims against parties seeking redress or the time limit within which claims may have to be made.

The sixth section is about the execution of the contract, more particularly in terms of the discharge of a contract, failed contracts and what remedies a party may expect in the event of a breakdown of relationships.

The final section is on Web-based contracts, given the prevalence of new generation GI system products and services. Here electronic transaction regulations are discussed first, before examining the general topic of electronic contracts. After this, there is a discussion of the law of contract formation, scope, jurisdiction, and terms and conditions in an electronic environment. A summary of the legal issues in GI contracts and a couple of contract precedents are given before a concluding section. The conclusion reiterates that contractual rights and obligations cannot be taken in isolation. There are interactions between the various legal theories including tort, intellectual property, and statutory provisions.

Chapter 6 Geographic Information and Liability Standards

Liability in law is a broad term that includes almost every type of duty, obligation, debt, responsibility or risk arising by way of a contract, a tort or a statute. While contract may regulate the extent of liability, common law tort also governs who should be legally responsible for an act or activity that has caused harm to a person. The amount and extent of liability may also be determined by reference to a statute. While criteria may be found in pre-established legal standards, liability in the use of GI systems and services is not fully understood and is in the process of evolving, but slowly. The complexity and diversity of environments in which GI is used presents broad, undefined liability concerns, and the legal standards that the industry is using is only now being developed.

This chapter explores what those legal standards are, whether in contract, tort or statute and to evaluate each of these theories in so far as it impinges on the provision of GI either as a product, service, or a combination of product and service. The basis of litigation and the defences that may be raised against claims is examined with a view to limit liability, but also to develop industry-standard better practice in the provision of information and services.

The first part of this chapter underscores the importance of legal standards and guidelines. These standards have been evolving and changing over time in the face of technological progress. In particular the five case studies demonstrate the classic issues of standards to be attained if liability is to be avoided. The legal standards include the duty to take care, the responsibility of due diligence to those who may be affected by a lack of care, the reliance on information to one's detriment, and the subsequent injury, damage and loss that occurs.

Part two interweaves the examination of contract, tort and statute liability exposure and risks in the context of the provision of GI. In the main liability is imposed on the provider of GI systems, databases, and reports; but this liability is also moderated in part by the user's role in contributing to the fault-based equation. A sub-section discusses contract and strict product liability because of the inherent nature of GI in which the goods–service distinction is paramount. Tortious liability may arise as a result of a wrongful act causing harm. The duty of care determines the class of persons to whom a defendant will be liable in negligence. The scope of this duty depends on reasonable foreseeability of the harm and the proximity as between the defendant and plaintiff. However, in terms of GI professionals that standard may be higher. Liability may arise also from an omission to act and one which causes purely economic loss. Other causes of liability include negligent misstatements, misrepresentation, and those imposed by statutes. There are also interactions with other legal theories such as privacy, IPR, defamation, and criminal law which may have liability risks.

Part three evaluates the various liability risk management strategies in order to minimise liability risks as well as the adoption of tactics aimed at eliminating such liabilities altogether. While disclaimers are a common method of reducing liability risks, pro-active strategies are also as important. Ensuring that the quality of the data is of the highest standards possible may eliminate liability. One area where it is not possible to guarantee the quality of data is in the use of global positioning systems (GPS) because of the reliance on instruments and third party providers. But here again, where map data quality elements are checked off, where map accuracy standards are maintained and published on the finished product, arguably these *prima facie* are a defence against negligence and liability. Employing legal risk management strategies will serve to minimise, transfer, and insure against legal liability.

In summary, this chapter suggests an agenda for users and providers of GI to raise standards in order to avoid legal liability. Such standards are to be made within the parameters of recent developments in GSDI and distributed GI.

Chapter 1
Geographic Information Science: Legal and Policy Issues

Learning Objectives

After reading this chapter you will:

- Understand the significance of geographic information as a resource that feeds into a technology, a discipline of study and a science of knowledge as well as a tool for various kinds of spatial analysis.

- Be familiar with law, legal systems and legal theories that reflect the customs and cultures of various jurisdictions in which you find it.

- Appreciate that the use of geospatial data and information will need various legal protection as well as legal defences for liabilities arising from this use.

- Understand how different factors influence geographic information policy and how the law has influenced its development. In turn you will observe how the use of geographic information has influenced geospatial policy development.

- Be aware of the legal theories of privacy, liability, and property issues, including that of intellectual property.

- Recognise that, even in an ageographic borderless environment, geography does really matter.

Geographic Information Science: Mastering the Legal Issues   George Cho
© 2005 John Wiley & Sons, Ltd   ISBNs: 0-470-85009-4 (HB); 0-470-85010-8 (PB)

1.1 Introduction

Geographic information (GI) is an intrinsically valuable resource that feeds in various ways into a technology, a discipline of study, and a science of knowledge. Geographic information and geospatial data are regularly used in urban and regional planning, in the identification of plant species, in agriculture, in environmental management, salinity, water resources management, transport and communication, and in oil and mineral exploration. GI is vital for making sound decisions at the local, regional, and global levels. At a socio-economic level, many organisations, including governments, see great potential in the use of the technology and are investing in GI systems for a wide range of applications. The value of GI systems has been shown in many projects in different parts of the world from micro-scale applications at a village level in poor countries at one point in time, through to macro- and meso-scale undertakings over several decades. The Mekong River Project is an example of development assistance to several countries—Vietnam, Laos, Cambodia, and Thailand—as well as involving private sector multinational corporations.

In January 2003 serious bush fires swept through the southwestern suburbs of Canberra, and resulted in the loss of four lives and over 500 homes. The bush fires got out of control and arguably the reason for this was that the authorities did not maintain sufficient detailed geospatial information on the build-up of fuel loads in the forest plantations and bushland. Geospatial information systems could have provided the tools to show where fuel loads had built up. Geoinformation logistics could have enabled the strategic positioning of fire fighting resources where they might be needed most and become most effective. Geographic information as digital maps and graphical documents could also have provided firefighters that had come from interstate to help in the firefighting efforts with a synopsis of local knowledge to ensure that everyone involved knew precisely the state of affairs and the geography of the terrain. The House of Representatives Report (2003) *A Nation Charred* on evaluating the Canberra fires seems to have appreciated the ramifications of poor spatial data.[1] In the aftermath, the Report recommended the acceleration of the national mapping program for the 1:100 000 series topographic maps and support for the development of a national spatial data policy. Such a policy

---

[1] House of Representatives 2003 *A Nation Charred: Report on the Inquiry into Bushfires*, Canberra: AGPS. Also at http://www.aph.gov/house/committee/bushfires/inquiry/report.htm.

is to be made a part of a national and holistic approach to emergency planning and management. The report also concluded that a functioning spatial data infrastructure could become the very foundation of an arsenal of strategic tools for disaster and emergency management. Such an infrastructure becomes fully functional only through continuous updating, on-going collection, collation and redistribution of geospatial data long before it is ever needed on the ground.

One may appreciate the complexities and difficulties in formulating such a policy given that GI systems are characterised by many actors in the collection and distribution of the data. There is also a proliferation of GI applications. Where datasets have to be specially tailored for particular uses there may be a duplication of effort in collecting the data. Difficulties in accessing data and the increasing diverse data formats and standards may hinder easy exchange and use of the data produced by different organisations. The use of metadata standards—that is, those standards describing the geospatial data in conjunction with data catalogues to make the data discoverable, and deliverable would infinitely add value to the asset. The applications may also leverage off other advances in the technology such as Web mapping, electronic commerce (e-commerce) and data warehousing and data mining. Such developments would allow a broader participation of industry in permitting value-added suppliers to create new data products and services.

In its transition to maturity, GI science has witnessed the growth and extension of its tools in the age of the Internet where e-commerce is in the ascendancy. E-commerce by its very nature is concerned with business activities where one aspect of its operations is electronic whilst maintaining traditional 'bricks and mortar' as the main business form. This is in contrast to e-business where the entire business is electronic. Fingar (1999) has identified what he calls the 'three pillars of electronic commerce'.[2] These pillars: electronic information, electronic relationships, and electronic transactions, provide a useful model of emerging paradigms in a digital age. Already the beginnings in the use of the terms 'M-commerce', 'M-banking', and 'M-investing' are becoming evident. There is a growing dependence on the use of mobile hand-held wireless devices such as the mobile-, hand- or cell-phone or digital personal assistants for transacting business. As these forms of business transactions grow there is already infrastructure for 'G-commerce' or geo-commerce where location-based

---

[2] Fingar, A 1999 A CEO's Guide to e-Commerce using Intergalactic Object-Oriented Intelligent Agents at http://home1.gte.net/pfingar/eba.htm.

services are delivered electronically through mobile hand-held devices, including GI system applications in navigation, asset inventories, and data collection.

However, one of the conundrums that courts of the future will have to unravel is whether or not the geospatial information is a product. When we go to a doctor for advice, if that is wrong and we suffer further harm the law allows us to sue the doctor for negligence. But what about the inanimate provider of information, such as the Internet or some Web-based GI system? What relief may we expect if the advice and information given were wrong and an injury resulted as a consequence? The issue is that when people take professional guidance from the Internet or some Web-based GI system, have they received professional advice directly or have they merely obtained an 'information product'? Present-day Australian law is ambivalent on the matter and the answer will dictate which set of standards is being used to ascertain liability—consumer protection, strict product liability, negligence, and contract. The point to be emphasised is that as modern technology becomes more widely accepted the law is struggling to keep pace with the rapidity of the changes taking place.

The economic value of the GI industry is difficult to estimate. However, it may be possible to provide 'orders of magnitude' so as to give an indication of the net worth of this industry to an economy. An estimate provided by the Price Waterhouse study in 1995 suggested that the benefits of the spatial information industry to the Australian economy was in the region of AUD\$4.5 million per annum over the period 1989–1994.[3] More significantly, however, the 'multiplier' attributed to this industry was a benefit–cost ratio of 4:1.

In the U.K. the OS is estimated to turn over £100 million per year with nearly three-quarters of this being generated by quasi-commercial business, that is, from other public sector agencies.[4] In an attempt to quantify the economic potential of public sector information (PSI) in Europe, the PIRA (2000) study estimated that the investment value of PSI for the entire E.U. was €9.5 billion per year while the economic value was estimated at €68 billion per year.[5] This compares with the estimates

---

[3] See Productivity Commission 2001 *Cost Recovery by Government Agencies*. Inquiry Report No. 15, Canberra: AusInfo. Also at http://www.pc.gov.au/inquiry/costrecovery/.

[4] Barr, R 2002 'Choosing the best route for Ordnance Survey. An old friend looks to a new life – the Ordnance Survey Quinquennial Review' at http://www.ginews.co.uk/0402_35.htm.

[5] Commission for the European Communities, 30 October 2000 *Commercial Exploitation of Europe's Public Sector Information: Final Report*, PIRA International Ltd. ftp://ftp.cordis.lu/pub/econtent/docs/commercial_final_report.pdf.

for investment and economic values for the U.S. at €19 billion per year and €750 billion per year, respectively.

A Statistics Canada survey in 1998 indicated that for mapping and survey services, 1614 establishments generated CAN$615 million per annum.[6] Using a broader definition for the geomatics industry to include GI software, hardware, navigation and other services the value was close to CAN$2 billion employing an about 20 000 people. More significantly, the study also suggested that worldwide the geomatics industry in 2001 was worth US$24 billion with an annual growth rate of 20% per annum.

An industry report suggests that GIS revenues for the U.S. is forecast to grow by 8% to US$1.7 billion in 2003.[7] This is compared with a 2.4% growth to US$1.6 billion in core business revenue in 2002 over the previous year. Core business revenues include software, hardware, services and other products. In 2002, software accounted for 67% for GIS vendors, reaching US$1.1 billion, while services accounted for 24% (US$393 million), and hardware declined to 5% (US$88 million).

The above gives an estimate of both the social and economic benefits of GI to national economies, to the spatial information industry as well as the people who are the direct beneficiaries of such use. Given this economic importance of the industry, legal and policy issues have also progressed in tandem with technological developments, first in developing socio-economic policies and then with legal instruments to put these policies into practice. After an outline of this chapter, we turn our attention to focus on legal instruments and GI policies.

Organisation of this Chapter

There are four parts to this chapter. The first is an introduction to law, legal systems, and legal theories. In considering the law we may ask why there is a need for the legal protection of geospatial data and information. Answers to this query pose questions about what possible legal mechanisms exist for the protection of geospatial data and information. These mechanisms may reflect the policy that lie behind the legal instruments and hence requiring separate treatment.

A second part deals with the key policy issues in so far as GI is concerned. In addressing these issues, regard is given to the factors that

---

[6] Industry Canada 2001 'Geomatics Industry' at http://www.strategis.ic.gc.ca.

[7] Daratech 2003 Press Release 'Daratech: GIS Revenues forecast to grow 8% to $1.75 billion in 2003: Utilities and Government Increase Spending' at http://www.daratech.com.press/2003/030807/ and also at GIS Monitor http://www.gismonitor.com/news/pr/2003/080803_Daratech.php.

influence GI policy, the existing policies, and policy-making processes. Following this discussion is an appraisal of the different data pricing policy regimes and an evaluation of the policies on the legal protection of data and GI. A short discussion on data preservation is also given, followed by a summary. This section concludes with a description of geospatial information policy developments in Australia.

In the third part the discussion focuses on the nexus between GI on the one hand and law on the other. This nexus is an important one because there are aspects of the law that determine how GI systems and science have been developed and implemented. There is a direct causal relationship between what is prescribed by the law and how information systems have responded to accommodate such prescriptions. At the same time, there is the obverse situation in which GI systems and scientific knowledge and evidence have influenced how a policy should be framed and consequently how the law is to be drafted in order to support the policy. There is ample evidence to suggest that GI practitioners and scientists have a major and important role to play in this regard. This section takes as its framework important legal theories to demonstrate how close the nexus is in the world of practice. Personal information privacy and security, contractors and service providers of information systems, liability and duty of care in the use of information systems, proprietary issues in information products and services, and access to public databases and the public nature of GI either singly or in concert raise problematic dilemmas for everyone—users, providers, brokers, value-adders, and scholars.

Examples of dilemmas are given in five sub-sections, beginning with database issues, followed by data sharing, maps, global positioning systems, and concluding with aerial photographs in general and more specifically with photographs and images. These applications raise a whole raft of emergent legal questions that as yet may defy and defer definitive answers.

The fourth part reiterates that geography does really matter and that maps can be used either as evidence or as an indispensable tool, even in an 'ageographic' borderless electronic environment. Indeed the ability to graphically portray three-dimensional spaces as well as hyperspaces of several dimensions in magnitude demonstrates the greater utility of GI, but also the integrative aspects of the geographer's craft encapsulated in the 'geographic', 'information', 'system' neologism GIS. Unfortunately geographic information is also a neutral arbiter in overseeing both positive applications as well as its negative uses that may threaten national, organisational, and personal security.

A conclusion summarises the main issues canvassed in this chapter and provides the background and foundations for the substantive chapters of this book.

1.2 An Introduction to Law

This short introduction to the law, legal systems, and legal traditions is reflective of the custom and culture of various jurisdictions whether it is civil law, common law or statute. A reference to the *law* is simply a reference to the body of enacted or customary rules recognised by a community as binding. It also represents the law of a social system or subject of study, whether it is in a democratic, socialist, religious or customary law society. The defining words denote one of the branches or theories of the study of law, for example, commercial law, contract law. *Jurisdiction* is the term used to denote the power of a court or judge to entertain an action or other proceedings. Generally jurisdiction gives the limits within which the judgements or orders of a court can be enforced or executed. Sometimes laws are codified. Here the *code* refers to the systematic collection of statutes and the whole body of laws so arranged as to avoid inconsistency and overlapping, for example, the Civil Code, the Criminal Code. A code can also be a set of rules on any subject or branch of law, such as the *Sale of Goods Act*, that collect and state the whole of the law as it stood at the time the statute was passed.

In general, all legal systems serve three interrelated functions. The first function is in social control and regulating the behaviour of its citizens by way of the statutes and legislation. A second function of the law is to help in resolving disputes among parties. A final function of the law is in the so-called executive or 'social engineering' function in which a community is encouraged and given incentives to act in a certain manner. These disparate functions of the law have together produced different structures and rules depending on the culture and traditions of the community. The various laws, remedies and court systems have thus evolved to be supportive of the particular legal system. Some of the aspects of legal systems are discussed next.

1.2.1 Common Law, Statutes and Civil Law

The common law is a general body of law built up through time and which preceded written legislation. This body of law is expounded and interpreted by judges in courts and is a feature of countries with an 'English' legal tradition such as Australia, New Zealand, and Canada. The U.S. is also considered a common law country. The law of contract is largely common law and only recently have statutes been formulated to better define contracts in narrow and very specialised ways. The common law is characterised by individual rights, an equality before the law, and the universality of the law. The common law strives for certainty and clearly defined universally applied rights. It is partly for this reason that the common law has endured over time in which the common law both supplements and is supplemented by statute law.

The common law is usually found in the judgements of the courts that interpret previous judgements in similar fact situations. These are referred to as 'precedents' and legal conventions dictate that rulings of 'higher' courts, for example, a Supreme Court, are binding on 'lower' courts, such as a District Court or Magistrates Court. The precedent provides guidance to judges to make decisions in similar situations. This alone provides some degree of consistency in judgements and would be unjust if this were not the case.

The common law is the basis of the substantive law in the laws of contract, tort, and evidence. However, the common law is also the source of 'procedural law', that is, law governing the procedures in court and the granting of damages, injunctions and other remedies.

Sometimes the common law becomes the basis of a law code, especially where Parliaments pass laws on particular issues. In the U.S. for example, the law may be formulated as a legal code. The Uniform Commercial Code (UCC) (1987) or the U.S. Code Title 17 *Copyright Act* (1976) are codified public laws. These codes have statutory force when adopted by legislation within a jurisdiction.

In parallel with the common law are published statutes, ordinances, regulations and bylaws. These both formalise and enrich the common law or change the common law in a particular way, creating a new law. Thus, changes to the common law may come about with the enactment of consumer protection law, law of contract and the law of evidence.

In court, judges apply both the common law and statute law. But this is not always a straightforward task because statute law is sometimes written in very general terms so that both a 'wide' and a 'narrow' interpretation are possible. In this way, according to the prevailing social and economic policy climate of the day, judges have some flexibility in putting into effect the subtle nuances of policy that may have been difficult to express in written form. Expert opinion and commonsense approach are *sometimes* employed to resolve difficult cases. The *Acts Interpretation Act* 1901 (Cwlth) in Australia and the *Interpretation Act* 1978 (UK) in England help in the interpretation of legislation.

Similarly there are also rules governing how the laws of precedent may be followed and applied, and in cases where the precedent may be distinguished. As there is only a finite set of common law and statute law, there is no remedy for every infringement of the law. It is here that lawyers play an important role in using and arguing the law on behalf of their clients. Both parties may have a claim, but one may have a better right to the remedy.

Civil law refers to Roman law, the *Corpus Juris Civilis*, for example, of those countries such as Italy which have based their legal system on its rules. More generally civil law pertains to the law regulating conduct between private individuals such as the law of trespass, negligence, and defamation. Thus, a civil action is a proceeding by way of court litigation, in contrast to criminal law, which is a proceeding in court in which the State attempts to regulate and punish the behaviour of its citizens for criminal activity such as murder, theft, and conspiracy.

In all countries law making is accomplished either through the parliamentary process or by means of court decisions. The Parliament of the people makes laws through legislation. The Executive implements the laws and the Judiciary puts law into practice and adjudicates in disputes between parties as well as interpreting the legislation and makes law where there is none.

A Constitution is the basic political and legal document that provides for the law-making functions of a parliament. The Australian Constitution, for example, vests what powers the Commonwealth may have and bestows the residual powers on the States. The High Court of Australia serves to interpret the Constitution and adjudicates in disputes between the Commonwealth and the States over constitutional matters. Most of Australian constitutional law is unwritten and is based on convention and practice. The Constitution may be amended, but only rarely. Constitutional amendment takes place through a referendum that requires a majority of all electors in Australia (including the Territories) plus a majority of electors in a majority of States (not including the Territories). Since 1900 of the 43 referenda that have been held, only eight have been successful.

1.2.2 Court System

Most court systems are organised hierarchically and it is important to bring one's action to the appropriate 'court of first instance'. While the structure and levels of courts vary between different jurisdictions, in general, lower courts hear 'simple' cases and higher courts are courts of appeal, which decide subtle points of law brought up in difficult cases. Appeal courts usually hear arguments and review points of law, except where new evidence has been uncovered.

The highest court in Australia is the High Court, which consists of seven justices, though it is unusual that the court sits *in banc*.[8] The High Court hears appeals from the State Supreme Courts and is the final court of appeal in the Australian legal system. The High Court has original jurisdiction in a number of areas such as indictable Commonwealth offences. The Federal Court of Australia has original jurisdiction in other federal matters such as industrial law, patents, trademarks, bankruptcy, and antitrust. Appeal from the Federal Court is to the Full Court of the Federal Court and to the High Court. Each of the States and Territories has its own court system and is comprised of various administrative tribunals, inferior courts, a court of general jurisdiction, and a court of appeal.

1.2.3 Alternative Dispute Resolution (ADR)

Sometimes parties to a dispute may not wish to go to court, either because of the expense or the expected delay in protracted litigation. Also parties may wish to engage in non-adversarial proceedings so as not to sour relationships and future dealings. Alternative dispute resolution (ADR) techniques include facilitation, negotiation, mediation (conciliation) and arbitration.

In using ADR techniques, an independent decision-maker decides on the merits of the case as well as lays down the procedures to be followed, the evidence that may be admitted and the extent to which experts and lawyers may be involved in the proceedings. Parties may also agree to ADR as a first course of action and only on a failure to resolve the dispute go to court as a last resort.

Facilitation involves a neutral third party that meets with the disputants in order to assist in determining an agreed course of action regarding the dispute. This third party acts as an impartial facilitator and is not there to resolve the problem, but rather to bring an objective perspective so that the parties themselves may determine their best course of conduct to resolving the dispute.

Negotiation is an attempt by parties to reach an agreement. In some contexts this involves two parties sorting out their differences whereas in others an independent authority may act as negotiator between the parties.

---

[8] Until 1986 it was possible to have an appeal from a State Supreme Court to the Privy Council in England. This avenue of appeal ended with the passage of the *Australia Act* 1986 (Cwlth). *In banc*: When sitting *in banc* as a Court of Appeal the justices deal only with questions of law as opposed to sitting at *nisi prius* [on circuit] where only questions of fact are dealt with.

Mediation is a process in which a neutral third party, the mediator, assists the parties to a dispute, to negotiate their own resolution of the dispute. The process also called conciliation often involves an identification of the disputed issues, the development of options, a consideration of alternatives and the endeavour to reach an agreement. A mediator has no advisory or determinative role in regard to the content of the dispute or the outcome of its resolution, but may advise on or determine the process of mediation whereby resolution is attempted.

Arbitration is a more formal process, similar to the adversary system of the courts except that it is generally quicker, less formal, and conducted in private. The arbitrator, a neutral third party, is empowered by the parties to resolve the dispute by making a binding award on the parties. The arbitrator hears submissions and takes evidence on behalf other parties in order to determine an appropriate outcome.

Arbitration is perceived as an alternative to litigation. In some senses, arbitration is private litigation, but others may see it as part of a commercial transaction. A number of codes have been developed by national and international bodies such as the International Chamber of Commerce (ICC), UN Commission on International Trade Law (UNCITRAL) and the London Court of International Arbitration (LCIA) that have published rules of international arbitration.

1.2.4 Remedies

Remedies are court orders in order to restore a loss or to correct a wrong. These may be of two types: damages and monetary orders or injunctions and other restraining and mandatory orders. In common law jurisdictions, courts made orders for payment of money or damages for an 'injury' or loss sustained from a breach of contract, negligence, a criminal act, and sometimes for a breach of statutory duty. The damages awarded by the courts are to compensate for actual and demonstrable financial loss arising from the wrong.

In general, damages are not punitive since it is argued that if it were meant to punish, then the person wronged would have profited from the wrong suffered. Damages are strictly limited to loss flowing directly from the wrong. Thus, at common law, losses arising out of a breach of contract will be compensated only if it may be shown that the loses are regarded as a natural consequence normally arising from a breach of contract.

Sometimes 'exceptional losses' will not be recoverable unless these were expected when the contract was entered into. For example, in international trade there are *force majeure* [irresistible compulsion] clauses—an act of God, storm, earthquake or wild fires. The term *vis major* is also used and denotes one of the 'expected perils' usually found in marine insurance policies.

In tort law on the other hand, legal wrongs are the result of negligence and in the absence of contract, the losses attract much greater damage claims. This is because in tort, damages will extend to all those losses of a kind reasonably foreseeable at the time of the injury, and this includes exceptional and special losses. In defamation cases, for example, 'aggravated damages' may be granted where the monetary loss does not reflect the special circumstances of the victim. On rare occasions, punitive or exemplary damages may be available where a person has been found guilty of malice or a 'contumelious disregard' of the rights of another.

For loss arising from deceit, defamation, and assault, some jurisdictions have given statutory recognition such as the law of copyright and damages in U.S. anti-trust law. Some U.S. courts have awarded substantial amounts for aggravated and exemplary damages.

Sometimes where monetary recompense given as damages awards is insufficient there may be other avenues that may be pursued. In circumstances where damages are inadequate, courts have devised 'equitable' remedies such as injunctions and mandatory orders. An injunction is an order or decree by which a party to an action is required to do, or refrain from doing, a particular thing. Injunctions can either be restrictive (preventative) or mandatory (or interim) or perpetual. In general, an interim injunction is one which is granted before the trial for the purpose of maintaining the *status quo* until the question in dispute between the parties can properly be argued and a final decision made by the court.

However, note that in an information services context, injunctions do not cover the provision of personal services because the provision of such services is contractual in nature and other legal remedies may be more effective. Should injunctive orders be disobeyed, the persons responsible are regarded as being guilty of contempt of court and may be punishable either by imprisonment or a fine at the court's discretion, or both. Other types of injunction include interim or interlocutory orders issued to relieve an urgent apprehension of damage or loss until such time that a final hearing of a claim may be heard in full in court. However, persons seeking such orders must be prepared to underwrite the costs should the action ultimately fail in court. The remedies are described here as equitable because it is up to the discretion of a court to so order.

1.2.5 International Law

International law is a body of rules of law, which regulate the conduct of independent states between themselves. There are generally two sources of international law: one consisting of general principles derived from custom and the other consisting of treaties among nations. While custom and treaties operate primarily in the international sphere, international law can and does have domestic application. More particularly reference is made to the system of law known as 'public international law' and 'private international law'. Public international law is made up of all the multilateral and bilateral treaties and conventions signed by sovereign nations and other entities accorded an international personality. These treaties become domestic laws when adopted by enabling legislation in national parliaments.

Private international law is in reality hardly international law at all since it deals primarily with what is described as 'conflict of laws'. For instance, when a local resident enters into a contract with a foreign national, various questions arise. Questions such as: in which jurisdiction will the contract will be enforced? In the event of a breach, what type of damages would apply especially where one jurisdiction provides punitive damages and the other does not contemplate such an eventuality? What if the contract is perfectly legal in one jurisdiction, but illegal in the other? These questions suggest that the law in this area is very complex and extremely difficult. In general, however, the law of the country in which the action is taken will determine legal procedures, damages, and other remedies. Knowing the jurisdiction, the law and its pitfalls beforehand can help avoid problems.

1.3 Key Policy Issues

The policies governing the use, dissemination and access to geographic information are equally as important as the technical issues of employing GI. Any GI becomes useful, not only because of its availability, but also the technology to access and manipulate the data. GI policies are frequently to be found in disparate statements, locations and form and few are widely circulated outside the policy group. To be able to operate GI systems and data there is a need for a clear understanding and appreciation of the rationale of the policy. This appreciation is also needed in order for there to be widespread adoption and support of the policy. Sometimes the policy impinging on GI is indirect in the sense that the broader policy issue may address a parallel theme and GI is only a part of that theme.

Apart from all other general uses in the mapping community, GI is deployed to satisfy three broad groups of users: public sector socio-economic planners, scientific researchers, and commercial users. While each of these users may present different perspectives on GI policy, a common thread running through all three groups relates to data access and data pricing. Public sector agencies produce and use GI for strategic purposes and for public benefit, for example, the Australian Bureau of Statistics. Scientists use GI for modelling applications in health and welfare planning, economic development, environmental management, mineral exploration, and geospatial infrastructure. The objectives include the discovery of facts about the environment, but also an understanding of the physical processes and ecosystems of the planet. Commercial entities build upon the geospatial infrastructure and provide value-added products and services for end-user clients. The objective is to grow the market and make the geospatial industry sustainable and profitable.

The key policy issues for GI therefore are access, economics, and protection. Access to GI is concerned with the fair and equitable treatment to all whether it is another public sector agency, impoverished scientists, or cash-strapped private sector companies. Given that a majority of GI is collected by publicly funded agencies there is a concern for fair treatment for all in areas where 'data stinginess' might prevail. Any term and condition for access need to be transparent and applied fairly to all so that no one group is disadvantaged. While the growth in the volume of GI and types of information are important factors, the more widespread interest is in the sharing and maximising of the use of GI. As will be noted in later chapters, the more the data are shared, the more it is going to be used and valued. A vibrant GI industry will be one where it is sustainable and where there are incentives for strategic investment plans to be made in order to ensure that quality data are timely, up-to-date, and commonly useable by all over a long-term horizon.

1.3.1 Factors Influencing Geographic Information Policy

There are several factors influencing GI policy, including government objectives, growth in the volume of geospatial data, digitisation, and the growth in the use of public sector information and data. A major factor influencing GI policy stems from governmental objectives in producing geospatial information. The investments in the collecting of geospatial data could be interpreted as part of the social investment in the data infrastructure so that the data are readily in place when it is required. Social objectives are tightly coupled with environmental benefits so that the data are available for modelling, mapping and descriptive purposes. The major beneficiaries would be scientists who would use the data for further research and applications. In the corporate sector, data brokers, value-adders as well as the commercial users themselves would be highly interested parties. Unfortunately, the differing needs of each of the above categories of users would bring about conflicting policies—indeed even within public sector agencies who are both major investors in, and users of, GI. Similarly, within the scientific community, GI is seen as a public good and should be available free of cost, if only because the tax payer has already funded its initial collection. At worst GI should be freely available, but at a low cost.

The next factor that may be seen as influencing GI policy is the growth in volume, quality, and use of the data that are being collected. As fundamental datasets, particularly those that do not change with time, such as physiography, continue to accumulate, the data are also further being refined and replaced those with a greater degree of accuracy and recency. Better quality datasets would exert a strong influence on the extent to which the data would be used. Where pricing and access policies are correctly balanced, it is expected that a greater use of the data would result. Here public–private cooperation and collaboration in policy formulation would ensure that the policies so developed would be the most efficacious ones ahead of other policies that may unintentionally stifle innovation, usage, and growth of the industry.

The digitisation of geographic data is a further factor because it is a product that can now be made more widely available than before and which arguably can be easier to use where it conforms to international standards. Moreover, information products may be regarded as public goods that are non-rivalrous and non-excludable. Non-rivalry is taken to mean that the consumption of information does not diminish the capacity of another to use it. Non-exclusion is understood to be where every user can have access to the information, even though it is being used by another (for example, a navigation beacon). The seller of the information still retains the dataset, even after its sale. A public policy on GI would thus ensure that access conditions are easy to fulfil and helpful to end-users.

In regard to the commercial use of publicly owned GI, it is recognised that the market would develop not just from data availability, but also from easy access to it at little or no cost. A correct balance in sale and access is imperative to ensure that commercial applications are encouraged, but at the same time that scientific and other uses of GI are still possible and within reach despite the costs of the information.

1.3.2 Existing Policy and Policy-making Processes

Existing policy and policy-making in the private and public sectors are variable. Private and public organisations that produce the data and GI create their own policies. Users and others who bear the brunt of the policies can and do act in a collective manner to identify policy issues and serve to influence changes and formulate about new policy. Many national policies governing GI spell out the principles, conditions of access, and other guidelines of its use. The standards for data storage, transmission, metadata, sharing, and legal instruments are part of the policy statements. With the growth in demand for GI in the range and variety of applications, and easier electronic dissemination, there is an impetus for a new approach to the policy and changes to it. Greater flexibility in data access is required so that those who have either purchased or are given access may do so more easily. The only control would be on deciding who may be given a right to have access to the data. Such policies would encourage the innovative use of the Internet and electronic dissemination, and help grow the industry with new uses and applications.

1.3.3 Data Pricing Policy

A pricing policy for GI could lay the framework for data access, distribution, dissemination, and use. Such a policy could also reflect the compromise in the tensions between the data producer and users—public, private, and others—as well as other competing suppliers and substitutes. With the growth of the private sector and the availability of re-purposed and re-aggregated data, pricing policies for public sector data will play an important role in determining the extent and degree of use of GI and data. A pricing policy that is supportive of particular uses will have significant impacts on the successful exploitation of the GI in the public, commercial, and research sectors. But what is the 'right' policy is a difficult question. Part of the answer to such a policy will be one where there is regard for the different categories of use, access conditions, and pricing options.

Several different kinds of pricing options have been proposed and applied in different jurisdictions. One includes *free data* for all users while another uses *marginal cost price* to all users, that is, a price beyond the cost of basic infrastructure collection costs. Other pricing options include *market price* for all users where the price is the 'realisable price' that may be obtained in the market; *full price* that includes investment costs plus all other costs; *two-tier price* to allow preferential treatment in financial terms where for example researchers pay a lower price than the market; and, *re-balancing government funding price* where even researchers pay what the others are paying. In re-balancing government funding price, a government would provide researchers access to grants to offset the extra expenses while at the same time the provider GI agency can recoup its investments in data collection and production.[9]

Whatever GI pricing model is adopted the issue is a sensitive one because of the need to satisfy all types of users. Pricing policy can also have significant implications for the greater exploitation of the GI and growth and maturity of the GI industry.

1.3.4 Policy on the Legal Protection of Data

A review of legal issues will show the need to protect data as an element of an asset in its raw form and information in its processed form, for example, as intellectual property. A data policy will have to take this observation into consideration. Such protection is also necessary where the data are distributed free of charge because of the legal circumstances surrounding issues of liability and infringement of regulations. In usual circumstances where the data are transformed into property and a creator seeking to recoup expenses for its creation will seek a protection of such assets. Intellectual property protection in the form of contracts and licenses where restrictions may be imposed on further distribution and on-sales are available.

A market-based means of protection is the adoption of differential pricing for different types of users of the GI. For example, commercial users may be charged more than say, educational users and researchers. However, this may raise cognate legal issues such as anti-competitiveness and price discrimination that are against public policy. Both practices may be interpreted as an abuse of market power. In Australia, competitive neutrality is a cornerstone of the government's competition policy. The Commonwealth and all States and Territories have implemented such a policy as part of the Competition Principles Agreement (1996).[10] Under this policy, the

---

[9] See Harris, R 1997 *Earth Observation Data Policy*, Chichester: John Wiley & Sons, pp. 110–125.

[10] Commonwealth of Australia 1996 *Commonwealth Competitive Neutrality Policy Statement*, Canberra: AGPS.

prices charged by government businesses are to be adjusted to reflect the advantages and disadvantages of public ownership.

The protection of GI may also be effected by physical and software access controls. In addition, inserting digital watermarks in the data may help make legal protection easier to enforce. In a maturing technology like GI systems, the trend is towards the need to respect rights, licence conditions, and security of the data and GI among users. Such an attitude could become the norm and would only be to the overall benefit of the spatial information industry as a whole.

1.3.5 Data Preservation

The preservation of GI, especially fundamental datasets that do not change over time, is as yet unaddressed as a policy question. In Australia, for example, while there may be pockets of information agencies such as Geoscience Australia who have been given the task by specific legislation to act as custodians and archivists of the geophysical and mineral data, there is no specific policy on preserving GI. Some of the more important considerations in the formulation of a data preservation policy are the duration for which the data are to be kept, the media, and the agency that should take responsibility for the archiving.

With the growth in the number and range of GI applications as well as the number of datasets being generated and developed, finding datasets fit for a particular purpose will become increasingly difficult. Fortunately, among the GI community, plans are already being implemented to ensure that all datasets contain the necessary metadata—data and information that describe the dataset, so that finding the data is made easier.[11] Clearinghouses and directory services have been set up in the U.S., E.U. and Australasia so that electronic search engines on the Internet can direct users to appropriate data archives.[12] Here too, access is to be facilitated by standards and protocols, with 'open' computer architectures and interoperability as an ultimate objective.

---

[11] See for instance the Australian Spatial Data Directory (ASDD), a fully distributed metadata directory launched in 1998 and has since achieved 19 separate nodes including over 40 000 individual metadata entries. Metadata tools and guidelines are found at http://www.auslig.gov.au/asdd/tech/tools.html.

[12] See National Research Council 1999 *Distributed Geolibraries: Spatial Information Resources*, Washington, D.C.: National Academies Press; and see also the U.S. Federal Geographic Data Committee site at http://www.fgdc.gov/clearinghouse/clearinghouse.htm.

1.3.6 Conclusions

A policy on GI and its use features prominently in public discussions among policy makers in government information agencies. While there may have been conflicts in the past, the present attitude seems to be an open supply of GI funded by the taxpayer and intended for the public good. But at the same time, there is the desire to seek growth in the geospatial information industry and to be internationally competitive. In the past GI has previously been dominated by the public sector producing a majority of the fundamental datasets. Invariably, as owners of datasets this sector is effectively responsible for developing access policies. The private sector, on the other hand, has acted as a data broker, value-adder, and developer of information products and services.

1.3.7 Policy Developments in Australia

In Australia, a landmark spatial data policy has been formulated and is being implemented. In September 2001 Senator Nick Minchin, the Minister for Industry, Science and Resources, announced that the government aims to maximise the benefits to the community from the government's investment in spatial data.[13] Under the policy, the Commonwealth will immediately provide free access to its fundamental spatial data that is readily available on-line. Fundamental spatial data that are not available on-line will be supplied at the marginal cost of transfer. All restrictions on the use of this fundamental spatial data have been removed. A single point of Internet access or a portal will be established to ensure users are provided with easy access to this fundamental data. This portal will be linked to the Australian Spatial Data Directory (ASDD) administered by ANZLIC, the Australia–New Zealand Land Information Council. A high-level policy executive, the Office of Spatial Data Management, with dedicated resources and replacing the Commonwealth Spatial Data Committee, will administer the policy. In this new arrangement, the private sector will have greater access to publicly funded data from which to develop innovative new products and services. The community on the other hand will reap maximum economic and social benefits from its investment in the data.

Simultaneous with the announcement of the above policy, the Spatial Information Industry Action Agenda *Positioning for Growth* was

---

[13] Media Release Senator Nick Minchin 25th September 2001 at http://www.minister.industry. gov.au/minchin/releases/2001/September/cmr485-01.doc.

also published.[14] This Agenda sets out a vision for the spatial information industry and identifies strategies to remove impediments to industry growth and participation in the global information economy.[15] The Action Agenda is a joint effort of all key stakeholders: businesses, educational and research institutions, and Commonwealth, State and Territory governments. One of the outcomes of this Agenda is the formation of a single organisation to represent business interests in the spatial information industry, the Australian Spatial Information Business Association (ASIBA). ASIBA will have much of the responsibility for implementing the recommendations of the Action Agenda. A grant to a consortium of spatial industry firms of AUD$2 million under the Technology Diffusion Program will be used to increase the effectiveness with which spatial information is used in emergency management.

AUSLIG, the national mapping agency has put its new Global Map Data Australia 1:1 million scale data on-line for free download for *all users* on the day of the announcement of the new policy.[16] Key datasets such as maritime boundaries and GEODATA TOPO-10M, as well as data on critical aeronautical heights, dams and storages, minerals, digital map index, and Australian surface water management areas are available on-line. Existing prices and conditions will continue to apply until the planned expiry dates, and new prices and licensing conditions are being developed. The policy will not affect prices and copyright on hard copy products such as topographic NATMAPs, which are already priced at 'cost of distribution'.

Demonstrably, the Australian spatial data policy is an important step for maximising the use of geospatial data by facilitating easy access with as few restrictions as possible.

1.4 The Geographic Information and Law Nexus

In presenting his course on 'First Readings in GIS Law' at the University of Maine, Harlan Onsrud (2001) characterises the pervasive use of GI technologies as one that generates conflicts and one where a balancing of

---

[14] Department of Industry, Science and Resources 2001 *Positioning for Growth. Spatial Information Industry Action Agenda*, September. Canberra: InfoProducts.

[15] Details of the Spatial Industry Action Agenda can be found at http://www.isr.gov.au/agendas/Sectors/siiaa/index.htm.

[16] To view and download Global Map Data Australia 1M visit http://www.auslig.gov.au.

competing interests is necessary.[17] Conflicts also confront those using GI systems and implementing such systems and those designing the next generation of spatial information technology. The webcast lectures for the course touch on six major areas of law and GIS:

- conflicts in regard to personal information privacy both in terms of its range and also the volume of issues arising in the use of GIS;
- conflict and societal harm in relation to the balance between protecting the rights of creators as against society's need to know and share in the intellectual capital of a nation moderated by copyright law, other intellectual property rights, additional ownership claims, contractual agreements or database legislation;
- liability issues and conflicts over the responsibility for damage and the lack of a duty of care causing loss, injury, and harm through the use of spatial information technology and databases;
- resolving conflicts in the access to public data generally and the complex set of tensions that this use generates between private users, government officials, non-profit groups and the commercial sector;
- conflict and tensions that arise when treating works in geolibraries as commodities and the harm this causes to other valuable societal functions of information;
- the contentious issue of the sale of taxpayer funded geospatial data by government agencies, and the imposition of restrictions on its use.

This reading guide may be contrasted to the National Science Foundation funded National Center for Geographic Information and Analysis (NCGIA) Core Curriculum (1990) proposed over a decade ago where the idea of information as a legal and economic entity was presented.[18] The syllabus focuses on the protection of private and public property rights. The material also highlights the 'responsibilities' of users of GIS in terms of contract law, liability and a duty of care, as well as positive acts to keep private and confidential information secure, but public information open and 'free' with the help of the Freedom of Information legislation. Also

---

[17] Onsrud, HJ 2001 SIE 525 *Information Systems Law. First Readings in GIS Law*, at http://www.maine.edu/~onsrud/GISlaw.htm.

[18] The NCGIA Core Curriculum in GIS 1990 Unit 70 (Legal Issues) gives suggestions as to how GIS courses might be organised. See http://www.ncgia.ucsb.edu/pubs/core.html.

the curriculum includes the presentation of information as evidence and information as intellectual property.

Comparing the two syllabi, Onsrud and NCGIA, one may be struck both by the commonality of legal topics presented, but also by the relative emphasis on conflict on the one hand and rights and responsibilities on the other. It seems that our perception then and our current understanding has changed much with advances in application, in the nature of the technology, as well as the changing roles of the major participants. The legal theories may remain the same, but the solutions have shifted ground. In part this may be because we have become more aware by being drawn into litigation or having to pay damages. More generally another explanation could be that our understanding and sophistication in the use of technology has awakened a need to protect and to assert our rights, together with the economic imperatives that accompany such assertions.

There are legal implications in all that we do. But more importantly, we need to be aware of the strong linkages between the law, geography, and GI. For example, legislation imposes certain challenges for the management of geospatial data and databases, not only in terms of the design of a GI system, but also in terms of how the information is presented and even the amount presented.

In New Zealand, for instance, local governments employ District Plans/Schemes as comprehensive documents and as key sources of public information. Such plans are 'fixed' as to the content with a planning horizon of ten years or more. In addition, local governments produce Project Information Memorandums (PIM) and Land Information Memorandums (LIM) as part of the effort to capture the ever changing *cadastre*—register of properties, and providing a timely 'snapshot' view of planning.[19] The aim of local councils is to create a single geographical database that is maintained as current as possible and which may be accessible to a range of uses. The dilemma, however, is how to manage this geospatial database while at the same time fulfilling the requirements of the legislation. While GIS has been used to produce the District Plan maps, the legal copy is the paper copy accepted and sealed by Council. Such copies must be exact replicas of the sealed copy. The reasons for this practice is that digital data are easily altered and manipulated and therefore arguably have no legal standing as part of the District

---

[19] PIMs are produced in accordance with the *Building Act* 1991 (NZ) and LIMs are produced in accordance with the *Local Government Meetings and Official Information Act* 1997 (NZ).

Plan. This alone raises interesting issues when presenting the maps as evidence at court hearings when there is a dispute between parties. The implications for customer services, database management as well as the design and management of geographical data contained in GIS can be serious indeed.[20]

The iconic Statue of Liberty and the lesser-known Ellis Island are symbols of a new life for many of America's immigrants. Despite the Statue's welcome message as Liberty enlightening the world and the invitation to a shining new life, the twelve million immigrants processed through Ellis Island between 1892 and 1954 were already on ambiguous 'ground'. When the immigrants stepped off the boats onto the island were they in New York or in New Jersey?[21] This territorial conflict has been brewing for over 160 years and the case involves sovereignty and jurisdiction, but not ownership. The Federal government owns Ellis Island. In 1993, the two states asked the U.S. Supreme Court to decide whether Ellis Island was subject to either New York or New Jersey sovereignty and jurisdiction. GI system technology played a vital role in settling this long-standing dispute.[22]

Ellis Island was originally slightly less than 3 acres, but grew to 27.5 acres after the federal government filled-in the tidal waters around the island to house and process the immigrants. New Jersey claimed jurisdiction over the filled-in areas while New York insisted that the entire island belonged to her. The New Jersey Department of Environmental Protection (NJDEP) used GIS technology to determine the actual boundary line. As this case is one which is concerned with original jurisdiction, a constitutional provision gives the Federal Supreme Court exclusive (original) jurisdiction over lawsuits between states. The New York legal team objected to the use of GIS to determine both the size of the island and the boundary lines. However, the Special Master (a Justice of the Federal Supreme Court) overruled the objection since the GIS-generated data had already been explained and accepted in pre-trial documents and all parties had access to the data then. But, the Special Master disagreed with NJDEP's use of the mean high-waterline on the island as the state

---

[20] See presentation in Lindley, M 1995 'GIS Customer Service vs Legal Status', AURISA/SIRC'95—7th Colloquium of the Spatial Information Research Centre, University of Otago, AURISA NZ and Massey University, 26–28 April.

[21] See commentary by Castagna, RG, Thornton, LL and Tyrawski, JM 1999 'Where's Ellis Island? GIS and Coastal Boundary Disputes' *ArcUser*, v. 2(4), October–December, pp. 67–69.

[22] *New Jersey v New York* 523 U.S. 767 (1998).

boundary. It was argued and the court accepted that the low-waterline would be a more accurate boundary.[23]

In May 1999 the U.S. Supreme Court issued its final decree and approved the boundary line as mapped by the NJDEP. The Supreme Court decreed the following:

> It Is Hereby Ordered, Adjudged, and Decreed as Follows: The State of New Jersey's prayer that she be declared to be sovereign over the land-filled portions of Ellis Island added by the Federal Government after 1834 is granted and the State of New York is enjoined from enforcing her laws or asserting sovereignty over the portions of Ellis Island that lie within the State of New Jersey's sovereign boundary as set forth in paragraph 4 of this decree.

This could be the first use of GIS to present and implement a boundary dispute before the Supreme Court, but certainly will not be the last. The success and power of GIS in this case demonstrates its coming of age in legal adjudication since it is now considered a 'tool' for evidentiary purposes and to which its credentials have been accepted, much as speed radar guns have been accepted as legitimate 'expert' tools that record the speed of errant vehicles.

1.4.1 Databases

Many people are familiar with 'horror stories' of intrusions into electronic credit card databases that make headlines in the media. The horror is because such intrusions could relate to us personally and more importantly to financial liability through no fault of ours. For example, a computer hacker in the U.S. broke into a computer database containing about 8 million Visa, MasterCard and American Express credit card numbers, prompting an investigation by the Federal Bureau of Investigation (FBI).[24] The intruder had cracked the computer security of a firm that processes credit card transactions for merchants. While the credit providers have required merchants to encrypt cardholder information, one reassured its customers that they would be automatically credited for any unauthorised purchases while another found it unnecessary to inform its customers of

---

[23] The lack of 'metes-and-bounds descriptions' in the 1834 compact between the two states indicated that it merely applied to Ellis Island as it existed then and did not contemplate ownership of the expanded portions. A metes-and-bounds survey describes a land parcel with a closed traverse of courses. Also the common law doctrine of avulsion—sudden shoreline changes having no effect on boundaries—was applied in this case.

[24] Krim, J 2003a '8 million credit accounts exposed. FBI to investigate hacking of database' 18 February, p. E01 at http://www.washingtonpost.com/ac2/wp-dyn/A27334-2003Feb18?language=printer.

the security breach. Although it may be difficult to gather further personal information from credit card numbers alone, a criminal could use the information to impersonate the cardholder in the classic case of an identity theft.

A study of database security in the U.S. has observed that an overwhelming majority of states have failed to require insurance companies to protect their computerised data from hacking and other attacks.[25] Fourteen states have complied with Federal mandates to ensure the protection of computer systems that hold confidential information, twenty other states have no such policies, while fifteen states and the District of Columbia have pending regulation. There is no claim in the report that the insurance company's data are insecure, but it is noted that there is an alarming lack of a strategic plan to respond to concerted attacks on the security of the data.

In another example, Equifax Canada, a database aggregator, informed more than 1400 people that detailed personal data: social security number, bank accounts, credit histories, home addresses and job descriptions, may have been compromised in an attack on their computer systems.[26] Those affected were mostly located in Alberta and British Columbia, with a few in Ontario. It seems that criminals are very aware of the value of locational metadata, although the question may also be asked as to whether the attack was funded?

'Victims' whose personal details in databases have been compromised have also retaliated. The decision by the Australian Privacy Commissioner in the case between the Tenants' Union of Queensland (TUQ) and TICA Default Tenancy Control Pty Ltd represents the first successful class action in Australia against a company under the *Privacy Act* 1988 (Cwlth). The Commissioner had found that TICA had breached the *Privacy Act* thirteen times.

TICA is a private national register and operates much like a credit reporting bureau except that it collates and provides information on so-called bad tenants based on reports it receives from landlords and real estate agents.[27] TUQ complained that many people were disadvantaged by incorrect or unfair listings on the blacklist and suffered financial hardship

---

[25] Krim, J 2003b 'States seen as lax on database security. Study faults efforts to police insurers', 26 March, p. E05 at http://www.washingtonpost.com/wp-srv/resetcookie/front.htm.

[26] Suppa, C 2004 'Credit agency reports security breach', 17 March at http://www.computerworld.com/printthis/2004/0,4814,91319,00.html.

[27] Dearne, K 2004 'Black listed. A landmark decision has exposed databases to expensive litigation', *The Australian IT Business*, 27 April, pp. 1, 4.

and loss of access to rental property. TICA was allegedly overcharging tenants for accessing their own information and was lax in its approach to the accuracy and recency of the information, and slow to correct errors. The Commissioner ordered TICA to cease repeating practices outlined in four determinations and also issued a number of recommendations that TICA implement to remedy its information-handling processes.[28] While the ruling does not have the weight of law, such as a Federal Court decision, it is a public statement to be seriously considered by business and privacy interest groups.

On a humanitarian note, a database is being developed to chronicle war crimes during the Khmer Rouge era in Cambodia and is expected to underpin the prosecution of senior leaders of the Pol Pot's genocidal regime.[29] The Yale University Cambodian Genocide Database Project established in 1994 has created four databases of images, bibliographic, biographical and geographic information that show how and where 2 million Cambodians or 20% of the country's entire population died between 1975 and 1979.[30] The Cambodian government passed a law in August 2001 to establish a special court with international participants to bring to trial senior leaders of the Khmer Rouge.[31] While the data are to be primarily used as evidence in the trial, the aim of the Database Project is also to write a factual historical record and to help Cambodians learn of the fate of their family and friends and assist in a search for legal accountability. The methodology developed in this project is also being used in other parts of the world that have experienced the terrors of warfare, genocide and other modern tragedies.

1.4.2 Data Sharing

Conceptions about data sharing and sale, the proprietary nature and protection of data, and database cultures in the corporate sector differ throughout the world. A culture of sharing is pervasive if it is perceived that it is for the common good, whilst secrecy, confidentiality, and competitiveness prevail where there are economic imperatives such as profits and account-ability to shareholders. Sometimes such imperatives encourage cooperative behaviour because this subscribes to the 'private common good', that is,

---

[28] Findlaw, 2004 'Tenancy Database Operator Breaches Privacy Act', 20 April at http://www.findlaw. com.au/news/default.asp?task = real&id = 19703&newstype = L&site = NE.

[29] Brown, P 2002 'Khmer genocide in total detail', *The Australian*, 16 January, p. 20.

[30] Cambodian Genocide Database Project is at http://www.yale.edu/cgp.

[31] A draft agreement between the United Nations and the government of Cambodia concerning the prosecution under Cambodian law of crimes committed during the period of Democratic Kampuchea is at http://www.yale.edu/cgp/Cambodia_Draft_Agreement_17-03-03.doc.

benefiting only those who participate in the endeavour. At other times, supranational goals will dictate the fate of data where the creators are forced to deposit these to a common pool for all comers. With data, while it is of little use in its own right, when processed, value added to, and re-purposed, become an invaluable resource and asset.

The Canadian Oil and Gas GIS (CANOGGIS) consortium was formed in the early 1990s as a result of attempting to resolve the problem of searching for data.

Studies in the oil and gas sector in Western Canada in the late 1980s had indicated that exploration companies were spending about 60% of their time searching for data and only about 20% doing something useful with it. Since the formation of the consortium the cost of access to data had been reduced by a factor of 10 and the number of participants in the consortium had expanded from the original ten to approximately 50 in three years. Groot and Georgiadou (2001) have observed that this is a particularly interesting development given that the oil and gas industry is notoriously secretive about its data and information resources. Yet, in this case it moved decisively in the creation of an information infrastructure to support the sharing of data assets.[32] The decision to share data resources was decidedly more advantageous than to keep it secret and confidential.[33]

Under a system unique to Australia all seismic data obtained by private companies belong to them for just three years, after which the data are in the public domain. The data and reports must be submitted to the Designated Authority, Geoscience Australia (GA), to be available to all comers. Since 2001 all mineral, coal and petroleum exploration reports and data have to be submitted under the provisions of the *Petroleum (Submerged Lands) Act* 1967 (Cwlth) and equivalent state/territory legislation.[34] The information must be submitted in digital form and remains confidential until eligible for public release, currently after a period of three years (s 150 of the Act). Submission of data according to agreed standards of digital data formats makes the sharing across the subcontinent a cheaper and simpler process, with the data becoming available in a more timely manner.

While the systems for data sharing differ, the ultimate objective is still the realisation that GI is too important a resource to remain in private

---

[32] Groot, R and Georgiadou, Y 2001 'Advancing the concept National Geospatial Data Infrastructure: Reflections on the "bottom line"' at http://www.gisdevelopment.net/application/gii/planning/giip1001 3pf.htm.

[33] See a commentary at http://www.geoconnections.org/iacg/gis/ind/n235n.htm.

[34] The guidelines for the submission of data are at http://www.ga.gov.au/oceans/projects/psla_guidelines.jsp.

hands and that it may have more value in the public arena. The examples cited also show that data infrastructures can develop either privately in a cooperative manner or one which is imposed by policy and legislation. Also, if data are to be shared and easily accessible there must be clear industry standards used in the collection, maintenance and deposit of the data.

1.4.3 Maps

The nexus between maps and the law can be easy to establish if one were to accept the humble property plan as legal title. But in practice this is not as simple because to attain this legal status the property would have had to be surveyed by a licensed surveyor. The property plan would need to be lodged with and certified by the local Land Titles Office. To protect against possible tortious claims the surveyor would have some form of liability insurance covering the work. However, with other kinds of plans and maps no such legal liability exists because anyone can draw a map and its reliability and accuracy cannot be guaranteed. The mapmaker may make all efforts to ensure its accuracy, but no reliance may be placed on it. Even topographical maps produced by National Mapping Agencies carry a disclaimer, especially with reference to boundaries with the statement that the features shown on the maps are indicative only.

Consumer maps, such as those used for navigating cities and suburbs, tourist maps, special purpose maps showing location of facilities, may sacrifice accuracy for graphics and aesthetics for commercial reasons. One genre of maps that has to have a semblance of high accuracy is the street directory, whether in paper or electronic form. These directories have to be accurate because people rely on them. However, the publishers of such maps have also inserted 'traps' to catch the unwary map copier. These traps could be nonexistent streets or street names embedded in obscure regions of the maps and could be on every map. The practice is to enable instant identification of maps reproduced without permission. So the question now turns on proprietary interests and the law's support of it, rather than a document having force of law in terms of its content.[35]

Inserting indistinct 'watermarks' to signify provenance is a further method of laying a claim to ownership. In the digital environment, altering pixel (picture element) values or eliminating random selections of it on a raster geospatial database are positive acts that leave an audit trail for

---

[35] Macey, R 2001 'Nightmare on bogus street: A work of fiction in your glovebox', *The Sydney Morning Herald*, July 7–8, p. 1.

possible use in court. With image-based datasets it is possible to hide messages in a way that is not apparent to the observer. The technique known as steganography embeds secret messages either in other messages or in images.[36] This technique is used for copyright protection and to assist in keeping track of original images.[37]

Even with property plans there has been litigation to establish proprietary rights in the plans themselves. As there is an absence of case law in the Australasian context an example from a U.S. jurisdiction may prove instructive. Traditionally judicial decisions from the U.S. have limited or no persuasive value in like cases in other jurisdictions because of different laws and regulations. The case is used here for illustrative purposes, to demonstrate how the law has been applied.

An infringement of proprietary rights has been litigated such as *Albert R Sparaco v Lawler, Matusky, Seklly Engineers*.[38] In this case a surveyor who prepared an original site plan sued the owner, architect, builder and subsequently hired surveyor who prepared the revised site plan, alleging copyright infringement and breach of contract. The court had to decide whether the surveyor had a copyrightable interest in the site plan and whether the owner had breached contractual prohibitions on modification of the original site plan. The court held that the plaintiff's only right was a *copyright*—that is, a right to keep others from making copies of the site plan; and that the owner had breached that contractual prohibition. More generally, the case is a salutary message for those who believe that tracing and digitising from published maps will result in an original digital work. The reality is that the very act of digitising from copyright material violates the proprietary and moral rights of the owner. The only legitimate way is either to get permission or obtain a licence from the copyright owner to do so.[39]

---

[36] *Stego* means roof or cover and is used in the same context as in stegosaur—the roof lizard, because of the large bony plates that decorate its back; *graphy* means writing, thus, steganography means covert writing. See Cho, G 2002 'Now u c it, now you don't' *E-law Practice*, Issue 10 (October), pp. 45–46.

[37] Several types of steganography software are available 'free' on the Internet such as from http://www.petitcolas.net/fabien/steganography/stego_soft.html; and http://www.stego.com/index.html; hiding Jpeg images at http://www.cryptobola.com; e-mail at http://www.cryptohaven.com; digital watermarks at http://www.jjtc.com; and, photographs at http://www.clickok.co.uk. Most software are distributed 'copyleft' under the GNU General Public License (GPL) published by the Free Software Foundation. For GPL see http://www.gnu.org/licenses/gpl.html.

[38] U.S. District Court, S.D.N.Y. 60 F.Supp.2d 247 (1999).

[39] For limits on the copyright of maps see *Streetwise Maps Inc. v Vandam Inc.* 159 F.3d 739 (2d Cir. 1998) and compare this for new arrangements or presentation of facts in *Rockford Map Publishers Inc. v Directory Service Company of Colorado Inc.*, 768 F.2d 145 (7th Cir. 1985).

1.4.4 Global Positioning Systems (GPS)

One of the wonders of navigation in the 21st century is global positioning systems (GPS).[40] The system is based on a constellation of 30 NAVSTAR satellites circling the globe. The satellites give location, synchronise the world's mobile phone networks, help keep accurate time on atomic clocks and generally transmit other scientific information and Internet traffic to ground stations. Use of GPS receivers has now become commonplace among consumers such as on-board vehicle navigation systems in cars and on the wrists of athletes in the sport of orientation and rogaining—bush running. But such services giving one's position cannot be guaranteed. Originally, only military personnel had access to the highest levels of positional accuracy and 'selective availability' could be switched on to degrade signals over specified regions if necessary. However, in May 2000 selective availability was removed and everyone can now get accuracy of better than 10 m, even with inexpensive equipment. The E.U. GALILEO system costing more than €3 billion to build and launch will provide advanced positional technology and give accuracy readings of up to within a centimetre. As a civilian system, in theory the accuracy of readings cannot be degraded in any way and a key objective of this project is to provide a guarantee of a continuous service, uninterrupted by military needs.[41] An agreement has been reached June 2004 between the U.S. and the E.U. aimed at making Galileo compatible with the U.S. Navstar GPS. Under the agreement both systems will use the same open signal, allowing access by GPS users to both satellite networks with a single device. Both systems will retain the facility for encrypted military usage.[42]

One of the weaknesses of GPS is that a unit will fail to give accurate positional readings if it is unable to 'see' enough satellites—a minimum of three is required to triangulate position. Trees block out signals, as do buildings and atmospheric interferences. If the units are to be attached for tracking purposes on either people or vehicles, there are sensitive issues of privacy and the potential for infringing human rights.

---

[40] GPS strictly refers only to the U.S. Department of Defense system, while GLONASS is the Russian equivalent offering similar coverage and accuracy as does GALILEO—the E.U. system. See also Bartlett, D 2001 *A Practical Guide to GPS-UTM* at http://members.rogers.com/don.bartlett/gpsutm.htm [30 June 2004].

[41] See Editorial 2002 'Location, location, location. Global positioning is too vital to be left in the hands of the Pentagon', *New Scientist*, 30 March, p. 5.

[42] See Pocket GPS World 2004 'US and EU agree to link Navstar and Galileo GPS Systems' at http://pocketgpsworld.com/modules.php?name = News&file = article&sid = 370.

The use of GPS may also get people into trouble, for example, in connection with protecting locational information of private resources. On the Alabama Gulf Coast of the U.S. charter boat captains forbid fishers bringing GPS units on board. This is to protect the known locations of private reefs that the charter operators have built as secret spots for their fishing charter expeditions. Charter captains use GPS to pinpoint these reefs, as may unscrupulous people who use GPS units to steal the locations of the reefs for later use.[43]

There is a case of a hunter from Douglas, Wyoming who was recently acquitted from a charge of trespassing for 'corner jumping' to get from one piece of public land to another.[44] Corner jumping is the term officials use to describe stepping over the corner created where four sections of land meet in order to reach a cater-corner parcel of land (across a diagonal) without touching the other two parcels. The hunter had used his GPS unit to locate a surveyor's pin that marked the official corner of land parcels. He did not step in or physically touch adjacent private lands when he stepped over the corner to get to the other side. But, the landowner and rangers from the Game and Fish Department maintained that it was against the law in Wyoming and the hunter was cited for trespass for the purpose of hunting. The citation was challenged in court where the judge agreed with the hunter. This decision is likely to challenge the long-standing assumption that corner jumping is illegal. In reality this is also a challenge to Wyoming and U.S. common law that grants 'at least as much of the space above the ground as they can occupy or use in connection with the land'; as well as the ancient civil trespass that says that 'he who owns the soil owns upward unto heaven'.

The examples cited above illustrate the constant evolution of the law to keep it in step with technological developments and the challenges the technology may pose to initiate policy changes. As businesses reliant on positional accuracy require real-time round the clock access for telephony and other Internet services, any loss of service can be catastrophic. Hence, the move away from the U.S. military GPS to private systems such as GALILEO may be justified on commercial and policy grounds. But there are also limits to the use of GPS to preserve privacy especially where these are used as tracking devices.

---

[43] See *The Birmingham News* 2004 'To catch a thief, charter captains outlaw the GPS' 25 April, at http://www.al.com/printer/printer.ssf?/base/sports/10828.html

[44] Luckett, B 2004 'Case could open public access' at http://www.casperstartribune.net/articles/2004/04/11/news/wyoming/df0fc4b4ae49db6287256e73001aeff2.prt. See Chapter Four *infra*. subsection on Privacy Risks: Location and Tracking Technologies.

1.4.5 Aerial Photographs and Images

The issue of privacy, informational and personal, has been raised several times before. Geography in all its manifestations adds further imbalance to the tension between private and public interests. Aerial photography as a tool of photogrammetry and remote sensing has provided pictorial geographies of the environment with data to be interpreted, measured and analysed. Sometimes these images can merely be a photographic record of the landscape.[45] One example is the California Coastal Records Project. This is a website that provides an aerial photographic survey of the California coast for scientific and other researchers.[46] It is a non-profit organisation and has captured 12 200 images for anyone to view and use. The entire California coastline has been photographed from a small helicopter—one picture every 500 feet (160 m)—from the Golden Gate Bridge to the Hearst Castle. The intention of the project is to provide a baseline for conservation and other land use researchers interested in a detailed record of the coastline. However, even with such altruistic objectives, there have been perceptions that personal privacy may have been violated. A prominent example is that of the Hollywood actress Barbra Streisand.

Ms Streisand sued the photographer and two other defendants for US$10 million, claiming that the pictures they provide to others of her Malibu home and estate violate her right to privacy. The lawsuit filed in May 2003 alleges five counts of privacy intrusion, and violation of the state's anti-paparazzi Act.[47] The suit seeks to stop the defendant from disseminating the photographs which use 'enhanced technology' and deprive her 'of the economic value of the use of the images of her property and residence'. In December 2003 a Los Angeles Superior Court decision reaffirmed the public's First Amendment right to participate in matters of public significance.[48] The court also held that Ms Streisand had abused the judicial process by filing the lawsuit and rejected her request for an injunction to remove the panoramic photographic frame of her bluff-top home and property from the Coastal Records Project.

---

[45] See Lillesand, TM and Keifer, RW 1999 *Remote Sensing and Image Interpretation* (4th edn), New York: John Wiley & Sons.

[46] See http://www.californiacoastline.org.

[47] See documents at http://www.californiacoast.org/streisand/lawsuit.html. California Anti-paparazzi Act is codified by the California Civic Code § 1708.8.

[48] *Streisand v Adelman* Case No. SC 077 257. Cal. W.D. 31 December 2003. At http://www.californiacoastline.org/streisand/slapp-ruling.pdf.

The U.S. First Amendment free speech protection and the private–public interest tension was drawn into this case. The defence argued that the photographs were taken in a public place in which Ms Streisand did not have a reasonable expectation of privacy. Also, the law of privacy, even with the paparazzi extensions of it (in California at least), is not about taking pictures of structures, it is about people. It is ironic that before the case, very few people would ever have known that the picture to Ms Streisand's home existed. After this litigation many will become aware of it, if only because of its 'celebrity' status.

In the U.K. the law has been tested in the continuing tension between a right to privacy (Art. 8) and freedom of expression (Art. 10)—articles spelt out in the *Convention for the Protection of Human Rights and Fundamental Freedoms* and included as a Schedule to the *Human Rights Act* 1998 (UK). In April 2003 the High Court decided in *Douglas v Hello! Ltd*[49] that there was no freestanding right to privacy. This litigation concerned the unauthorised publication of wedding photographs of Catherine Zeta-Jones and Michael Douglas. The court held that the law of confidence was sufficient to protect people in the Douglas' position. The court used the analogy of people who traded on their image rights and that of a manufacturer trying to protect confidential trade secrets.

Further judicial clarification on protection of confidences and privacy has been given in the Naomi Campbell and *Daily Mirror* case. The *Daily Mirror* had published a photograph of the supermodel's attendance at a narcotics support group. Ms Campbell sued for breach of confidence. MORLAND J at first instance in the High Court ruled in favour of the super-model.[50] On appeal, the Court of Appeal found that the disclosure was not in breach of an obligation of confidentiality.[51] The Appeal Court found that Ms Campbell's Art. 8 right was overridden by the newspaper's Art. 10 right, since disclosure of her drug abuse problem was in the public interest. However, the House of Lords overturned the Court of Appeal judgement by 3:2 and ruled that the *Daily Mirror* had violated Ms Campbell's right to privacy. LORD HOPE of CRAIGHEAD said that '[d]espite the weight that must be given to the right to the freedom of expression that the Press needs if it is to play its role effectively, I would hold that there was

---

[49] [2003] All ER (D) 209 (April).
[50] *Campbell v Mirror Group Newspapers* [2002] All ER (D) 448 (March); [2003] QB 633. See also http://www.lawreports.co.uk/qbmarb0.3.htm.
[51] *Campbell v Mirror Group Newspapers* [2002] All ER (D) 177 (October).

here an infringement of Miss Campbell's right to privacy that cannot be justified'.[52]

In the U.K. at least, the debate as to whether the law of breach of confidence meets the demands of the 21st century or whether a new law of privacy should be developed continues. The Campbell case may, however, be decided differently in Australasia and the U.S. where there may be laws and interpretations that may produce unanticipated results. Nevertheless, the example underlies the right of people to maintain important elements of their privacy, particularly in relation to therapy and those requiring treatment.

Copyright issues also feature in photographic images. Buying one set of photographs and reproducing these, either by photocopying or re-scanning/digitising would violate fair use principles and an owner's intellectual property rights. In *Images Audio Visual Productions Inc. v Perini Building Company Inc.*[53] a holder of copyright on aerial photographs showing a construction site's progress sued a contractor for infringing copyright. The contractor had made duplicates of the aerial photographs by photocopying, and distributed these to participants in an arbitration dispute rather than ordering extra copies from the photographer. The District Court agreed with the owner and held that the contractor did not make fair use of the photographs. In a similar case in *Tiffany Design Inc. v Reno-Tahoe Specialty Inc.*[54] the holder of copyright on aerial photographs of a city's entertainment centre sued a competitor in the souvenir business for alleged violation of copyright in a derivative work. The alleged infringer had engaged in copying by scanning copyright photographs and loading the result into a computer for further manipulation. The District Court held that the alleged infringer had not made fair use of the scanned elements.

As in the case of maps, the creation of intermediate copies of images by digitisation, that is, digitising of copyright images in order to manipulate or modify them, could be infringing of copyright. The reason is because the first step would involve a reproduction right that is exclusively in the hands of the owner. U.S. intellectual property law would protect pre-existing original expression.[55]

---

[52] Agence France Presse (AFP) 2004 'Court finds against tabloid. Campbell privacy case win', *The Canberra Times*, 8 May, p. 20; and also *The Guardian* 2004 'Tabloid's fear Naomi's court win will spell end for exposés', *The Sydney Morning Herald*, 8 May, p. 15.

[53] 91 F.Supp.2d 1075 (12 April 2000, U.S. District Court, E.D. Michigan S.D.)

[54] 51 U.S.P.Q.2d 1651 (12 July 1999, U.S. District Court, D. Nevada.)

[55] 'A publication of a derivative work would also constitute the publication of a pre-existing work upon which it is based.' Nimmer, MB and Nimmer, D (eds) (1993) *Nimmer on Copyright* § 4.12(A) Bethesda, MD: Matthew Bender Publishers and Lexis Nexis.

1.5 Geography Really Does Matter

At the start of the Internet boom the hype was the 'death of distance' and that geography no longer mattered once you were connected to the Net. The Internet was 'ageographic'. These thoughts may have brought great comfort to Australians and others living in vast uninhabited territories where the 'tyranny of distance' has reigned.[56] But Australians would call this a 'furphy'—a rumour without foundation.[57] Geography does matter and geography is important even on the Internet.[58]

There would be no workable law if there were no jurisdiction. The jurisdiction defines both the geographical reach of the law and as the authority that gives the law legitimacy. Those boundaries that are drawn are dependant on geography, and this is one reason why in the cyber-world there needs to be supranational bodies that oversee the policing and the enforcement of regulations on the Internet. The difficulties of enforcement are perhaps the result of ambiguities in the boundaries that are used in cyberspace. The situation is no different in the real world.

International boundaries are important for establishing an authority and a national identity.[59] In most cases the symbolic significance in terms of nationhood and international legitimacy is of the highest order. In some cases, however, boundaries are a source of friction between the states which they separate.[60] Treaties define most boundaries and maps play a prominent role in showing where these boundaries actually lie. But, conflicts arise where the boundaries are ambiguous and open to interpretation. Expertise in using geographic evidence, either by the disputants or by the legal fraternity involved, can often lead to protracted arguments and delays in the resolution of conflicts. Moreover, boundaries are complex entities and require a multidisciplinary approach, both in definition and interpretation. In treaties, maps showing agreed lines form an integral part of the legal definition of a boundary. Sometimes the map itself is *the* legal

---

[56] A phrase made popular by the eminent Australian historian Geoffrey Blainey 1967 in his book *The Tyranny of Distance*, Melbourne: Pan Macmillan Australia.

[57] Wilkes, GA 1978 *A Dictionary of Australian Colloquialisms*, Sydney: Fontana Books, p. 150.

[58] See Dabson, J 2002 'The "G" in GIS—What's New about GIS?' at http://www.geoplace.com/gw/ 2002/0203/0203gngs.asp [23 February 2002]. *The Economist 2003* 'The revenge of geography' March 13. Also at http://www.economist.com/science/tq/PrinterFriendly.cfm?story_ID=1620794 [29 March 2003].

[59] See discussion in Pratt, M and Donaldson, J 2003 'Drawing the Line: Mapmakers and International Boundaries', 2003 Cambridge Conference, Southampton: Ordnance Survey, Paper 4D.3.

[60] For some examples and a discussion see Monmonier, M 1995 *Drawing the Line. Tales of Maps and Cartocontroversy*, New York, NY: Henry Holt & Coy Pub., Ch. 4 Boundary litigation and the map as evidence, pp. 105–147.

document while at other times the maps serve only as an illustration of what is intended with the result that different types of maps have different evidential value.[61]

A mapping of the geography of the cyberworld too can help us understand the virtual world at a glance and to navigate our way through the electronic mail (e-mail) that arrive daily in our in-boxes. Dodge and Kitchen's (2001) atlas provides a literal view of the geography of the Internet, based on the premise that the infrastructure, the servers and the people using it are 'somewhere' on the globe.[62] Such an atlas helps us make sense of the world since the electronic space and Internet are another kind of unchartered territory, ripe for mapping. An appreciation of the topology might help engineers design a better network for the future, it may help facilitate statistical analyses of the Internet, visualise it as a map, and even develop commercial applications. The term 'spatialisation' is used to describe such a process where spatial map-like structures are imposed onto data, even if the data do not have any inherent structure or where the structure is not obvious and devoid of reference to geographical space. Two- and three-dimensional spatialisations may produce more legible and intelligent views of the data.

One application of this idea of mapping spatialisations in two and three dimensions are the data mining techniques employed by NetMap Analytics.[63] The algorithms automate a link analysis process through which business intelligence is gained. Business intelligence is the complete and timely understanding of each and every factor that drives business and how these factors interrelate. NetMap Analytics attempts to find subtle, hidden connections required to reveal true insights by *showing* the analyst what the interconnections are, why they are connected and what they mean. Gaining insight in time to take appropriate action is effective business intelligence. NetMap Analytics' algorithms cull large amounts of data to find relevant connections between seemingly unrelated entities. No matter how subtle, indirect or hidden, the resultant patterns, trends and chain of events that emerge are displayed in a graphical format. These charts are similar to the 1970s Spirograph toy where geometric, regular patterns can be drawn. However, on NetMap charts, if something is off-centre or pronounced, it is immediately obvious to the naked eye and the analyst can

---

[61] Rushworth, D 1999 'Geographic support to courts and states involved in boundary dispute settlement' in Dahlitz, J (ed.), *Peaceful Resolution of Major International Disputes*, New York, Geneva: United Nations, p. 172–173.

[62] Dodge, M and Kitchen, R 2001 *Atlas of Cyberspace*, Reading, MA: Addison Wesley.

[63] See http://www.netmapanalytics.com/other.asp?heading = 1.

focus on a particular node to examine why the links to it are particularly strong. This technology is being used in international intelligence, law enforcement agencies, insurance companies, fraud detection agencies, targeted advertising and marketing, and the airline industry.

Summary

Geographic information, as an intrinsically valuable resource, feeds in various ways into an information system technology. This technology is a discipline of study in its own right and a science of knowledge. In Australia, for example, there is no overarching policy other than the idea of a national spatial data policy after shortcomings were uncovered following a natural disaster. The complexities in formulating such a policy may be appreciated given the diversity of GI systems that is characterised by a multiplicity of actors in the collection, distribution and use of GI data. This chapter has laid out the major policy issues directly relating to GI to serve as a starting point for debate, analysis, and re-examination.

The key policy issues for GI therefore are access, economics, and protection. In detail, these would depend on government objectives for a spatial data policy as well as the need to manage the growth in volume, quality, and use of the data that are being collected. A further factor is the digitisation of the data, making it a product that has become more widely available than ever before and which arguably is easier to use where the data conform to international standards. A data pricing policy may lay the framework for data access, distribution, dissemination, and use. Such a policy reflects the compromise in the tensions between the data producer, and users—public, private, and others—as well as other competing suppliers and substitutes. A review of the legal issues that have data policy implications will show the need to protect data as an element of an asset and property.

It is conceded that the law touches all our activities and that there is a geography to the law. While the law may affect the way GI systems are developed, GI systems have also redirected application of the law. As discussed, databases might be an intrinsically good thing, but when used by those with evil intentions, protection of the content is necessary. Data use practices such as sharing and sale may be determined both by legislation as well as by cooperative agreements. The issue of intellectual property rights may yet have no easy resolution from the simple digitising of published maps that may infringe proprietary and moral rights to the use of licences. This is also the case of using aerial photographs without licence.

In the final analysis, geography does really matter. There would be no workable law if there were no jurisdiction. The jurisdiction both defines the geographical reach of the law as well as the authority that gives the law legitimacy. International boundaries are dependant on geography and in the cyber-world there is now a need for supranational bodies to police and enforce international regulations on the Internet. The difficulties are many because in cyberspace 'there is no there there'.

Chapter 2
Sharing Geographic Information and Data

Learning Objectives

After reading this chapter you will:

- Be able to describe how public sector managers of geographic information (GI) deal with intellectual property, legal liability, access, and marketing issues in their work.

- Know some of the main ways in which the sharing and commercialisation of geographic information may be achieved by the use of exchange standards and frameworks.

- Be able to compare how geographic information is disseminated in Australia with international practice.

- Obtain an outline of the international framework for the development of access policies, including standards and metadata.

- Appreciate the nature of the various proposals for a development of a global information infrastructure (GII) and the national spatial data infrastructure (NSDI).

Geographic Information Science: Mastering the Legal Issues   George Cho
© 2005 John Wiley & Sons, Ltd   ISBNs: 0-470-85009-4 (HB); 0-470-85010-8 (PB)

2.1 Introduction

Public sector information (PSI) resource managers would prefer nothing better than to show that the information resources that they provide are accurate, reliable and of the highest quality possible. However, with the evolution and growing maturity of the spatial information industry, these managers have now to turn their attention to deal with issues of the ownership and legal responsibilities arising from the provision of information resources. Key proprietary questions arising from licensing agreements, intellectual property, copyright, legal liability, pricing, access, and marketing have brought new challenges and imperatives to the operations of an information agency. These come about when spatial information products and services are shared, exchanged, bought, and sold in any resource network.

This chapter is about the sharing and commercialisation of GI and data. There is a certain degree of altruism in suggesting the sharing of GI data since there may be benefits to both parties involved. Equally, the commercialisation of GI by way of sales for cost recovery purposes and to generate profitable revenue streams brings benefits to all, the seller as well as the buyer. Whatever model of disseminating GI and data is adopted there is a reflection of both the underlying data policies extant in that jurisdiction and the legal regimes governing all such interactions and transactions. However, before any of these can take place there may be other considerations such as exchange standards and frameworks. Standards, for instance, provide a measure of quality assurance, whereas metadata provide information of what is contained in the data and descriptions of the data and content. Data discovery services tell where to find the data and information. The framework of exchange relations may be governed by way of mutual agreement, contracts and licences, or simply as a commercial transaction. Traditions differ in various parts of the world as do philosophies and rationales for sharing and commercialisation of GI and data. Ultimately, of course, in a globally interconnected world, the ideal would be one where there is universal sharing and use of GI data and an infrastructure supporting these activities, and where commodification may no longer be an issue.

There are four parts in this chapter. In the first part, GI as a resource, asset, commodity, and infrastructure is explored in its various guises. How GI is viewed may provide an insight into the developing and evolving policies that surround its use, sharing and sale as a good in the data marketplace. As expected, the access to and commercialisation of spatial information will depend largely on its quality and availability. Associated with the quality of the information and data are issues of liability on the one hand and its protection as an asset on the other. These need to be clarified before examining other substantive issues. The second part addresses the data policies and legal frameworks for accessing data. Australian experiences in data dissemination for public geospatial data are used as the basis for discussing the general topic of whether to share data or to sell data to recover costs. International comparisons are then made with the U.S. open records policy for federal data and other policies adopted by some of the states. The European Union's public sector information (PSI) policies then provide alternative viewpoints to the topic. A third part outlines the international framework for developing access policies by way of standards and the use of metadata, clearinghouse and registries to facilitate exchange, sharing, sale and use of GI. The framework is seen as a means of promoting the interoperability of GI worldwide. The final part discusses the possible scenarios for the development of a global information infrastructure (GII) and where a global spatial data infrastructure (GSDI) sits within this conceptualisation. In this discussion the development of national spatial data infrastructures (NSDI) beginning in the U.S., and then followed by Australia and New Zealand and the E.U. provide useful pointers for a vision in 2020.

2.2 Sharing Geographic Information and Data

A 1998 survey of the geomatics industry by Statistics Canada ascertained that there were 1614 mapping and surveying service companies in Canada with total annual revenue of CAN\$615 million. A broader definition, including GI system software and navigation and positioning applications, accounts for about CAN\$2 billion in revenues and employing 22000 people. Geomatics covers the disciplines of surveying, mapping, remote sensing and GI processing. Also included are global positioning systems, geodetic, cadastral, engineering and marine surveying and mapping activities as well as the creation and maintenance of spatial and GI systems. The world market for geomatics in 2001 was estimated to be US\$24 billion and the industry is said to be growing at a rate of approximately 20% per annum.[1] An accurate statistical definition of the industry is difficult because of the vast array of companies involved and the different degrees of participation of each.

---

[1] Industry Canada 2001 'Geomatics Industry' at http://www.strategis.ic.gc.ca.

A study for the E.U. in 2000 estimated the economic value of PSI in Europe at €60–70 billion per annum, of which over half was accounted for by GI that included mapping, land and property, meteorological services, and environmental data.[2] A study in the previous year by OXERA (1999) indicated that whilst the turnover of the national mapping agency (NMA) Ordnance Survey in the U.K. was approximately £100 million, the gross value added of the businesses it underpins in the country was approximately £100 billion.[3]

The statistics suggest that there is a growing and vibrant GI industry contributing to the economies in different parts of the world. Actual and perceived needs of users and producers alike and the availability and access to relevant GI and data account for this vibrancy. To this may be added the sharing of GI data and where possible of publicly available data at little or no cost. Most of such data are available freely, at the cost of transfer, from the public sector in some countries. GI as information *per se* is intrinsically valuable in terms of its usefulness and functionality. To know where one is in regard to other frames of references such as the home, the workplace, the business centre is useful enough in itself. To be able to use this knowledge of location and link it to activities and a time frame makes the information eminently functional because one can then operate effectively and efficiently within it. The value of geospatial data is realised through its usage. The more it is distributed the more it comes to be used. The more use that is made of the data, the more value is given to it.[4]

The *sharing of data as a resource* reduces the cost of data collection, maintenance and updates to all participants. Each participant may undertake a small part of the cost of data activity and the whole burden is proportionately shared. Each participant can contribute to the updates and make corrections to the common data source. This means that the original data source may be maintained more effectively, and, where the activity is properly organised, can be self-sustaining and grows in proportion to contributions and other value adding activities. Sharing data also ensures

---

[2] PIRA International Ltd, University of East Anglia, and Knowledge Ltd, 2000. *Commercial Exploitation of Europe's Public Sector Information*, Final Report for the European Commission, Directorate General for the Information Society, Luxembourg: EC DG INFSO. See also ftp://ftp.cordis.lu/pub/econtent/docs/commercial_final_report.pdf.

[3] OXERA 1999 *The Economic Contribution of Ordnance Survey GB*, Oxford Economic Research Associates Ltd at http://www.ordnancesurvey.co.uk/literatu/external/oxera99/contents.htm.

[4] Onsrud, HJ and Rushton, G (eds) 1995 *Sharing Geographic Information*, News Brunswick, NJ: Rutgers, The State University of New Jersey, Center for Urban Policy Research, pp. xiv and 502.

that the data will be more comprehensive and beyond that which may be available from anyone or any agency.

It is instructive to note also that the U.S. National Academy of Science (NAS) has been grappling with the thorny issue of sharing published works and its supporting data. The National Research Council Committee on Responsibilities of Authorship in the Biological Sciences has offered a draft resolution which says that it is the author's responsibility to take reasonable efforts to make data and materials integral to a publication reasonably and promptly available in a manner that furthers science. The proposed guideline not surprisingly has sparked immediate debate among scientists, researchers and other academics.[5]

In discussing the role of law in either impeding or facilitating the sharing of GI, Harlan Onsrud (1995) has said that while information and the knowledge it brings is a source of power, that 'power which information provides is antipathetic to sharing'.[6] Moreover this desire to control information is in direct conflict with technological developments that make it easy to copy, disseminate and share information inexpensively. Other barriers to sharing and the re-use of information include cultural, institutional and legal matters that either prevent or delay the use of existing spatial data. Policy dictates such as full cost recovery may be in place and hence precludes any sharing of public data. Technological requirements may mean that there is a lack of interoperability and hence difficulties in sharing, even if there were no policy impediments. The behaviour of people and institutions may also discourage the sharing and reaping of the collective benefits from the data, even if it is of little or no economic advantage not to do so. Thus, the factors and strategies that appear to assist the greater sharing of data and those that inhibit such sharing have to be further explored.

In business and market terms, a commodity, as an article of trade and an intrinsically useful thing, is one where everyone in the market is selling the same good. With this understanding, there is a view that the *commodification of GI data* has yet to occur. Unlike electronic games, music and other commodities, the market for GI data is not sufficiently large, as the trade statistics given previously have shown. There is no general market for GI data where every consumer needs it and neither is it

---

[5] Garretson, C 2002 'Whose data is it, anyway?' at http://www.itworld.com/Tech/2987/020409whosedata/pfindex.htm.

[6] Onsrud, HJ 1995 'Role of Law in impeding and facilitating the sharing of Geographic Information', in Onsrud, HJ and Rushton, G (eds) 1995 *Sharing Geographic Information*, News Brunswick, NJ: Rutgers, The State University of New Jersey, Center for Urban Policy Research, pp. 292–306 at 293.

required everywhere. But, recent trends indicate that the use of GI data for mapping purposes as a consumer product is slowly creeping into general usage, especially on the Internet where it is widely available. At present while cost considerations may prevent the widespread use of GI data, for instance, in on-board car navigation systems, a maturing product cycle may see its use in every road vehicle in the near future.

Cost considerations aside, it has been argued that *GI as a commodity* is like no other asset or resource and its special characteristics make it significantly different from other commodities. The 'public goods' dimension of GI make it a unique commodity in that it can be beneficial to everyone. This feature means that the information can be easily shared with one and all and can have multiple uses in various applications. Some kinds of GI remain static and do not change over the short term, for example, soil profiles or contours showing elevation, but others may use this same information in different contexts and in a dynamic fashion. For example, the changing volumes of water flowing in a river system that is shown as a static feature on a map.

GI is almost everywhere and unlike other commodities the information remains in the hands of the seller even after sale. However, GI 'leaks' and from a proprietary point of view ownership and its protection can be a problem. While GI may be compressed it is equally difficult to define what a unit of information refers to. On the other hand, GI may be infinitely expanded and this raises the opposite problem of how to gauge its value. Some information becomes obsolete and thus lose value while gaining an historical one instead after archiving. Barr and Masser (1967: 243) give the example of the non-substitutability of GI in the case of a geographic referencing standard to establish a master address file in the U.K. (BS 7666). Since only one authority can be responsible for the operation of this master address file a natural monopoly thus exists in the creation of geographic references. Such a monopoly may also be present in topographic mapping.[7]

There may, however, be a commercial market for GI and data, the former by way of information systems products and the latter in the form of derivative, value-added data. Value adding may be an important niché market that may generate significant revenues in the GI market. But this market is heterogenous in that the needs of the consumers differ by type,

---

[7] Barr, R and Masser, I 1997 'Geographic information: a resource, a commodity, an asset or an infrastructure?' in Kemp, Z (ed.) *Innovations in GIS 4. Selected Papers from the Fourth National Conference on GIS Research UK (GISRUK)*, London: Taylor & Francis Ltd, pp. 234–248.

by scale, and by quality. Different applications may require different types of data, depending on its use and application. Fine-grained data may be required for planning suburban streets while region-wide catchment area studies could do with medium-grained data. For generating maps on the Internet data of a lower quality may suffice, if only to show national spatial trends, say for weather charting purposes. These together mean that the GI data provider may have to differentiate the product to meet specific needs—one size would necessarily not fit all.

Nevertheless, the conclusion is that there is a market for GI and data in the commercial sector, and the major players include government data providers as traders as well as the private sector as buyers. But unlike traditional economic resources such as land, labour, capital and entrepreneurship where market forces may dictate their exchange, the uniqueness of GI requires a different set of controls such as those identified by Cleveland (1985).[8] The very nature of GI cries out for attention in its own right when it is used, shared and traded. These unique qualities of GI—as a resource, as a commodity, as an asset and an infrastructure are explored further in Barr and Masser (1997).[9]

That information is an *asset* is unarguable whether it is to an individual, a company, a government agency or to society as a whole. What might be more problematic is when information is withheld with reason, for example personal information and what responsibilities flow when private information is in the hands of an information officer. Issues that these stewards and custodians of information have to contend with include the rights of access to individuals, the maintenance and upkeep of the information, and how selected information may be manipulated for use without divulging details of individuals. The idea of custodianship in the Australasian context is that the corporate entity and not the separate parts comprising the entity owns the information, and that information collected, produced and maintained by the part is available to the whole. In protecting government investments in commercially valuable products the custodianship and *public trust* argument is used. In this circumstance the data and information are not owned by independent government agencies, but rather the agency become custodians for and on behalf of the state and

---

[8] Cleveland, H 1985 has identified six unique qualities of information that make it unlike other economic resources in terms of its expandability, compressibility, substitutability, transportability, diffusiveability and shareability. See Cleveland, H 1985 'The twilight of hierarchy: Speculations on the Global Information Society', *Public Administration Review*, January–February, pp. 185–195.

[9] *op. cit.* Barr, R and Masser, I 1997.

held in trust for the benefit of the entire public. Any *trust corpus* may not be given away or sold by the custodian.

An *infrastructure* may be understood to mean the basic facilities, services and installations that underlie the functioning of a society. Longhorn (2001) suggests that closer examination will show that virtually all infrastructures created by society have some elements in common.[10] These commonalities include:

- high-level policies that set the overall goals and objectives for creating the infrastructure;
- implementation technologies;
- standards that guarantee various levels of interoperability for the components within a single type of infrastructure as well as across related infrastructures;
- rules and regulations;
- resources to create the infrastructure, and to operate, maintain, and enhance it over time.

We may readily identify roads, railways, telecommunication systems, power, water, and school systems as part of the infrastructure of a modern society. To this list may now be added GI as part of the infrastructure that is used by public agencies for governance and social good. Geographic reference information is vital for mapping purposes but its intrinsic value comes into its own when integrated with other information that is then used for planning and other decision-making to improve the social fabric. This 'public good' feature of GI therefore means that it has to be maintained and nurtured as a public responsibility despite the high costs of development and maintenance of fundamental data sets and where the private sector and marketplace may have little or no role to play. It also implies that there will be a certain degree of regulation as well as standardisation when using GI in building the infrastructure.

When incorporated within a spatial data infrastructure (SDI) the common elements of GI noted previously imply that one is able to access the spatial data via existing or planned infrastructures. This is in addition to the ease with which that access is permitted, not only in terms of locating the information, but also in acquiring the information as well as using and re-using the information. SDIs are also the coordinating and control structures that develop and maintain datasets and make the data and information

---

[10] Longhorn, RA 2001 'The impact of data access policies on regional spatial data infrastructure' available at http://wwwlmu.jrc.it/Workshops/7ec-gis/papers/html/longhorn/longhorn.htm.

accessible through the system. SDIs will also have dependencies upon other infrastructures—telecommunications, commercial, information technology (IT), and legal—and may at the same time underpin other infrastructures, for example navigation and transport systems. Any definition of SDI therefore depends on the context and the frame of reference of the person or agency proffering the definition. The policy and strategies that follow subsequently depend on particular national contexts.

We should conclude this section with a quotation from Barr and Masser (1997: 247) who have observed that the formulation of national GI strategies 'will vary from country to country because of the different institutional contexts that govern information and geographic information policy making. This will be reflected in the choices that are made about the mix between public and private sector involvement and between public interest and cost recovery in each case...'.[11] The decisions made in regard to the mix of involvement, cost recovery, and the balancing of interests by governments in Australia, the U.S. and the E.U. are examined in the next section.

2.3 Policies on Access to Public Sector Information (PSI)

This section presents the debates advocating free access to government information as against a cost recovery regime and the positions in between. This section is also concerned with the funding and charging mechanisms relative to the dissemination and use of GI. Blakemore (2001) has paraphrased these funding and financing 'religions' as either of privatisation, nationalisation, or commercialisation; or that liberation is good. The mantra is that for some, markets are best, and for others facts should be free, whereas strident critics claim that the 'commons' are tragic. Proponents of these positions reside on opposite sides of the Atlantic. As shown in Table 2.1 there are large differences in the implementation of general policies on cost recovery and the different philosophies that advocate open records and user pays—U.S. and E.U. respectively, together with an emergent mixed model from the Antipodes. But such a characterisation may be overly simplistic and hence careful description of the different pricing, access and marketing of public sector GI is necessary. This section addresses the different

---

[11] *op. cit.* Barr, R and Masser, I 1997, p. 247.

Table 2.1 Guidelines for cost recovery in selected jurisdictions

Jurisdiction	Policy
International	
Canada	*Cost Recovery and Charging Policy*, Treasury Board of Canada, 1997
New Zealand	*Guidelines for Setting Charges in the Public Sector*, NZ Treasury 2002
OECD	*User Charging for Government Services*, OECD 1998
UK	*The Fees and Charges Guide*, Treasury UK 1992
US	*Circular A-25 Revised*, Office of Management and Budget 1993
Australia	
New South Wales	*Guidelines for Pricing User Charges*, NSW Treasury 2001
Victoria	*Guidelines for Setting Fees and Charges Imposed by Departments and Budget Sector Agencies*, Victoria Treasury and Finance 2000
South Australia	*A Guide to the Implementation of Cost Reflective Pricing*, SA Treasury and Finance 1998
Tasmania	*Costing Fees and Charges: Guidelines for Use by Agencies*, Tasmania Treasury and Finance 1998
Western Australia	*Costing and Pricing Government Outputs*, WA Treasury 1998
Queensland	*Full Cost Pricing Policy*, Queensland Treasury 1996

regimes for access, cost recovery, and cooperative ventures in Australasia, the U.S. and the E.U. A comparison of the different philosophies in particular circumstances will be illuminating. Such circumstances as the purpose, the legal authority, fiscal and economic framework, and the nature of the product itself may explain the divergent solutions, paths and models.

To begin with, two common objectives of a cost recovery regime may be readily observable. The first objective is a promotion of more efficient allocation of resources. User charges will gauge market demands as well as eliminate frivolous requests at the same time empowering consumers—'user pays, user says'. Secondly, cost recovery may promote greater equity in that the beneficiaries of the products and services pay for the privilege instead of costs becoming an impost on the general public and not contribute to revenue raising. But practices and policies differ among the jurisdictions and hence a separate treatment for each is necessary. Figure 2.1 provides an illustration of the various kinds of cost recovery models that have been proposed for sharing public geographic information and data.[12]

---

[12] The advantages and disadvantages of the various funding models are evaluated in greater detail in Harris, R 1997 *Earth Observation Data Policy*, Chichester: John Wiley & Sons, pp. 110–125.

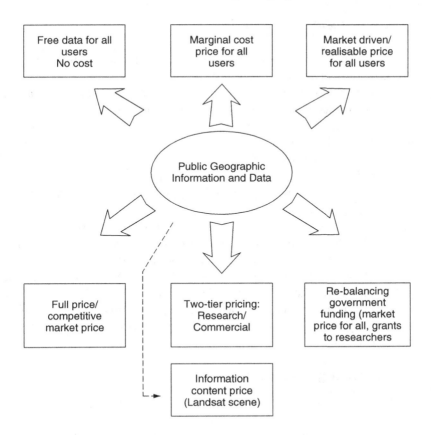

Figure 2.1 Models for sharing public geographic information and data. Source: developed from Harris 1997: 110–125

At the conclusion of this discussion on the sharing of public GI and data we should be able to provide some answers to the following set of questions.

- What is the legal authority and purpose of providing data access to the public?
- Which information service activities are mandatory and which are discretionary?
- What is the fiscal and economic framework for such activities?
- Should public agencies merely cover costs, make a profit, or even charge a fee?
- What is the role and competitive behaviour of the government in the marketplace?
- What is the role of the information industry and private companies?

2.3.1 Australia–New Zealand

As elsewhere there are important economic, social, and political reasons for the government to be involved in the provision of information and information products. The public good characteristic of many information products and services being non-rivalrous and non-excludable means that only governments are best placed to provide such products.[13] For some types of information products there are the so-called spill-over effects, since it is unlikely that the marketplace is in any position to provide these sorts of products. Then, there may be natural economies of scale and scope for such products to be provided by governments and other public bodies.

The rationale for cost recovery by information agencies is based on the fact that the consumption of information products is usually discretionary.[14] Only those who have a need for the data and information may seek these out from a public agency. While the cost of collection, assembly and compilation of a dataset can be very high, the cost of dissemination is low, other than the cost of reproduction. In such circumstances, to impose cost recovery charges could impede the desirable use of the information.

Purpose and Legal Authority

In Australia a legal and fiscal framework underpins the design and operation of cost recovery arrangements. The Commonwealth Constitution places legal constraints on the nature of charges that may be made by government agencies. For instance there is a distinction between taxation and fee-for-service. Taxation is generally 'a compulsory exaction of money by a public authority for public purposes enforceable by law, and . . . not a payment for services rendered'.[15] A fee-for-service refers to a direct charge for the provision of a good or service. As a general principle, a fee should bear a direct relationship to the cost of providing the good or

---

[13] Non-rivalry in which the consumption by one person of a good or service will not diminish the amount available to others; and non-exclusion in the sense that once it is provided to one person, others cannot be prevented from also consuming it. These concepts were also discussed in Chapter 1.

[14] Information agencies are those whose primary function is the collection, compilation, analysis and dissemination of information to the public. Included under this is the Australian Bureau of Statistics (ABS), Bureau of Meteorology, AUSLIG—the national mapping agency, AGSO—the Australian Geological Science Organisation, National Library of Australia and Screen Sound Australia. AUSLIG and AGSO has since merged to form Geoscience Australia (GA).

[15] *Matthews v Chichory Marketing Board* [1938] 60 CLR 263.

service, or could be open to legal challenge as amounting to a tax. The separation of 'powers' between the Commonwealth and the States means that the design of revenue collecting and revenue generating arrangements are already prescribed. The Commonwealth powers on tax are to be found in ss 51(ii) and 55 of the Constitution. Sections 90 and 96 further define the limited ability of the states to raise revenues independently. International obligations also place a cap on an agency's ability to recover costs.

In 2001 the Productivity Commission undertook a review of the nature and extent of cost recovery arrangements across Commonwealth regulatory, administrative, and information agencies. The scope of the inquiry included the identification of the activities of those agencies for which cost recovery is undertaken, who benefits, the impact on business and users, and the consistency of such arrangements within government agencies.[16] Cost recovery is taken to mean the recovery of some or all of the costs of a particular government activity or product and is distinguished from mere taxation. In its most direct form, a user is charged a fee, based on the cost of providing the government product that is consumed. Cost recovery is usually not undertaken with a view to generate a profit as opposed to a profit-making government business enterprise.[17] Full cost recovery is where the value of all resources used or consumed in the provision of an output. The cost recovery includes direct costs, indirect costs and imputed costs (in order to comply with competitive neutrality rules).[18]

Financing—Fiscal and Economic Framework

For information agencies, cost recovery is inappropriate where the information products have a high degree of 'public good' character or where there are significant positive spill-over effects. Information products that meet these tests would be budget-funded as part of a basic product set. Such information products may be categorised into three types and priced accordingly:

---

[16] See Productivity Commission 2001 *Cost Recovery by Government Agencies*. Inquiry Report No. 15, Canberra: AusInfo. Also at http://www.pc.gov.au/inquiry/costrecovery/.

[17] *ibid*. Productivity Commission 2001: xxxii.

[18] Competitive neutrality policy principles require that the prices charged by government businesses in actual or potential competition with the private sector are to be adjusted to reflect the advantages and disadvantages of public ownership. Prices should at least cover costs (including a return on capital invested and all relevant taxes and charges).

1. dissemination of existing products at marginal cost;
2. incremental products (like those involving additional data collection or compilation) at an incremental cost;
3. commercial (contestable) products that are priced according to competitive neutrality principles.[19]

Commonwealth agencies and authorities operate under the *Commonwealth Authorities and Companies Act* 1997 (Cwlth) as well as the *Financial Management and Accountability Act* 1997 (Cwlth). The principal working fund of the Commonwealth is the Consolidated Revenue Fund where all public monies are deposited and drawn from. Section 83 of the Constitution requires parliamentary approval for expenditures and allocations. The Department of Finance and Administration (DOFA) provides pricing review guidelines as well as the costing of government activities—full cost, marginal cost pricing and staff, labour on-costs.[20]

Pricing

The issue of how one goes about establishing a sale price of the data is a topic of considerable uncertainty because for some data and information it may be impossible to establish a market value. In general, data for the GI industry may be priced either at *true cost*, that is, all the costs and over-heads that have been spent to create and produce the data, or *fair market value*. This is sometimes described as charging what the market will bear. The pricing for GI can thus be very complicated, as shown below.

GI data and databases have two unique differentiating characteristics. First, data and databases have very high initial production costs. But while the cost for the first database may be very high, the marginal cost of each additional sale is low, albeit one which has high marginal returns. Secondly, public agencies have been collecting data, not only for decision making, but also for policy formation. This implies that some data are collected with no immediate and apparent use. It is arguable that the costs to the community may be less when data are gathered in anticipation of need rather than left to an eventuality when decisions need to be made. On the other hand, it may be impossible to know in advance what information

---

[19] *op. cit.* Productivity Commission 2001 p. xlii.

[20] The Constitutions of New Zealand, Canada, and the U.K. are similar to that of Australia in that taxes are implemented on the basis of legislation. The same may be said of charges for cost recovery: New Zealand 2002 *Guidelines for Setting Charges in the Public Sector*, NZ Treasury; Canada 1997 *Cost Recovery and Charging Policy*, Treasury Board Canada Secretariat; U.K. 1992 *The Fees and Charges Guide*, UK Treasury.

should be collected, stored and analysed for some unknown future decision or application. Thus, it seems that only when the private sector has discovered a use for such data that both the public and private sectors pay special attention to the quantity and quality of the data collected and the way the data are analysed and reported. In such circumstances the value of data depends on its use and the demand for it.

As noted previously the price for public data can range from nil to the market price, depending on the nature of the product and the custodial agency. The mood among some government circles in Australia is that the data should be 'sent out the door as cheaply as possible'. However, 'economic policy based on public benefit considerations is steering the price policy [into a] marginal cost regime with extra charges for value adding performed by the custodial agencies' (Millhouse 1994: 10).[21]

In the pricing of government data it is uncertain if the historical or 'sunk' costs should be recovered from a sale. The reasoning here is that costs of data collection have been funded by taxpayers and as such should not be passed on to users. Information as a public good may have been collected as part of the government's 'community service obligation' (CSO), similar to providing power and telecommunications to remote communities at a subsidy (Grant and Krogh 1995: 11).[22] An extension to the argument is that attempts to recover sunk costs in an information project should be avoided because these have been expended in the project for a specific purpose. Any additional application should be considered a collateral benefit—a bonus.

A comparison of the pricing policies of each jurisdiction within Australasia by Nairn and Holland (2001) shows significant variations in both prices and access conditions.[23] In some jurisdictions full cost pricing applies while in other jurisdictions pricing is left to individual agencies. The New Zealand policy of access to government data is currently the least restrictive in terms of low prices for access and the absence of royalty payments for those wishing to value-add and on-sell the data. At the other end of the spectrum some agencies charge a high price for data and require licences and royalties from

---

[21] Millhouse, D 1994 'A merchant banker's view of GIS', *AURISA News*, 55, pp. 1, 6–11.

[22] Grant, DM and Krogh, B 1995 'Partners in the spatial information systems industry—an "open marriage" between the public and private sectors', *GIS Law*, 2(4), pp. 9–17.

[23] Nairn, AD and Holland, P 2001 'The NGDI of Australia—Achievements and Challenges from a Federal Perspective', paper presented at a Workshop on NGDI—Towards a Road Map for India, 5–6 February, New Delhi, India available at http://www.gisdevelopment.net/policy/international/interna007.htm.

users. In between the extremes are agencies that have policies that provide data at minimal cost to other government agencies for internal use. Information agencies such as Geoscience Australia (GA) use a fee-for-service to cost recovery rather than through taxes or levies. Copyright licences and royalties are also used, but these are 'exceptions rather than the rule'. Such variations in charges may change as a result of the review by the Productivity Commission (2001), the implementation of the government's Spatial Information Industry Action Agenda (Department of Industry, Science and Resources 2001) and other recent developments. To provide some orders of magnitude, Table 2.2 shows variation in prices of state and territory cadastral databases in Australia.

One conclusion to be made here on pricing is that a public policy strategy should include a pricing regime that is tailored to cater for different classes of users. This flexibility could be on the basis of need, potential market for the information product, and commercial profitability (see also discussion in the next sub-section on governance below). As an example, the output of geoscience information on Australia's natural resources and environment is determined by government policy and funded by appropriation through parliament. However, there are also co-funded and collaborative undertakings with external bodies in which the costs are shared. A further alternative is where the work is commissioned or fully funded in situations where it complements the strategic programs of an agency, for example GA. There is no charge for general and reference information, but charges arise when the information is packaged and customised for particular purposes such as catalogues and geodetic controls, map data and customised products such as satellite imagery.

Table 2.2 Prices for cadastral databases in Australia, November 1999

State/Territory	Price (AUD$)
Northern Territory	2 000
Victoria	5 500
South Australia	10 000
Australian Capital Territory	26 150
Queensland	87 500
Tasmania	189 000
Western Australia	200 000
New South Wales	4 725 000

Source: Nairn and Holland (2001)

Governance

The Australia New Zealand Land Information Council (ANZLIC) is the peak coordinating body for geographic information in Australasia.[24] The ANZLIC Guiding Principles for Spatial Data Access and Pricing Policy (2004) are aimed at providing easy, efficient and equitable access to fundamental spatial data. Under this policy each jurisdiction is responsible for determining its own access conditions and arrangements as well as data pricing and access policies.

More recently, the organisation and structure for the delivery of geospatial data has been streamlined with the establishment of the Public Sector Mapping Agencies (PSMA) Australia Ltd consortium.[25] The PSMA consortium was originally created in 1993 as an unincorporated joint venture between the nine mapping agencies of the Commonwealth, states and territories. It came into being as a response to an Australian Bureau of Statistics (ABS) tender for the provision of mapping services for the 1996 Census of Population and Housing. In winning the contract PSMA broke new ground in the delivery of the national topographic dataset that augmented the Australia's cadastral framework. Since 1997 PSMA has focused on the assembly and framework for national geospatial datasets. A review of the future options for PSMA concluded that the organisation should be transformed into an unlisted public company limited by shares and owned by the governments of Australia. The advantages of such a structure were seen to be the separate legal personality independence that it would have, financial transparency and accountability, an efficient management structure while permitting each jurisdiction to protect its own interests; and limited liability to shareholders. PSMA was incorporated in June 2001. Figure 2.2 shows PSMA as the crucial link between the supply and demand sides of the market.

The vision of PSMA is 'the creation of a national asset of comprehensive, quality and accessible spatial knowledge'. The company determines the datasets to be assembled and defines the detailed specifications, including outsourcing through competitive tendering. Revenue is generated through data licensing and royalties using data supply agreements. PSMA

---

[24] ANZLIC has ten members one from each Australian State and Territory, a Commonwealth representative and a New Zealand representative. The Council directs its activity in industry development, policy development and spatial data infrastructure. See http://www.anzlic.org.au.

[25] See Hedberg, O, Paull, D and Bower, M 2003 'Spatially enabling Australia through collaboration and innovation' paper presented to the Cambridge Conference 2003, Southampton: Ordnance Survey, paper 7.1.

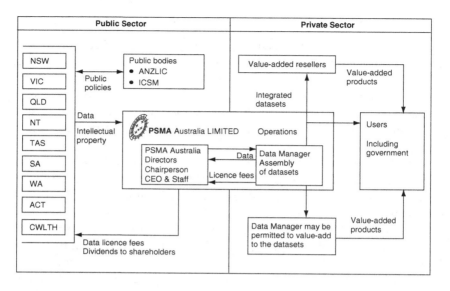

Figure 2.2 PSMA Australia Ltd: supply chain positioning. Source: with permission, ©PSMA Australia Ltd 2003. http://www.psma.com.au

does not receive any government funding and is required to ensure that its activities are fully funded from data licensing and data supply contracts. PSMA is a government owned, but not government funded organisation.

Pricing of PSMA data reflects the value of each dataset for different applications. Thus, high-value products are shown up in the pricing matrix while also facilitating opportunities to have the data used ubiquitously in low-price digital products. The pricing of the products is therefore based on a matrix of data type, data volume, user applications, and the number of users.

In Australia it is now not necessary to negotiate data access and pricing with the nine governments involved. The complex and time-consuming tasks have now been eliminated and the 'bottleneck' to geospatial information supply avoided. PSMA is also a crucial link between the supply and demand sides of the market. Also it does not deal with end-users, but rather with organisations that develop products and services for end-users through a process of value adding. A recent initiative is the launch of a single authoritative geocoded database for street address data called G-NAF (Geocoded National Address File). This database is used for cross-checking multiple address sources and provides an authoritative reference for such information.

Much of the developments in Australia may be traced to the government's Spatial Information Industry Action Agenda (SIIAA) of September 2001.[26] The Action Agenda provides a common vision and implementation plan for all sectors of the spatial industry. Under this strategy all levels of governments will adopt data policies that increase the creation and use of spatial information. High priority is given to the development of a common approach between government and industry towards spatial access, pricing and copyright policy. A key initiative in the Agenda was the incorporation in July 2001 of a single association known as the Australian Spatial Information Business Association (ASIBA) to represent private business interests in the spatial information industry. This Association will create formal linkages between business and government agencies and is positioned to provide advice and policy recommendations to government.

Since 2002 a new Commonwealth spatial data access and pricing policy has been implemented that will improve access to fundamental spatial data held by Commonwealth government agencies and which will ensure a whole-of-government approach to its distribution.

The basic access model proposed under this policy is described in Figure 2.3 and features the following:

- a choice of access methods for the user;
- a 'single point of entry' to Commonwealth spatial data;
- a metadata search engine, the Australian Spatial Data Directory (ASDD);
- links to all relevant AGIMO (Australian Government Information Management Office, previously NOIE, National Office for the Information Economy) endorsed portals;
- distributed spatial data holdings, managed by relevant Commonwealth custodians;
- support for e-commerce and value-added services.

Under the policy, there is free on-line access for specified spatial data where technology is available. Products unavailable immediately on the Internet will be supplied at a marginal cost of transfer. The policy also removes restrictions on commercial use or value-added activities by private companies and individuals. There are more than 80 separate datasets produced by eleven Commonwealth government departments that have

---

[26] Department of Industry, Science and Resources 2001 *Positioning for Growth: Spatial Information Industry Action Agenda*, September, Canberra: InfoProducts. See also http://www.isr.gov.au/agendas/sectors/siiaa/aa_presentation.pdf.

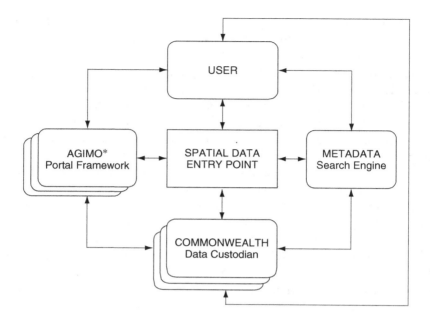

*AGIMO : Australian Government Information Management Office previously
the National Office for the Information Economy (NOIE).

Figure 2.3 Australia: basic spatial data access model. Source: CSDC 2001

been classified as 'fundamental'. This excludes non-spatial datasets, for
example, statistical collections or those that are part of an existing commissioned service arrangement and commercial contractual agreement
with external agencies. The policy is focussed on activities that are
directed at growing the spatial information industry in Australasia.[27]

2.3.2 United States

Purpose and Legal Authority

The 'open records' regime in the U.S. means that cost recovery and sale of
public data is not possible other than by exception. The regime is mandated

---

[27] See Commonwealth Spatial Data Committee (CSDC) 2001 A proposal for a Commonwealth Policy
on Spatial Data Access and Pricing, Belconnen, ACT: CSDC. Also at http://osdm.gov.au/osdm/docs/
Commonwealth_Policy_on_Spatial_Data_Access_and_Pricing.pdf.

by virtue of the *Freedom of Information Act* 1966, 5 U.S.C. § 552 and the *Electronic Freedom of Information Act* 1996. These instruments assure free public access for all federal data. Most states have laws that complement this principle of open access for both state and local government data. In an open access regime all data collected by public agencies are deemed to be 'public data' or records and are therefore be freely available to the citizen at large. As public access to government data is a constitutional right the data are generally provided at marginal cost, that is, for the cost of reproduction of existing data. This policy is to ensure that all government activity is 'transparent' and public officials accountable and open to scrutiny. Intellectual property law has no application to public information. Dando (1994) surveyed and analysed open records laws of 50 states and the District of Columbia in relation to the ability of agencies to recover database development costs.[28] The commentary focused on legislation to balance the right of the public's access to government information and the desire of local governments to realise returns from the private sector in the use of a valuable asset developed at substantial public expense.

While in theory every State has a GIS data distribution statute of one form or another, the approaches adopted by the states have been based on different justifications. Some provide access rights on the basis of an exception to open records law, others depend on the nature of the request that is made. There are yet others that make no distinction between geographic information and any other type of digital database while some states have removed the issue from open records law by treating any request as an administrative matter.

Iowa legislation has amended its open records law and added geographic information to the list of exceptions. In North Carolina, geographic information has been made an exception to the law in one county. These two examples raise the difficult question of whether a government agency has a right to charge fees for data and why geographic information data is distinguishable from other types of automated data. In Kentucky no such distinctions are made. The cost depends on a number of elements such as how the request is to be serviced, namely whether the information is to be customised, whether access is by electronic means and whether the information is for commercial or non-commercial purposes. In Alaska, the approach attempts to strike a balance between cost recovery, open records and privacy.

---

[28] Dando, LP 1994 'A survey of open records laws in relation to recovery of database development costs: An end in search of a means' in *Marketing Government Geographic Information*, Washington, D.C.: URISA, pp. 5–22.

Geographic information is no different from any other kind of automated data, and the provision of all data contain some cost element with specific fee waivers for certain groups of users. Wisconsin has adopted a pragmatic approach where geographic information is separate from open records law. Here the information is included as part of the Land Information System (LIS) of the state and any fee collected is put towards the development of that system. This scheme is seen as a more equitable form of cost recovery and real estate transactions provide the revenue for the upkeep and development of the Wisconsin LIS (see Dansby 1992b: 10).[29]

The guiding principle and legal authority for cost recovery and charging for Federal public data and information is contained in the *Paperwork Reduction Act* 1995 and the accompanying Office of Management and Budget (OMB) Circular A-130.[30] Under the Act, s 3506 stipulates a federal agency's responsibilities in regard to ensuring that the public has timely and equitable access to the agency's public information. The Act encourages affirmative dissemination of information and requires agencies to provide timely and equitable access to public information. Government agencies should not, except where specifically authorised by statute:

(a) 'establish an exclusive, restricted, or other distribution arrangements that interferes with timely and equitable availability of public information to the public;
(b) restrict or regulate the use, resale, or re-dissemination of public information by the public;
(c) charge fees or royalties for resale or re-dissemination of public information; or
(d) establish user fees for public information that exceed the cost of dissemination'.[31]

In the U.S. the controversy is over the appropriateness, legality, and effectiveness of public agencies selling digital geodata. The tension is between a public's right to public data as against a public agency's need to fund its GI operations.

From a U.S. constitutional point of view it is unclear what limitation there is on the application of user charges. The OMB views cost recovery as an alternative to general taxation. In the U.S. Supreme Court case of

---

[29] Dansby, H 1992b 'Public records and government liability. Part II', *GIS Law*, 1(1), pp. 7–13.
[30] OMB 1996 Circular A-130 *Management of Federal Information Resources*. Washington, D.C.: OMB also at http://www.whitehouse.gov/omb/circulars/a130/a130.html.
[31] *Paperwork Reduction Act*, 1995, s 3506.

Skinner v Mid-America Pipeline Co. 1989 user charges were not considered as taxes.[32] However in *U.S. v U.S. Shoe Corp.* 1998 the court held that user charges could be considered as a tax and hence unconstitutional.[33] The reasoning behind the decision was that the revenue source and the revenue use were too closely linked, so as to give the charge the character of a tax.

Financing—Fiscal and Economic Framework

According to the OMB Circular A-25 Revised full cost recovery applies to identified recipients of government activities, irrespective of whether all or some of the benefits are passed on to others, including the public in general (OMB 1993: 4).[34] This rule does not apply to all government activities, as some are exempt. The OMB Circular A-130, states that user charges for government information products should be set 'at a level sufficient to recover the cost of dissemination but no higher' (OMB 1996: 9).[35]

Within the parameters of the Open Records regime, Onsrud (1992) has previously argued in favour of the need for cost recovery for publicly held geographic information.[36] The distribution of GIS products and services should be cost recovered on the basis that these are similar to the costs of operating schools and maintaining streets. So, even though GIS data have been acquired at taxpayer expense and therefore should be free to anyone who has an interest in them, the counter-argument is that it takes additional time and expense to provide such a public service. It has also been argued that the government agency should recover marginal costs incurred when responding to citizen requests. This includes staff time, pro-rated to hardware and software maintenance costs as well as clerical time to process the requests. Onsrud has further argued that an agency could also recover greater than marginal costs so that GI products and services may become revenue generators.

Pricing

Following the passage of the *Uniform Freedom of Information Act* 1986 the OMB issued a Fee Schedule and Guidelines that dealt amongst others

---

[32] *Skinner v Mid-America Pipeline Co.*, 490 U.S. 212 (1989).

[33] *United States v United States Shoe Corp.* 523 U.S. 360 (1998).

[34] OMB 1993 *Circular A-25 Revised*. Washington, D.C.: OMB also at http://www.whitehouse.gov/omb/circulars/a025/a025.html.

[35] *op. cit.* OMB 1996 Circular A-130.

[36] Onsrud, HJ 1992 'In support of open access for publicly held geographic information', *GIS Law*, 1(1), pp. 3–6.

with pricing issues.[37] The Guidelines provides charging for search, dupli-cation and review costs, but not for value adding by the public sector to the raw data. This pricing policy is one where the public sector value adds to the data only as a tool for its own efficiency purposes and not as a means for profit making. The task of enhancing the data is left to the private sector to make a commercially viable product or service.

Governance

In general, the OMB has oversight functions over all federal cost recovery activities in the U.S.

Experiences

A study by KPMG Consulting Inc. in 2001 reported that 'U.S. agencies reporting data income had revenues equal to 2 per cent of their expenses'.[38] The small income stream is also collaborated elsewhere. An Open Data Consortium (ODC) Project 2003 funded by the U.S. Geological Survey (USGS) through the GeoData Alliance provides similar findings.[39] The study was undertaken to develop a model data distribution policy for local governments throughout the country. This project found that most local government agencies that sell public data have not realised revenues, with many having operated in deficit. Indeed, it was discovered that there were better ways of raising funds to support GIS operations within local government. These conclusions are not new, given that there have been numerous examples of failed cost recovery experiments in the U.S. at both the federal and state levels.

In the early 1980s the USGS adopted cost recovery policies by increasing prices for its data and map products. As a result demand for these fell sharply, forcing the USGS to reduce prices in order to recapture its previous market. It took a few more years before sales returned to earlier levels.[40] The agency is still looking to find a balanced method to recover dissemination costs and only recently has it recovered close to 100% of such costs.[41]

---

[37] *op. cit.* OMB 1993 Circular A-25.

[38] Sears, G 2001 *Geospatial Data Policy Study*, Ottawa, Ontario, Canada: KPMG Consulting Inc., March 28, p. 18.

[39] Joffe, BA 2003 *Open Data Consortium Project: Model Data Distribution Policy*, Oakland, CA: GIS Consultants. Also see http://www.joffes.com/.

[40] Donato, DI 1985 A review of pricing issues and alternatives for 1:100 000 scale digital cartographic data. Reston, VA: USGS.

[41] Blakemore, M and Singh, G 1992 *Cost Recovery Charging for Government Information: A False Economy?* London: GSA Ltd, pp. 30–34.

In 1992 the U.S. Federal Maritime Commission (FMC) created the 'Automated Tariff Filing and Information System' (ATFI) to collect, manage, and disseminate data on tariffs filed by common carriers that included information on cargo types, shipping destinations, and service contract terms. The *High Seas Driftnet Fisheries Enforcement Act* 1992 Public Law 102–582 included a requirement that the FMC collect user fees from anyone who directly or indirectly accessed ATFI data. The goal was to raise US$810 million over three years by charging 46 cents per minute to retrieve the information directly or indirectly. The upshot was that actual user fees collected were US$438 800 approximately 0.05% of the original mandate.[42] This failure to achieve targets has been attributed to the overoptimistic assumptions about the perceived inelasticity of tariff data and the failure to consider that users could obtain tariff data from other sources other than from the FMC.

The ODC 2003 Report cited earlier reveals that at a state and local government level the lack of success from data sales operations among ODC participants has been even more dramatic. Few have made money, none has raised significant revenues compared with the costs of maintaining GIS and geodata assets, while some have lost money. The example from Ventura County, California is instructive. During a five year period when the county sold their data for US$1 per parcel they raised US$15 000 per year compared with an annual cost of nearly US$1 million to maintain a ten-person team to update geodata and create GI applications. The county has now lowered its price for the entire county GI database to US$3000, which includes quarterly updates, and have twenty annual subscribers, producing a revenue of US$60 000 per year.

2.3.3 European Union

The 'European' model is characterised by Treasuries and legislation that pressure government agencies to recover costs for PSI directly from users. Arguably, as a result of this cost recovery regime there is limited use of the PSI and thus limited financial resources for the government agency to collect and maintain the data. Direct funding and cross-subsidies are thus needed as agencies transfer taxpayer monies to originating agencies. There have been different E.U. funding models that have been used including total government funding (taxpayer pays), private sector funding

---

[42] See Weiss, P 2002 'Borders in Cyberspace: Conflicting Public Sector Information Policies and their Economic Impacts' p. 13 at ftp://ftp.cordis.lu/pub/content/docs/peter_weiss.pdf.

(user pays), public sector funding where fees are charged by public agencies, and indirect funding such as revenues from services, advertisements and other sources.

Purpose and Legal Authority

An overview of E.U. legal authority for cost recovery and user charges for PSI may be obtained from Table 2.3. Access to PSI in most E.U. countries is

Table 2.3 E.U.: Legislation and policy on access to public sector information (PSI) 2000

Jurisdiction	Legislation/policy	Distribution model
Austria	Constitutional law 285 of 1987	Free/MC*
Belgium	Federal laws 1994, Regional laws 1991	Free/MC
Denmark	Freedom of Information and General Right of Access to Government Documents Bills	MC
Finland	Publicity of Official Documents Act 83 of 1951; User Charging for Government Service Act 1992	Depends on type of information
France	General law on access to administrative documents 78–753 of 1978	Free
Germany	No general access law. Some Länder constitutional provisions	Depends on type of information
Greece	General law on access 1599 of 1986	No policy yet
Ireland	Freedom of Information Act 1997	Fee-for-service and CR**
Italy	General access law 241 of 1990	Free and CR for some information
Luxembourg	No general access law or rulings regarding exploitation of public sector information	Free/MC
Netherlands	Government Information Act 1980	No policy yet
Portugal	General access to public sector information law 65 of 1993	Lowest price possible
Spain	General access law 30 of 1992, but excludes computer information	Free to market prices
Sweden	Freedom of the Press Act 1766 (last amended 1994), Personal Data Act 1984; Government Information Technology Act 1996	Free/MC
United Kingdom	No general access law to government information; DTI Guidelines for Government-held Tradeable Information 1985; Code of Practice 1994; Freedom of Information Act 2000	CR

* MC = marginal cost; ** CR = cost recovery
Source: Adapted from European Union 1999a *Green Paper on Public Sector Information in the Information Society*, Annex 1

provided either free or on a marginal cost basis or full cost recovery basis. But there are variations in the charging of PSI. In general, most charging schemes would require some legislative mandate and user charges are devised to avoid the character of a tax. As discussed previously a user charge would not resemble a tax where the transaction is voluntary, where there is no link between the revenue source and its use and the amount charged, and whether direct and indirect costs are being recovered.

In Sweden the user charge is directly related to services rendered and is not to exceed the full cost of the service. Finland on the other hand gives agencies broad leeway in implementing user charges within guidelines. Agencies in Ireland, Sweden and the U.K. are unable to charge for services unless the legislation has given that agency a grant to do so. This legislative mandate is given on an agency-by-agency basis as there is a lack of a general legislation on cost recovery. Moreover, the terminology used can also be misleading since market prices may also be regarded as a full cost recovery scheme—compare Spain and the U.K. In the latter case the guidelines state that 'charges should normally be set to recover the full cost of the service' (UK Treasury 1992: 19).[43] Also, the retrieval of deficits is encouraged far more than the return of surpluses. It is the 'normal presumption' that charges should be set to recover past deficits, but surpluses should not 'normally be taken into account' when setting charges (UK Treasury 1992: 9).

Financing—Fiscal and Economic Framework

E.U. publicly funded agencies treat their information resources as a commodity to generate short-term revenue. By controlling access to certain types of information the agencies are able to recover the costs of collecting or creating the public data. This tends to prevent private entities from developing markets for the information or in disseminating the information. The economic framework is thus one of government commercialisation or cost recovery with the hint of a natural monopoly.

Pricing

According to the 1999 E.U. Green Paper on PSI, pricing as well as pricing models vary enormously between different Member States and between

---

[43] In the U.K. full cost recovery is defined as including a 6% annual cost of capital. This is the amount of interest the government could earn on its capital if it were put into an alternative use, U.K. Treasury 1992 *The Fees and Charges Guide*, London: HMSO, p. 8.

different public sector bodies within the same state.[44] The French Prime Minister's *Circular of 14 February 1994* distinguishes between information collection and production costs for which there is no charge and costs such as printing, updating, data retrieval and transmission for which there is a charge.[45]

The types of information and use of the information are also distinguished for charging purposes. The 1985 U.K. *Guidelines on Government-held Tradeable Information* (revised in 1988) favoured a market approach.[46] The rule is that where an established market exists Departments may charge a reasonable market price. However, where the tradeable information has not previously been exploited, the charges will initially be on the basis of costs incurred over and above those that would be normally incurred in handling the data or information for their own purpose.

The 1989 European Commission *Synergy Guidelines* favoured a distribution cost approach where pricing policies varied depending on the nature of the information.[47] A price is established which reflects the costs of preparing and passing it to the private sector, but which does not necessarily include the full cost of routine administration. The price may be reduced if provision of the resulting information service is deemed to be necessary in the public interest.

Governance

There are different governance and accountability arrangements among the various E.U. states. There is no formal consultative process in Finland for example, whereas in neighbouring Sweden there is a requirement in the legislation for an annual consultation with the National Audit Office. This consultation is intended to ensure that user charges are uniform, relevant, and do not exceed the full cost of the service over the service's lifetime. In Finland, on the other hand, public agencies have to publish their pricing

---

[44] European Union 1999a *Green Paper on Public Sector Information in the Information Society*, COM(98) 585 Final, adopted on 20 January 1999. Available at: http://europa.eu.int/ISPO/docs/policy/docs/COM(98)585/.

[45] France 1994 *Circulaire du 14 février 1994 relative à la diffusion des données publiques*, Paris, also at http://www.environment.gov.fr/donnees/diffdon/balladur.htm.

[46] United Kingdom, Department of Trade and Industry 1988 *Government-held Tradeable Information: Guidelines for Government Departments Dealing with the Private Sector* (2nd edn), London: HMSO, 35 pp.

[47] See European Communities—Commission 1989 *Guidelines for Improving the Synergy between the Public and Private Sectors in the Information Market*, Luxembourg: Office for Official Publications of the European Communities.

decisions in the Official Gazette for transparency and accountability purposes.

There are also fair competition concerns when the public sector offers value-added information products on the market. Legislation dealing with competitive neutrality issues in the Member States exists. U.K. guidelines prohibit direct competition with the private sector. French rules hold that the public agency not directly intervene in the private sector and only provide value-added products if it is within its legal mission to do so. General competition rules, for example, under the European Commission (EC) *Single European Act* 1986 and the *Treaty of Rome* (Art. 85–94) are designed to avoid market distortions because of public subsidies, abuse of dominant position, agreements between companies, and monopolies.[48] In general where information is available only from one source, then the practice is to make it available with reasonable conditions attached. Where the information is available from different sources (the diversity principle) then market prices may apply to those public products.

Weiss (2002) suggests that recent experiences in the E.U. show that charging marginal cost of dissemination for PSI will lead to optimal economic growth and will far outweigh the immediate perceived benefits of an aggressive cost recovery strategy.[49] Accordingly, a pan-European fully commercial approach to PSI will be doomed to failure for several reasons. First, the small private user base cannot support full cost recovery for information services. Second, charging other government users merely shifts expenses from one agency to another rather than saving the Treasury any money. Third, the fundamental characteristics of information—its public good nature and the high elasticity of demand—make it well nigh impossible to raise revenues adequate to pay not only for its dissemination, but also for its creation in the first place. Finally, high prices for information will lead to predatory and anti-competitive practices such as price dumping and favouritism of private sector partners in joint ventures and other government-owned enterprises to the exclusion of others from the market.

Weiss (2002) further suggests that the most sensible solution is to separate commercial activities and endow these into a truly commercial entity. Such an entity is to be divorced from the government and to permit it to adopt open access policies. These steps are in accord with open market rules under European competition laws and will also ensure the setting up

---

[48] E.U. 1987 *Single European Act* OJ L 169, 29.6.1987 also at http://europa.eu.int/abc/obj/treaties/en/ entr14a.htm#41; and EEC 1957 *Treaty of Rome* at http://www.europa.eu.int/abc/treaties_en.htm.
[49] *op. cit.* Weiss 2002.

of market structures with maximum overall economic potential. Short-term losses as a result of liberalisation of data distribution policies and subsequent budgetary cuts could be more than offset by the creation of wealth and tax revenues in the long term.

In Europe it has been recognised that an open access policy to PSI is critical for the Information Society, for scientific endeavour, Research and Development (R&D), and for economic growth. But there are still pockets of resistance to more liberal, open access policies from Treasuries and civil servants in charge of commercialisation initiatives. The E.U. (1999a) *Green Paper* concluded that PSI is a key resource for Europe, and that E.U. nations should more closely follow the U.S. federal government model policies of a broader access to government databases.[50]

Experiences

Rather than review the experiences of the each of the Member States in the E.U. the following focuses on the Ordnance Survey (OS) in the U.K. as an exemplar of cost recovery strategies and how the organisation demonstrates the workability of the principles, but yet fulfil its 'public' functions. This presentation is an historical description of the trials and travails of the OS in its recent past from a civil service organisation to becoming an Executive Agency with Trading Fund status and to an uncertain future as to whether it might remain to be so. The possibility of transforming the OS into a public limited company (Plc) has been mooted in various circles.

An early debate as to the status of OS as a revenue generator may be traced to 1979. The Serpell Committee Review of the Ordnance Survey argued against charging users for information on a cost recovery basis.[51] But in 1981 in a review of the U.K. government's statistical service by Sir Derek Rayner new cost recovery policies were put in place. The so-called Rayner doctrine decreed a general policy of charging for data, even between government agencies in some circumstances.[52]

Within the so-called tradeable information initiative, the OS adopted a cost recovery policy for its mapping products. This initiative encouraged government bodies and Executive Agencies to trade data for profit where

---

[50] *op. cit.* European Union 1999a *Green Paper*.

[51] Department of the Environment 1979 *Report of the Ordnance Survey Review Committee*, Chairman: Sir David Serpell, London: HMSO.

[52] HMSO 1981 *White Paper: Government Statistical Services*, Chairman: Sir Derek Rayner, London: HMSO.

the income was used to underwrite the cost of collection, analysis, and storage. Openshaw and Goddard (1987) provide a useful commentary on the impact of such policies on access to data.[53]

In 1990 the OS was chartered as a semi-independent Executive Agency, a form of privatised government body, designed to make a profit from is operations. Even so the OS has yet to achieve full cost recovery. The OS, for example, recovers only a percentage of its total costs of its operations from the sale of goods and services. Of the estimated £100 million annual OS revenues, only £32 million comes from commercial product sales.[54]

The OS is still dependent on parliamentary appropriations since the OS functions as the national mapping agency in the U.K. The OS has a natural monopoly because of the economies of scale and other factors. It has been estimated that the income of the OS is largely from the public sector. For example, the 23% of OS income is derived from the utilities, 16% each from local government and consumer sales, 20% from the central government, 14% from the National Interest Mapping Service Agreement (NIMSA), 10% from commerce and 1% from educational institutions.[55]

Until 1999 when the OS was granted Trading Fund status, it had placed no asset value on the National Digital Topographic Database. However, since then the National Audit Office (NAO) assessed the asset with a depreciated value of £2.2 million based on an estimated cost of £4.8 million. The NAO had concluded that the database should be more appropriately accounted for as a tangible fixed asset and made the assessment based on Treasury guidelines.

Apart from its status as a Trading Fund and the accompanying cost recovery mandate, the OS has fulfilled its 'public' functions and engaged in non-profitable areas of mapping through other means. In conjunction with the Department for Environment, Transport and the Regions (DETR) a formal National Interest Mapping Service Agreement (NIMSA) has ensured the provision of agreed mapping services required in the national interest will be available.[56]

---

[53] Openshaw, S and Goddard, JB 1987 'Some implications of the commodification of information and the emerging information economy for applied geographical analysis in the United Kingdom', *Environment and Planning A* 19: 1423–1439.

[54] *ibid.* Weiss 2002.

[55] Barr, R 2002 'Choosing the best route for Ordnance Survey. An old friend looks to a new life— the Ordnance Survey Quinquennial Review' at http://www.ginews.co.uk/0402_35.htm.

[56] DETR 1999 'National Interest Mapping Service Agreement', DETR. http://www.detr.gov.uk/annual99/12.htm.

A Quinquennial Review in 2001–2002 considered the future of the OS in terms of the different types of ownership models from abolition to privatisation and options in between. Widespread consultation with users in both the private and public sectors was undertaken as part of the process. At one stage the Review's preferred option was to recommend OS become a wholly state-owned public limited company (Plc) with the government owning 100% equity as opposed to a privatised Plc. However, after further consultation and research the government decided to keep OS as an Executive Agency with Trading Fund status with some extended freedoms to enable continued development.[57]

The Treasury's view was that, with OS as an information utility, the taxpayer would not have to bear the burden of producing national mapping products. As a commercial entity the OS will be exposed to the market discipline of meeting what the customer wants. To achieve this objective the way forward for the OS was for it to progress to either full privatisation or a Plc. The Quinquennial Review rejected full privatisation, leaving open the Plc route. In commenting on the OS Quinquennial Review, Barr (2002) proposed a different route that the OS could take.[58]

Barr suggested that the Plc could have non-tradeable shares owned by the users and organisations generating the revenues. Some regulation will ensure shareholders fulfil their obligations to maintain the national spatial data infrastructure whilst engaging in profitable commercial undertakings. As a Plc with non-tradeable shares, the OS will be even more competitive and will change the objective from simply cost recovery to usage maximisation. This mechanism will shift from the high-cost–low-volume market to a low-price–high-volume market for OS data and will change the structure and value of the income stream. However, no decision has been taken and to date the OS remains an Executive Agency with Trading Fund status.

In the new millennium, the U.K. government has accepted the general principle of providing government data at marginal costs. For instance, from 2000 national statistics in the U.K. have become available to the public free of charge on the Internet. However, Trading Funds such as OS and the Met Office are specifically excluded from this marginal cost principle. Paradoxically these Trading Funds have the most interesting and potentially useful datasets for the private sector R&D and the scientific community. This trend towards making data available has been further stimulated by

---

[57] Ordnance Survey Media Release 22 July 2002 at http://www.ordnancesurvey.co.uk.
[58] *ibid*. Barr 2002.

the passage of the *Freedom of Information Act* 2000 (U.K.) that could facilitate more open access.

2.3.4 Conclusions

The efficient allocation of resources and the need for equity in the use of government resources are some of the reasons for policies on access to government information. In the Australasian environment there is no overarching access or pricing policy for PSI. Constitutional constraints provide the limits to which charges may be made. The review by the Productivity Commission on cost recovery identified the 'public good' character of information products and the difficulty of pricing such goods at either the true cost or fair market value. ANZLIC as the peak public sector body provides the policy and framework for funding and pricing in the spatial information industry. The new pricing model in general is one where the information products are provided at a marginal cost and in time where all GI will be delivered on-line virtually at zero cost.

The Open Records regime in the U.S. to ensure government accountability in its activities has had its supporters and detractors. While every State has passed GIS data distribution statutes, the success at cost recovery is variable. Full cost recovery may only apply to some government activities. Case studies have demonstrated that full cost recovery does not work in the U.S. The Federal Office of Management and Budget oversees the principles and policies in cost recovery. It has been observed that 'many disparities exist regarding PSI access policies across agencies in the Federal government, across state governments, and even across agencies within a single state government' (Longhorn 2001).[59]

In general there is a cost recovery regime for PSI in most Member States in the E.U., but the models for charging differ. Information resources are treated as commodities, which may account for pricing differences among the states. Governance and accountability is through consultation with the customer base, but general national Audit Office reviews are also in place. The OS experience at cost recovery demonstrates that it can be profitable but agencies must also look after the 'public good' aspects of its operations so as not to forget the geography of places where no mapping may have taken place because there was no commercial imperative to do so. The OS cost recovery model is a well-developed counterpoint to U.S.

---

[59] *op. cit.* Longhorn 2001

open access FOI regime (Longhorn 1998).[60] In the U.K. the cost recovery regime has proven to be a useful means of funding the GI needs for all. However, in practice the regime has to be balanced with the needs of the community, for example, through NIMSA. All these endeavours may take time, money and political commitment at the administrative and policy levels.

Overall there is a growing maturity of the spatial information industry world-wide that is coupled with converging spatial technology derived from mainstream information management know-how. This maturity has also brought about an appreciation of the business benefits in traditional spatial information areas and has sparked a shift in industry dynamics.

2.4 Frameworks for Accessing Geospatial Information

One enduring feature of geospatial information in non-digital form is its diversity in terms of the scale, scope, coverage, quality and abundance or lack thereof. Such information may be in the hands of stewards, custodians and national mapping agencies in a variety of formats and vintages. The information is heterogeneous and is contained in manual data files that defy integration or the development of a truly national dataset that can be used by agencies and individuals within that country. The dataset may also not be truly portable and easily useable between government agencies let alone with and among private users or across jurisdictions. In economic terms there is a loss of value because of the high cost in terms of time and money in the collection, analysis and maintenance of the data. In the age of digitisation, the same criticism may be laid on digital geospatial information. The need for digital data standards is even more critical if GI is to be used across jurisdictions. Signposts showing what data are available, its accessibility and the possibility of its use are also required. The keyword in the digital age is interoperability for the interchange of geospatial resources. This section is concerned with the twin issues of metadata content standards and access and exchange standards for geospatial data and information. These together contribute towards the development of an interoperable *info-structure* for GI.

---

[60] Longhorn, RA 1998 'Strategic Initiatives for the Global Spatial Data Infrastructure (GSDI)', GSDI-3 Conference, 17–19 November, Canberra, Australia at http://www.gsdi.org/docs1998/canberra/ralstrat.html.

2.4.1 Metadata Content Standards

Metadata is the data about data. Metadata is used to organise and maintain an agency's investment in data; provide information about the data for use in catalogues and clearinghouses; and, provide information to assist data transfer. Metadata that accompanies a digital data file are usually found as a 'header' to this file. A metadata record lists several characteristics of the data theme that might be useful to a user. It may inform the user whether the datasets are appropriate for their needs. A standard geospatial metadata record could list the following information:

- *Footprint.* Spatial coverage of the dataset, its geographical extent, giving the latitudes and longitudes of the area, projection and coordinate system used.
- *Thematic coverage.* The geographical subject matter represented by the dataset.
- *Data format.* The formal structure of the data (SDTS, DXF, ArcInfo Export) and how it is organised.[61]
- *Lineage.* Source of the information represented in the dataset.
- *Data quality.* An indication of the level of positional and attribute accuracy of the elements in the dataset.
- *Contact information.* Information about people to contact and the agency responsible for the development and maintenance of the dataset.
- *Distribution.* Instructions as to how the dataset may be obtained, updated and revised.

A sample summary of the core elements of the ANZLIC metadata for high-level land and geographic data directories in Australia and New Zealand may be gleaned from Table 2.4. ANZLIC guidelines aim to promote consistency in the description of a small number of core metadata elements that are common to all types of data. It is designed to indicate what data exists, its contents, geographic extent, and how useful it might be for other purposes and where the data may be obtained. Such

---

[61] SDTS refers to the spatial data transfer standard, a U.S. Federal Information Processing Standard (FIPS 173) to define data formats called profiles. DXF is a proprietary drawing exchange format for vector data developed by Autodesk that is used to transfer cartographic data between software platforms. ArcInfo Export is a format for exporting data used in Environmental Systems Research Institute Inc. (ESRI) products.

Table 2.4 Summary of ANZLIC core metadata elements

Category	Element	Comment
Dataset	Title	The ordinary name of the dataset
	Custodian	The organisation responsible for the dataset
	Jurisdiction	The state or country of the Custodian
Description	Abstract	A short description of the contents of the dataset
	Search word(s)	Words likely to be used by a non-expert to look for the dataset
	Geographic extent name(s) or	A picklist of predefined geographic extents such as map sheets, local government areas, and catchments that reasonably indicate the spatial coverage of the dataset
	Geographic extent polygon(s)	An alternate way of describing geographic extent if no predefined area is satisfactory
Data currency	Beginning date	Earliest date of data in the dataset
	End date	Last date of information in the dataset
Dataset status	Progress	The status of the process of creation of the dataset
	Maintenance and update frequency	Frequency of changes or additions made to the dataset
Access	Stored data format	The format or formats in which the dataset is stored by the custodian
	Available format type	The formats in which the dataset is available, showing at least, whether the dataset is available in digital or non-digital form
	Access constraints	Any restrictions or legal prerequisites applying to the use of the dataset, e.g. for licence
Data quality	Lineage	A brief history of the source and processing steps used to produce the dataset
	Positional accuracy	A brief assessment of the closeness of the location of spatial objects in the dataset in relation to their true position on the Earth
	Attribute accuracy	A brief assessment of the reliability assigned to features in the dataset in relation to their real-world values
	Logical consistency	A brief assessment of the logical relationships between items in the dataset
	Completeness	A brief assessment of the completeness of coverage, classification and verification
Contact information	Contact organisation	Ordinary name of the organisation from which the dataset may be obtained
	Contact position	The relevant position in the contact organisation
	Mail address 1	Postal address of the contact position

	Mail address 2	Australia/New Zealand optional to mail address 1
	Suburb or place or locality	Suburb of the mail address
	Country	Country of the mail address
	Postcode	Australia—post code of mail address NZ—optional postcode for sorting
	Telephone	Telephone number of contact position
	Facsimile	Facsimile number of the contact position
	E-mail address	E-mail address of contact position
Metadata date	Metadata date	Date metadata record for dataset was created
Additional metadata	Additional metadata	Reference to other directories or systems containing further information about the dataset

Source: ANZLIC 1996 *Guidelines: Core Metadata Elements, Version 1*, Canberra: ANZLIC

information is made as freely available as possible so that the data can be reused for other purposes. For top-level directory systems, metadata may be summarised from more detailed levels held and maintained by data custodians.

ANZLIC metadata standards have been derived from the U.S. Federal Geographic Data Committee (FGDC) *Content Standards for Digital Geospatial Metadata* (CSDGM) first developed in 1993.[62] ANZLIC metadata standards are consistent with the Australia–New Zealand Standard on Spatial Data Transfer AS/NZS 4270. The Australian Spatial Data Directory (ASDD) is a fully distributed metadata directory launched in 1998 and has since achieved 19 separate nodes, including over 40 000 individual metadata entries. To participate as a node organisations need to meet several criteria including using the ANZMETA XML Document Type Definition (DTD) and run a server with Z39.50 search and retrieval protocol.[63]

The CSDGM is also the basis of an international metadata standard for geospatial data: ISO 19115 Geographic Information/Geomatics Metadata.

---

[62] See Federal Geographic Data Committee (FGDC) website at http://www.fgdc.gov/Metadata/metahome.html and at http://www.fgdc.gov/metadata/constan.html.

[63] See metadata tools and guidelines at http://www.auslig.gov.au/asdd/tech/tools.html. XML refers to eXtensible Markup Language, a flexible and powerful means of encoding data in text for programmatic manipulation. As plain text XML may be read by humans, but Web browsers with appropriate software can 'parse' or process XML structured text. Z39.50 an ISO standard is a network protocol which allows searching of remote heterogeneous databases and retrieval of data via one user interface. A protocol is a set of rules governing the exchange of information between computers. Since 1997 Z39.50 has become the standard protocol known as ISO 23950.

The CSDGM specifies exactly what elements should or may be included in a metadata record. In order that the metadata are useful, shared, and exchanged in any way, both the creators and users must understand the rules for creating metadata. There are more than 220 items in this metadata standard and is intended to describe digital geospatial datasets adequately for all purposes. As the creation of these metadata records can be a laborious task, software tools have been developed to create metadata records automatically.[64] The CSDGM has ten sections of which two sections are mandatory: section 1 on land information and section 7 on metadata definition. Apart from that, the standards do not specify how the content should be structured or in what form. This gives developers the freedom to implement the standards that suit particular software environments and applications.

Another metadata standard that must be mentioned here is the *Dublin Core Metadata Standard* (DC) which is an effort to find a minimum set of properties needed to support a search and discovery of datasets in general.[65] In its development the DC consists of a series of fifteen broad categories.[66] Each of these elements is *optional*, may be *repeated* as many times as required, and may be *refined* through the use of a set of sub-elements. The DC as a 'lite' or minimalist version of the CSDGM when used for geospatial data, however, can prove problematic. This is especially because the standard treats space and time as part of a single property and does not specify how that property is to be defined, for example, scale, projection and geodetic reference.

There are suggestions that the DC is of limited use to the GI community and this may be apparent by the preference of many users for the CSDGM and by implication the ISO standard. EuroGeographics, representing 40 national mapping agencies throughout Europe has rejected the DC in favour of 'discovery level' metadata emerging from the ISO standards.[67] Part of the reason for the rejection is that it contains too little information and handles spatial information poorly. But the DC metadata may have applications elsewhere, for example, in the discovery and retrieval of government services and information, as used by the Australian Standard

---

[64] See FGDC website for list of suggested tools and commercial vendors at http://www.fdgc.gov.

[65] For Dublin Core metadata standard see http://www.dublincore.org.

[66] The categories in the DC include Title, Creator, Subject, Description, Publisher, Contributors, Date, Type, Format, Identifier, Source, Language, Relation, Coverage and Rights. For practical implementation suggestions see Miller, P and Greenstein, D 1997 *Discovering On-line Resources Across the Humanities: A Practical Implementation of the Dublin Core*, Bath: UKOLN.

[67] See http://www.eurogeographics.org/eng/01_about.asp.

for Records Management AS4390.[68] The Open GIS Consortium (OGC) has recommended the use of catalogue services approach employing ISO metadata standards when these are ratified.[69] Indeed, there are plans to provide a pathway to extract DC metadata and to provide links to the relevant items in a DC metadata structure.

2.4.2 Clearinghouse and Geolibrary

A clearinghouse is intended to permit indexing and searching on-line metamedia entries that point to on-line data or mapping services. Clearinghouses support search for data based on geographic coordinates, time period content and other geospatial attributes. In some parts of the world clearinghouses are distinguished as either *catalogue services* or *registries* that help users or application software find information that exists anywhere in a distributed computing environment. Some registries interlink with other registries and where these conform with the OGC Catalogue Services Specification will mean that a query to one will be a query to all others. A clearinghouse helps a user find and evaluate data, but it does not usually provide the means for automatically viewing, downloading or operating on the data.[70]

The term *geolibrary* is used to describe a digital library on the Internet that can be searched for information about any user-defined geographical location. Just like a book catalogue in a library, it is possible to sort any GI collection by location and by date.[71] The term *collection level metadata* (CLM) describes the entire collection of datasets rather than individual datasets. It is as if it were a 'meta' metadata. Clearinghouses and digital libraries provide centralised interfaces that can search 'libraries' and 'supermarkets' on the Internet for the information that we may seek. A *geoportal* is an assembly of components that provides a community-wide, Web-based access point to distributed data and processing resources, including, for example, clearinghouses. A geoportal usually offers selected and evaluated links to other websites.

---

[68] See National Archives Australia (NAA) (1999) *Recordkeeping Metadata Standard for Commonwealth Agencies*, Ver. 1.0, May, 136 pp. Canberra: NAA.

[69] The OpenGIS Consortium (OGC) metadata specifications are found at http://www.opengis.org/techno/specs.htm.

[70] See introduction to building registries or catalogue services at http://wwwlmu.jrc.it/ginie.

[71] See National Research Council 1999 *Distributed Geolibraries: Spatial Information Resources*, Washington, D.C.: National Academies Press.

The U.S. National Geospatial Data Clearinghouse (NGDC) is a network of distributed data server nodes connected to the Internet. The NGDC will search all metadata catalogues according to title, keywords or bounding coordinates. In reality the Clearinghouse is a collection of over 250 spatial data servers with six gateways available in North America using Z39.50 search and retrieve protocols. The Clearinghouse permits custodians to make their data visible to users who access it and to search across multiple servers using a simple protocol.[72] This distributed service helps locate geospatial data based on characteristics identified in a metadata. It is like an Internet search engine for spatial metadata—a 'Google' for GI—but which allows a search to be performed by geographical attributes, words and phrases. To be recognised as a Clearinghouse a node would need to fulfil certain criteria. For example, the metadata must be provided in formatted text or marked-up in Standard Generalised Mark-up Language (SGML); indexed metadata must be available in various forms and connected to the Internet, and the ANSI Z39.50–1995 (ISO 23950–1997) search and retrieve protocol is used. Examples of Clearinghouses include the USGS facility based at the EROS Data Center, Sioux Falls, South Dakota; Space Imaging, Terraserver and GlobeXplorer.[73]

Clearinghouse sites are now beginning to provide hypertext linkages within their metadata entries to enable users to directly download the digital dataset. Where such datasets are too large to be made available through the Internet or where the data are for sale, linkage to an order form may be provided in lieu of the dataset. Most clearinghouses around the world are being updated to conform to OGC's OpenGIS Catalogue Services Specifications. This includes Australia (NSW Community Access to Natural Resource Information (CANRI) Natural Resources Data Directory), Canada (GeoConnections Discovery Portal), the U.K. (Association of Geographic Information's GIgateway), Germany (*Geodaten Infrastruktur Deutschland* GID NRW), Netherlands (The Dutch National Association for GI (RAVI) NCGI), South Africa (National Spatial Information Framework), Spain (IDEC project in Catalunya) and others.

---

[72] See http://www.fgdc.gov/clearinghouse/clearinghouse.htm.

[73] Space Imaging www.spaceimaging.com; Terraserver www.terraserver.com; GlobeXplorer www. globexplorer.com.

2.4.3 Access and Exchange Standards

The importance of standards for using and transferring GI as part of the spatial data infrastructure is well recognised. Standards permit the interoperability of agency systems across jurisdictional boundaries as well as within agencies themselves. Interoperability is the ability for autonomous, heterogeneous, distributed digital entities to communicate and interact and be used despite differences in systems (technical) or applications (semantic).[74] In general there are three types of geospatial standards:

- *Content standards*—including land use codes, surveyor codes, data dictionaries for cadastre, geographical place names, bathymetry;
- *Access standards*—including GDA94, ISO 19100 series (*Geographic information*), ISO 23950 (*Information Retrieval Z39.50*), most OpenGIS Consortium standards;
- *Exchange standards*—including Geography Mark-up Language (GML), Scalable Vector Graphics (SVG), Uniform Resource Identifiers (URIs also known as URLs).

The International Organization for Standardization (ISO) is responsible for the promotion and development of standards to facilitate the international exchange of goods and services. The ISO Technical Committee 211 (ISO TC211) is charged with developing and deploying standards relating to GI/Geomatics namely the ISO 19100 series.[75] The OGC is an international grouping of businesses, governments and universities that is developing publicly available geoprocessing standards known as OpenGIS® Specifications.[76] Working closely with ISO TC211 the OGC specifications are designed to support interoperable solutions that 'geo-enable' the Internet, wireless and location based services and mainstream information communications technology (ICT).

The U.S. FGDC actively provides international leadership in implementing spatial data standards through sponsorship of bodies such as the ISO and OGC. In addition the World Wide Web Consortium (W3C)

---

[74] *Technical interoperability* refers to the ability of different software systems to communicate and interact through a common interface. *Semantic interoperability* is the ability of people and software systems to find and use various types of spatial data that may or may not use standard naming or metadata schemas.

[75] See http://www.iso.org/tc211. Membership includes 29 participant (voting) nations, 27 observing nations and 22 external liaison organisations.

[76] The vision of the OpenGIS Consortium is '[a] world in which everyone benefits from geographic information and services made available across any network, application, or platform'. See http://www.opengis.org.

has developed standards for interoperable technologies enabling delivery of geospatial information on-line.[77] Standards Australia and Standards New Zealand jointly prepare and publish standards where appropriate. The Joint Technical Committee IT-004, consisting of experts from industry, government and other sectors, is responsible for GI/Geomatic standards in Australasia and is a representative at the ISO TC211. The Australian Government Information Management Office (AGIMO) successor to the National Office for the Information Economy (NOIE) and the New Zealand e-government initiatives promote an interoperability framework using open W3C standards for on-line government web services.[78] Figure 2.4 illustrates the relationship between the spheres of government, industry, and the wider community in the Australian context.

The U.K. National Geospatial Data Framework (NGDF) initiative was launched at the Association for Geographic Information (AGI) conference in 1995. NGDF aims to facilitate the unlocking of GI by enabling better awareness of data availability, improving access to the data and integrating data by using standards, for example, the 1998 Discovery Metadata Guidelines. The Inter-governmental Group on Geographic Information (IGGI) enables government departments to liaise effectively and exchange best practice in GI. The Geographic Information Charter Standard Statement (GICSS) adopted in 1998 is an industry agreement signed by organisations to conform to certain standards in the delivery of GI. Competitive funds obtained to promote the work of the NGDF together with the assistance of the OS resulted in projects like askGIraffe (launched 2000) and its later transfer to the metadata portal GIgateway.[79] With the latter spelling the demise of NGDF, GIgateway is NIMSA funded through the Office of the Deputy Prime Minister (ODPM).

This survey of geospatial standards has highlighted some important standards and specifications. Prominent among these are the ISO 19100 series standards, W3C recommendations and OGC specifications. Other

---

[77] See http://www.w3c.org. The mission of the W3C is 'to lead the World Wide Web to its full potential by developing common protocols that promote its evolution and ensure its interoperability'. A *Recommendation* is work that represents consensus within W3C and has the Director's stamp of approval. *Specifications* developed within W3C must be formally approved by the membership. Consensus is reached after a specification has proceeded through the review stages of *Working Draft, Proposed Recommendation,* and *Recommendation.*

[78] See information about NOIE at http://www.govonline.gov.au/projects/standards. See also NOIE 2003 *The Interoperability Technical Framework for the Australian Government,* Canberra: NOIE and at http://www.agimo.gov.au/publications/2003/08/framework/overview. Reference to the New Zealand e-government is at http://www.e-government.govt.nz.

[79] See http://www.gigateway.org/.

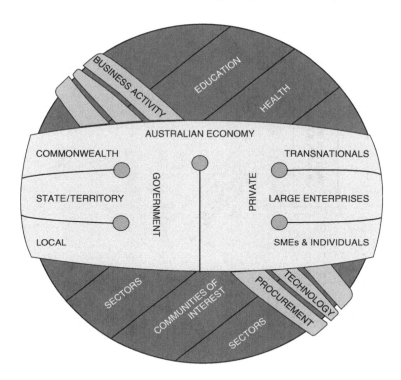

Figure 2.4 Australia: relations between sectors influencing interoperability. Source: with permission, ©AGIMO 2003. http://www.agimo.gov.au/publications/2003/08/framework/ overview

standards that deserve mentioning here are the Unified Modelling Language (UML), Simple Object Access Protocol (SOAP), Universal Description, Discovery, and Integration (UDDI) and the Web Services Definition Language (WSDL). UML is an industry standard for visualising, specifying, constructing and documenting artefacts of software intensive systems. It is a platform neutral environment for the abstract modelling of data and processes and is *the* conceptual schema language for ISO TC211. SOAP uses XML for sending messages over the Internet while the UDDI project is aimed at speeding up interoperability and adoption of Web services. The UDDI is seen among industry partners as a global business registry. Finally, the WSDL is an XML language for describing Web-based services and addresses the 'how' and 'where' of accessing a Web-based service. The relationship between of each of these standards and how each sits within the scheme of things is best depicted diagrammatically as shown in Figure 2.5.

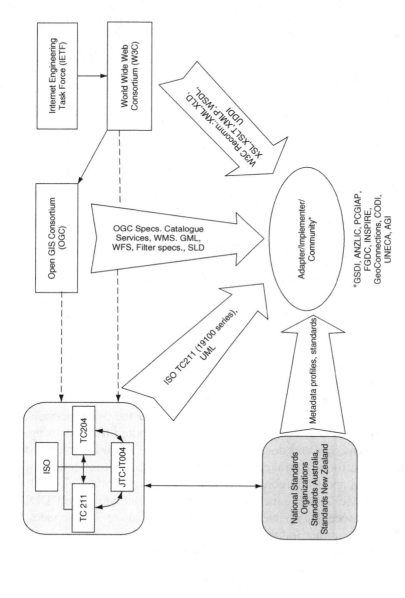

Figure 2.5 The Relationships and Interactions between international standards and national organisations

2.5 Towards a Global Information Infrastructure (GII)

The metaphor for an information infrastructure that most of us have come to know as the 'information super-highway', also known as the 'I-way' and the 'infobahn' has its roots in U.S. national policy, beginning with the passage of the *National Information Infrastructure Act* of 1993. A year later the Bangemann Report was released proposing a European version of the information infrastructure.[80] In the same year Vice-President Al Gore proposed a global information infrastructure (GII) at the International Telecom Union (ITU) *Buenos Aires Declaration for Global Telecommunications Developments for the 21st Century: An Agenda for Cooperation* (Brown *et al.* 1995).[81] The Group of 7 plus Russia agreed to collaborate in order 'to realize their common vision of the Global Information Society' and to work cooperatively to construct a Global Information Infrastructure.[82] The quotations from Vice-President Al Gore and the Bangemann Report eminently summarise the aims for such an information society.

> Let us build a global community in which the people of neighboring countries view each other not as potential enemies, but as potential partners, as members of the same family in the fast, increasingly interconnected human family. Vice-President Al Gore (1994).

> The information society has the potential to improve the quality of life of Europe's citizens, the efficiency of our social and economic organization and to reinforce cohesion. Bangemann Report (1994).

In parallel with developments at the world stage were those taking place within the U.S. and in particular President Clinton's decree in setting up the National Spatial Data Infrastructure (NSDI). In its launch President Clinton stated that 'Geographic information is critical to promote economic

---

[80] Bangemann Report 1994 *Europe and the Global Information Society: Recommendations to the European Council*, Brussels: European Council.

[81] Gore, A 1994 'Al Gore Speech on a U.S. vision for the Global Information Infrastructure', World Telecommunication Development Conference, Buenos Aires (March) quoted in Brown, RH, Irving, L, Prabhakar, A and Katzen, S 1995 *The Global Information Infrastructure: Agenda for Cooperation* (March) at http://www.iitf.nist.gov/documents/docs/gii/giiagenda.html.

[82] The G-7 nations include Canada, France, Germany, Italy, Japan, the United States, and the United Kingdom with Russia having participated in recent meetings. See G-7 Ministerial Conference on the Information Society (1995: 1–2). Available at http://europa.eu.int/ISPO/intcoop/g8/i_g8conference.html.

development, improve the stewardship of natural resources and protect the environment.'[83]

2.5.1 United States National Spatial Data Infrastructure (NSDI): Evolution and Growth

The evolution and formal launch of the NSDI began with a very strong political mandate given in President Clinton's Executive Order of 1994. The NSDI's role is primarily to facilitate the sharing of digital GI. In 1994 the Mapping and Science Committee (MSC) conceptualised the building blocks to the NSDI to include users, policies and procedures, institutional support, people, and GI, material and technology. Prior to this in 1990 the inter-agency FGDC, comprising of 19 cabinet level federal agencies, was set up as a result of the OMB Circular A-16. Revised in 2001 it requires coordination in the development, use, sharing, and dissemination of surveying, mapping, and related spatial data.[84] The four objectives of the FGDC include increasing the understanding of the NSDI through outreach and education; developing common solutions for discovery, access and use of GI; using community-based approaches; and, building relationships among organisations. These are to be undertaken through eight working groups and 13 thematic sub-committees.

The National Academy of Public Administration's 1998 publication gives unequivocal support for the development of the NSDI. In particular, the recommendation is for the drafting of a new statute in cooperation with state and local governments and other organisations to create an NSDI, establish a National Spatial Data Council, and better define federal agency roles and responsibilities for NSDI so as to meet the participating organisations' programmatic needs.[85]

In the U.S., several notable projects merit separate comments. First, since its inception the FGDC has awarded Cooperative Agreement Program (CAP) grants to help communities initiate NSDI. By 2001 the CAP had seeded some 270 NSDI resource-sharing projects across the U.S., involving more than 1300 organisations. The annual budget is in the

---

[83] Clinton, W 1994 Executive Order 12906—Coordinating Geographic Data Acquisition and Access: The National Spatial Data Infrastructure, *Federal Register*, v. 59(71), 11 April, pp. 17671–17674. Also at http://www.fgdc.gov/publications/documents/geninfo/execord.html.

[84] OMB 2001 Circular A-16 at http://www.whitehouse.gov/omb/circulars/a016/a016_rev.html.

[85] National Academy of Public Administration 1998 *Geographic Information for the 21st century— Building a Strategy for the Nation*, January. See publication summary at http://www.napawash.org/ napa/index.html; and http://38.217.229.6/NAPA/NAPAPubs.nsf/.

range US$1–2 million per year. Second, the I-Team Initiative addresses institutional and financial barriers to the development of the NSDI. An I-Team is a voluntary body of leaders representing all sectors of the geospatial community to plan, steward and implement the production, maintenance, and exchange of community information resources.[86] Already 39 states are actively involved in the I-team process and 6 have already submitted their I-plans.

A further project of note is the 2001 USGS 'The National Map' initiative with the goal that by 2010, the country will have current, accurate, and nationally consistent basic spatial data, including digital data and derived topographic maps.[87] In particular, basic spatial data will be made available for use by those carrying out programs and for re-use by organisations. As well licenses and other policies will allow for data sharing among partners to carry out federal activities.

The President's management agenda provides a pathway to a government that delivers results by making it easier, faster, cheaper for all levels of government and public to access geospatial data and information. The e-government initiative and funding of over US$350 million for the period 2003–2006 includes the use of digital technology to transform government operations to improve efficiency and effectiveness.[88] Hence with the launch of the Geospatial One-stop portal in 2002 the goal is to give federal and state agencies with a single point of access to GI, and to reduce and eliminate redundant data collection and archives. Milestones achieved so far include the identification of a federal inventory of framework data, a focus on standards, and the deployment of Geospatial One-stop portal in support of the NSDI.[89]

The federally driven NSDI involves stakeholders at the state and local levels as well. Altogether it is estimated that there are close to 80 000 agencies involved in the creation of GI in the U.S.

Apart from the complicated metadata requirements imposed on all creators of federal databases and information, the top-down approach has proven to be a disincentive to local government uptake of the NSDI. Federal agencies and state governors themselves are unsure how NSDI developments should take place and prefer new legislation and more

---

[86] See http://www.fgdc.gov/I-Team/index.html.

[87] USGS 2003 *The National Map*, US Geological Survey Report, 26 April at http://nationalmap.usgs.gov/nationalmap.pdf.

[88] U.S. 2002 *E-Government Act*, H.R. 2458/S. 803 signed 17 December 2002, effective 17 April 2003. See http://www.whitehouse.gov/omb/egov/pres_state2.htm.

[89] See http://www.geo-one-stop.gov/Intro.html and the portal at http://www.geodata.gov.

coordination between federal and stage agencies. Also while the research agenda needs to be supported into the future, the implementation and delivery of the NSDI promise is predicted to take place slowly. As an example, an attempt to privatise the Landsat imagery program was unsuccessful in the 1990s. It required Congress to pass the *Land Remote Sensing Policy Act 1992* that returned the program to the public sector and to more open and cheaper access to Landsat imagery. Moreover, agency budgetary cuts in future may reduce the scale and activities undertaken. In an open records environment like the U.S. there is no guarantee that at some stage users may be asked to pay for the cost of collecting and maintaining data when agency budgets are cut and priorities given to other areas of activity.

2.5.2 Canadian Geospatial Data Infrastructure (CGDI): Private Sector Leadership

Unlike other jurisdictions, the private sector plays a major role in the Canadian Geospatial Data Infrastructure (CGDI) through the Geomatics Industry Association of Canada (GIAC). The development of the CGDI has taken the form of a host of different projects, partnerships, and cooperative activities which together work towards a national infrastructure for accessing geospatial information in Canada. The target is 'a Canadian geospatial data infrastructure that is accessible to all communities, pervasive throughout the country, ubiquitous for its users, and self-sustaining, to support the protection and betterment of Canada's health, social, cultural, economic, and natural resources heritage and future'.[90]

The Canadian government funded GeoConnections to bring together all levels of government, the private sector and academia to work towards the establishment and implementation of the CGDI.[91] It has an allocation of CAN\$60 million, spread over six years from 1999. The five key policy thrusts implemented through the GeoConnections programs include fostering geospatial data access, providing a foundation of framework data, harmonisation of geospatial standards, encouraging private–public partnerships (GeoPartners, GeoInnovations), and supporting policies for the wider use of GI (GeoSkills, Sustainable Communities, The Atlas of Canada).[92] The lead agency for GeoConnections is Natural Resources Canada. The Canadian

---

[90] Canadian Geospatial Data Infrastructure 2001 'Canadian Geospatial Data Infrastructure Target Vision', *Geomatica*, 55, pp. 181–185.

[91] See CGDI GeoConnections site at http://cgdi.gc.ca/english/

[92] See The Atlas of Canada at http://atlas.gc.ca/site/english/learning_resources/ccatlas/index.html.

Clearinghouse set up under CGDI is interoperable with the U.S., E.U., and Australasian counterparts.

2.5.3 European Geographic Information Infrastructure (EGII): Balanced Representation

The European Umbrella Organisation for GI (EUROGI) defines GII as infrastructures that 'deal with matters that are much broader in scope and integrated in nature than GI policy matters. Typically the term "infrastructure" refers to all the materials, technologies, and people necessary to acquire, process, store and distribute such information to meet a variety of needs' (EUROGI 2000).[93]

The European Commission's draft Communication *GI2000* defines the European Geographic Information Infrastructure (EGII) as 'a stable, European-wide set of agreed rules, standards, procedures, guidelines and incentives for creating, collecting, exchanging and using geographic information, building upon and where necessary supplementing, existing Information Society frameworks' (EC 1998).[94] However, full support for the EGII throughout the E.U. never eventuated and the proposal lapsed and is being taken up by other general-purpose funded actions within the E.U. In attempting to explain the lack of success for the proposal, it has been suggested that different SDI models exist within each E.U. Member State and each is at a different level of sophistication in creating or promulgating a SDI at a national level (Longhorn 2001).[95] Indeed there is no mandate to create a European SDI of any kind.

GINIE, the Geographic Information Network in Europe is a project funded by the Information Society Technology Program of the E.U. (2001–2004).[96] It aims to develop a deeper understanding of key issues and actors affecting wider use of GI in Europe; to articulate a strategy to promote such use; and to pay closer attention to the role of GI in support of European policies with strong spatial impact. Part of the latter is the initiative to develop an Infrastructure for Spatial Information in Europe,

---

[93] EUROGI 2000 'Towards a Strategy for Geographic Information in Europe' ver. 1.0, October, Apeldoorn, Netherlands: EUROGI. Also at http://www.eurogi.org/geoinfo/eurogiprojects/strategy. pdf.

[94] European Commission 1998 *GI2000: Towards a European Policy Framework for Geographic Information*, Luxembourg: DG Information Society.

[95] *op. cit.* Longhorn 2001.

[96] GINIE partners include EUROGI, OpenGIS Consortium Europe, the Joint Research Centre (JRC) of the European Commission, and the University of Sheffield, U.K. as coordinator.

INSPIRE. The aim of INSPIRE is the 'making available of relevant harmonised, and quality geographic information to support formulation, implementation, monitoring and evaluation of community policies with a territorial dimension or impact'.[97] The message is further reiterated in the final GINIE Report (2004). In order to formulate a European GI strategy there is a need for 'strong leadership and a balanced representation of all these stakeholders... [as a] *conditio sine qua non* for instilling an implementing any strategy in Europe'.

While there have been pockets of activity at the regional level in different jurisdictions, the reality of a European SDI is hamstrung on a number of counts. Other than the lack of funding at a regional level, there is as yet no 'champion' overarching organisation to take the reins in fashioning a European-wide initiative. This stems also from a lack of a mandate, either at the Commission level or in an EC programme. The 'subsidiarity' principle of the E.U. is also unhelpful since it is intended to ensure that decisions are taken as closely as possible to those most affected by them. Thus, if the ESDI is to be a European level action it is to be initiated and paid for only in cases where Member States cannot undertake the required activities themselves. The commitment and willingness to put ESDI high on the political agenda is indispensable for its success.

The impasse may already be discerned from the different data access regimes—free access to full cost recovery. Unless GI is a commodity that is required everywhere for all sorts of applications, and unless accurate, detailed GI becomes freely available, the drive towards a truly ESDI could be hampered.[98]

In Europe, some commentators are not so optimistic, however. For example, the conclusions from a recent European study of the impact of data access policies have suggested that the wide variety of models have not been encouraging of a rapid formation of a pan-European SDI or EGII. Also, agreement on a truly harmonised national data access policy may never be achieved on a regional basis in Europe (Longhorn 2001).[99]

In an assessment of the GINIE Report entitled *Towards a European GI Strategy (2003)* Longhorn (2004) has suggested that the vision and

---

[97] GINIE 2004 *GINIE Final Report*, D-1.5.1 (January), 22 pp. available at http://wwwlmu.jrc.it/ginie/doc/GINIE_finalreport.pdf.

[98] Lemmens, MJPM 2001 'An European Perspective on Geo-Information Infrastructure Issues', at http://www.gisdevelopment.net/application/gii/global/giigp0007pf.htm.

[99] *op. cit.* Longhorn 2001. For an overview of SDI initiatives in Europe see GINIE: Geographic Information Network in Europe 2004 *A Compendium of European SDI Best Practice*, Ver. 1 (January) at http://wwwlmu.jrc.it/ginie/doc/d511_Book_v1.pdf.

strategy in that report should become the 'bible' for SDI implementers everywhere.[100] While the Report accepts that the 'overall picture in Europe is still one of considerable fragmentation [due to]...multi-cultural, multi-lingual and multi-national nature of Europe...the main challenges are organisational, institutional, and political in nature'. Longhorn laments the fact that 'unfortunately without continued strong political leadership and funding for necessary preliminary activities including either an Advisory Board for Geographic Information (ABGI) or a European Spatial Data Committee (ESDC) one wonders just how far we will get in fulfilling the dream of achieving a first fully functional regional (transnational) spatial data infrastructure in the world'.

2.5.4 Australian Spatial Data Infrastructure (ASDI) Developments

The vision of the Australian Spatial Data Infrastructure (ASDI) initiative is '[t]o provide better access for all Australians to essential spatial data'. ASDI has progressed from theory and organisation (1999) towards implementation and development of the component parts (2001). A consistent policy for development, access and pricing of GI has been achieved, together with the adoption of internationally compliant data directories and on-line atlases. Standards have also been developed so that more reliable framework datasets have become deliverable.

ANZLIC is the peak coordinating council for GI in Australasia.[101] Consistent with the government's initiatives with on-line access there is increasing pressure to provide fundamental spatial data on the Internet free and to improve signposts on the availability of information through metadata. Whereas in the past there had been little or no interest in GI in political circles, the new millennium has witnessed heightened interest and expectations politically and economically. First, the government's SIIAA (2001) involving both public and private sectors started the process. The incorporation of the industry association ASIBA (2001) also took place at about the same time as the incorporation of PSMA (2001) a government-owned entity. Then followed the formation of the Spatial Sciences Institute (SSI) (2003)[102] which brought together all professional 'spatial' associations.

---

[100] Longhorn, RA 2004 *GINIE: Geographic Information Network in Europe. Document Peer Review Report* (January), 6 pp. Available at http://wwwlmu.jrc.it/ginie/doc/ginie_peer_review.pdf.

[101] *op. cit.* see Nairn and Holland 2001.

[102] See http://www.spatialsciences.org.au.

These public and private developments have all been a part of the ASDI mission to deliver quality spatial reference information. At the same time there is an increased understanding of the importance of GI and its relevance to policy making. The challenge for the industry is to build on this momentum and to obtain further funding as well as involve areas of government that traditionally have little use of GI to help develop the ASDI.[103]

2.5.5 Asia-Pacific and Africa Spatial Data Infrastructure (SDI) Efforts

The vision for an Asia-Pacific SDI (APSDI) is one where there is 'a network of databases, located throughout the region, that together, provide the fundamental data needed by the region in achieving its objectives: economic, social, human resources development and environmental'.[104] The Permanent Committee on GIS Infrastructure for Asia and the Pacific (PCGIAP) made up of 57 member countries in the region is responsible for setting up principles and practice in the implementation of the APSDI. However, these principles are not binding in law or in any way enforceable against individual countries. Information access policies are also variable across all member countries.

In Africa, the development of a SDI is focused on the activities of the Southern African Development Community (SADC).[105] Here too, the resolution of important issues is yet to be achieved especially those concerned with data access. SDI delivery is hampered with the lack of telecommunications infrastructure for the Internet. There is as yet neither a national nor regional SDI policy and there are unresolved issues regarding data ownership and pricing.[106]

---

[103] Hobson, D 2001 NSDI Development in Asia and the Pacific. Report on Australian Activities, paper presented at the 7th PCGIAP Meeting, Tsukuba, Japan, 24–27 April. Available at: http://www.gsi.go.jp/PCPGIAP/tsukuba/seminar/paper_au.pdf.

[104] See PCGIAP website at http://www.permcom.apgis.gov.au/.

[105] For the African SDI see http://www.nsif.org.za/africasdi_main.htm.

[106] See Groot, R 2001 'Economic Issues in the Evolution of National Geospatial Data Infrastructure' Background paper for the 2nd meeting of the Committee on Development Information (CODI-2), 4–7 September, Addis Ababa, Ethopia. Available at http://www.uneca.org/eca_programmes/it_for_development/geoinfo/doc8EN(Economic%20issues%20in%20the%20evolution%20of%20Geo-information).pdf.

2.5.6 Global Spatial Data Infrastructure (GSDI) Strategic Plan

The first conference on a GSDI was in 1996 and in the second conference a definition of GSDI was formulated. A GSDI is one where '... [t]he policies, organizational remits, data, technologies, standards, delivery mechanisms, and financial and human resources necessary to ensure that those working at the global and regional scale are not impeded in meeting their objectives...' (GSDI 1997).[107] The definition has been further embellished with greater detail in the *SDI Cookbook*.[108] A GSDI Strategic Development Plan has also been developed where the vision is to 'support all societal needs for access to and use of spatial data'.[109]

The GSDI has neither formal organisational status nor funding base other than that which participants contribute.[110] The Secretariat operates out of offices at the U.S. FGDC in Reston, Virginia, U.S. The plan is to develop a more robust organisational structure that can interact with other related global organisations; one that actively promotes capacity building and research, and one that seeks and distributes funds to help develop fledgling SDIs around the world. Four permanent committees covering regions around the globe have now been established: PCGIAP and EUROGI, mentioned in the previous section; the Permanent Committee for the Americas (PCIDEA); and the recently supported UN Economic Commission for Africa (UNECA).

In attempting to implement a GSDI it seems that there are too many elements in the equation that are inconsistent across national boundaries and cultures that would serve to make the disparate components of the infrastructure 'work'. But one should not be discouraged—if one looks at how a phone link can be made from anywhere in the world to anywhere else in the world, the promise of a GSDI can be delivered in the not too distant future. Everyone in the geospatial industry must want it because it offers better decision-making, improved economic growth, social development

---

[107] GSDI 1997 *Conference Findings and Resolutions*, 2nd GSDI Conference, Chapel Hill, North Carolina, USA 19–21 October 1997, available at http://www.gsdi.org/docs/gsdi97r.html.

[108] GSDI Technical Working Group 2001 *Developing Spatial Data Infrastructures: SDI Cookbook* Ver. 1.1 (May) available at http://www.gsdi.org/pubs/cookbook/index.htm.

[109] GSDI 2004 *Global Spatial Data Infrastructure. Strategic Development Plan*, Ver. 0.9 (January) Draft at http://www.gsdi.org/DSP/GSDI_PlanD_v_0-9.htm.

[110] Masser, I and Stevens, AR 2003 'Global Spatial Data Infrastructure (GSDI): At the Crossroads, Moving Forward', paper presented to the Cambridge Conference 2003, Southampton: Ordnance Survey, paper 6.2. Borgman, CL 2000 The premise and promise of a global information infrastructure, *First Monday*, Issue 5(8) at http://www.firstmonday.org/issues/issue5_8/borgman/index.html [19 May 2004].

and environmental management. Hopefully, only then will the GSDI become a reality.

Longley *et al.* (2001: 426ff) have asked pertinent questions about the GSDI. They question whether the GSDI is a real global SDI or a federation of national or regional SDI? Is GSDI really about encouraging developing countries or is GSDI a process, a framework, or is it a particular product such as a world map or a comprehensive database? Who are the stakeholders and where is the demand for a GSDI? There may be no answers to these questions except to note that progress has been slow to date and that the real reason is not about inadequate technology, but about committed finance, expressed need, and politics.

Summary

This chapter has advocated the sharing of geospatial information and data at a local, regional, national and global level. However, as discussed in the first section, the sharing of data needs to have more than an altruistic motive because at some stage someone needs to pay for the data creation, compilation and maintenance. The ideal would be for that GI and data to be shared as a public good in the public commons; but the reality suggests otherwise. The sharing of this common good is not feasible because of the multitude of data access policies that exist around the world each with their own rationales and justifications. In examining the purpose and legal authority, the fiscal and economic frameworks, pricing policies and governance issues, the evidence suggests that there are as many models as there are participants in the sharing and sale of public information. The experiences and the successes also vary in width, depth and extent from zero cost to full cost recovery and variations between the extremes. However, what is undeniable is the growing maturity of the spatial information industry world-wide. This has also brought about an appreciation of the commercial benefits to be derived from GI products and services and the radical shift in industry dynamics that has included the public sector as an integral part of the business plan.

There are several routes and directions that an NSDI or GSDI can take, and each of these can be undertaken either individually or in tandem with all the others. The analyses and descriptions above suggest six ways in which to move forward. First the vision, concepts and benefits of the NSDI have to be sold both to the 'converted' as an exciting and innovative development, and to the rest of the professions that use GI and data. This selling of the

NSDI could be through demonstration projects, formal and informal education and training, but more importantly should be at a whole-of-government and private sector effort. As exemplified by the Australian Action Agenda for the spatial industry nearly everyone who has an interest in GI is involved. There is now a changed attitude towards public sector information in terms of its usefulness in functional terms, but also in a commercial sense. There are indeed new sectors and new resources for data production, integration and maintenance from non-traditional areas such as health care, tax, business and the law that have been awakened to the vast possibilities of the use of GI. The change of attitude is also in terms of identifying and supporting institutional and economic behaviours in policy and legal frameworks that promote development of NSDI.

A second way forward stems from the key word 'interoperability'. Efforts must be continued to finding common solutions for the discovery, access and use of GI. Through the use of metadata, the adoption of XML and GML that permits human and machine readability of metadata files, clearinghouse mechanisms, standards particularly open standards that together contribute to the interoperability of systems and the interconnectivity of agencies both public and private. Support has to be given to the development of non-proprietary tools such as open source for easy exchange, development, applications, information and results. There must also be research to develop and implement architectures and technologies that enable data trading and sharing.

A third way forward is to ensure that the efforts in creating the NSDI are community-based so that each participant has a stake in its development. The efforts of the NSW CANRI (community access to natural resource information) project, the community based IDEC project in Catalunya provide supreme examples that it is equally feasible to build systems from the bottom up. Similarly in Alberta, Canada, the Spatial Data Warehouse is a not-for-profit company is owned by utilities, municipalities and the provincial government. It was set up to maintain and build the digital mapping infrastructure for the province. Another Canadian corporation with the same structure is that of Service New Brunswick. These efforts demonstrate, not only an attitudinal change, but also one where the mindset has moved from data stinginess and a proprietary culture to one that is predisposed to giving and sharing.

Fourth, the building of relationships comes from a sense of community. The promotion of NSDI and indeed that of a GSDI should start in schools, tertiary institutions as well as in public and private agencies. People tend to be trusting and less wary in relationships when working towards a common good rather than in a competitive, adversarial environment. This could also be engendered in a competitive, commercial environment since a 'win' for one should also be a 'win' for the other.

Fifth, there is an old Scottish saying that 'many a mickle makes a muckle'. The idea being that one could build on a few achievable goals. The focus should be on practical activities that can be identified, funded and completed successfully in as short a time as possible. Together and over time, the successes will be greater than simply the sum of the parts. In tandem, there should be continued efforts at R&D between industry and public sector partners as well as encouraging pure and applied research in universities, quasi-public research organisations and institutions. In the Australasian context the idea of a Centre of Excellence in GIS, the Australian Cooperative Research Centre for Spatial Information, the Australian Spatial Industries Business Association and the professional Spatial Sciences Institute are all integral to the nation's Action Agenda.

Finally, organisations can and do influence policy. Such policy can be powerful forces of change where there is a political 'champion'. Political will is therefore imperative in bringing about rapid and significant change in the GI industry as witnessed in the mid-1990s. In the U.S. President Clinton and Vice-President Gore did much to bring about not only the GII but also the NSDI first in the G7 plus Russia nations and then generally to most developed parts of the world. There is a need for more such champions locally, regionally, nationally and globally. However, it is also to be accepted that there will be long lead times for plans to come to fruition and that the progress can be glacial.

Chapter 3
Geographic Information and Intellectual Property Rights

Learning Objectives

After reading this chapter you will:

- Develop a mastery of conventional intellectual property legal rights and contemporary issues and how these may be similar or different in an electronic environment.

- Be able to outline protective mechanisms for intellectual property rights.

- Understand how conventions, agreements, and treaties influence the international environment for intellectual property protection.

- Know the range of possibilities of protection for copyright in geographic information systems, science, and services, including the protection for maps and electronic databases.

- Be aware of atypical developments in intellectual property rights protection from the copyleft movement through to free open source proposals.

Geographic Information Science: Mastering the Legal Issues   George Cho
© 2005 John Wiley & Sons, Ltd   ISBNs: 0-470-85009-4 (HB); 0-470-85010-8 (PB)

- Know how to handle the ownership of intellectual property issues claimed by employers, and those rights generated by employees and contractors.

- Be able to anticipate the ways in which a geographic information professional will need to deal with intellectual property rights and analyse issues of practical importance from the experiences of Gigo, the GI professional.

3.1 Introduction

The storyline of the imagined life of Gigo below provides an introduction to intellectual property rights (IPR) and the pervasiveness of IPR in the information age generally and to geographic information in particular. This chapter addresses substantive IPR issues raised in the Gigo case as well questions that may be frequently asked and its implications for geographic information systems (GIS). Practice notes offer practical suggestions and timely reminders of the do's and don'ts to avoid litigation and damage awards for infringements. Finally, as many IPR issues were raised in the life of Gigo storyline a possible scenario of outcomes is sketched in the event of litigation to ascertain ownership, rights and damages to be awarded.

3.2 The Life of Gigo

Gigo is a practising geographic information professional working for the city-based firm Multinational Aerospace Systems (MAS) Pty Ltd in the Canberra suburb of Fern Tree Technology Park. Gigo created a class of codes, which he called Gigo's App Builder when he was a student at university, both to help with his assignments and to practise his code-writing skills. These codes comprised material which could be re-used in many applications and are similar to 'macros'—algorithms and codes for performing repetitive tasks. On graduation from university Gigo was employed by a spatial information company RESI Pty Ltd, at which time he used proprietary software for personal computers (PCs). Here he became familiar with proprietary software that used small macro language (SML) as well as other scripts such as Avenue and geographic mark-up languages (GML) for use in website development. After a year he left RESI Pty Ltd and joined MAS on the basis of his considerable skills. His job was to lead and support several domestic and international projects in spatial information systems, particularly mapping and geographic information systems (GIS).

In his new job Gigo used his library of codes in program conversions, developing new work and adapting on-going work. All these relied heavily on the library of code that he had developed at university and at RESI Pty Ltd. On his appointment at MAS there were no express arrangements regarding the use of Gigo's previous work. Gigo continued to work on and improve the library both during and outside working hours.

One of the major overseas projects that Gigo worked on was the creation of a GIS-based field implementation of precision farming, mapping and marketing software for smallholders in developing countries in Southeast Asia. The program has been code-named p-ArcMap. The objective was to write software for the Jhai PC to bring computers and Internet access to remote villages in Southeast Asia.[1] A parallel project was to support the Indian Simputer—the Simple Inexpensive Mobile People's Computer. The Simputer is a hand-held device designed as a low-cost portable alternative to standard PCs. Rather than being 'personal', the computer is owned by the community at large and may also be used as a mobile phone.[2]

This overseas project has many partners from non-government organisations (NGOs), local government and private consultants. Each of these partners has contributed their particular skills and expertise to the project. The creation of the software for smallholders has other spin-offs. Gigo was able to adapt some of the script for use in his butterfly hobby database which he developed for the local butterfly interest group, Ornithoptera. The multi-purpose database, open-source program named Omni-Base, is an easy to use, easy to adapt and shared among members of the group.

---

[1] The Jhai PC is the brainchild of the non-profit Jhai Foundation (http://www.jhai.org) in the U.S. At its base is the use of solar and muscle-power systems and low-cost wireless technology (wireless fidelity WIFI) for voice and text communication systems based on the Internet protocol (IP) specially designed for developing countries. Muscle-power generated from stationary bicycles provides five minutes of power for every one minute of pedalling to the 6-watt Jhai computer. This computer has access to the Web as well as the ability to make phone calls over the Internet—voice over IP (VoIP).

[2] Limited storage capacity means that it may have to be linked to a PC for the transfer and storage of information. An interface based on the Information Markup Language (IML) permits its use to be based on sight, touch and sound, thus enabling its use by illiterate people. The technology is based on GNU/Linux software and a chip that requires just three AAA batteries for its power needs. Simputer projects have been used to bring technology to schools, provide sources of micro-finance to farmers in rural areas and in e-governance in the automation of land records procurement. See http://www.picopeta.com/showcase.

On two evenings a week, Gigo also teaches Spatial Information Services (SIS) at the local Technical and Further Education (TAFE) Institute, sharing his experience and knowledge with acolytes wishing to get a qualification and find a job in the spatial information industry. The teaching materials and examples that Gigo uses include his library of code. Recently the TAFE Institute decided that Gigo should develop on-line teaching materials so that its flexible delivery maxim 'any time, any place, any where' could be actively implemented by all teaching staff.

Gigo resigned from MAS without notice, without leaving a copy of the source code of the library which App Builder and p-Arc Map needed to function. Gigo commenced a venture to commercialise the software programs that he had developed. A Canadian company, Map Le Serup Ltd, was incorporated for this purpose and Gigo assigned his interests in the two software programs to this company, but not the open source program Omni-Base. A nominal consideration of one dollar has been recorded for tax purposes. Map Le Serup later bought out Gigo's interests and he left the company and went on to develop a further class of libraries called Paradime which was also based on his library of codes.

The life of Gigo may contain much that is unsurprising as it appears to be what one would expect of a hard-working professional person. However, the facts underlie significant reminders of many legal issues that may have to be considered both by individuals, employers and companies in regard to matters pertaining to intellectual property rights (IPR) and other issues.

This chapter addresses IPR issues that impinge on GI systems, services, and science. There are nine sections in this chapter and the first section deals with the question of what are IPR in a general sense and whether such rights are any different in an electronic environment. In the second section the Australian legislative framework is examined for each of the rights and protective mechanisms. This section also gives a summary of the basic features and characteristics of IPR in Australia. In the third section IPR protection is characterised as a *quid pro quo* for maintaining a proper balance between protecting private rights and property on the one hand and for sharing knowledge, utility, and interests with the public on the other. This section also discusses the international environment for IPR protection in order to give a global view and how domestic laws have influenced and have been influenced by international conventions, agreements and treaties. The fourth section deals with copyright *per se* and its influence on GI—systems, services, and science. In particular the section addresses specific issues relating to maps, to electronic databases, the E.U. Database Directive, moral rights, and *sui generis* schemas that have been proposed, business methods/systems patents that relate to GI, and the implications of the Digital Agenda on GI. Then the fifth section addresses other legal issues and atypical developments that are intended to provide IPR protection in other ways—the 'copyleft' movement, for example. This section also discusses other IP such as geographical indications, photographs and fonts. The sixth section discusses infringements, defences to infringements of IPR, and suggests some remedies to these. In the seventh section IPR generated by employees is described within the context of IPR agreements with employees. As most GI professionals would be engaged in or involved in some form of consultancies, both domestically and internationally, there is a discussion of multi-participant international GI projects and IPR issues. This is because there will inevitably arise questions of the ownership of the project data, the economic protection of IPR, the resolution of disputes, questions of jurisdiction and law, and access to the information post-project. Finally, in the eighth section the fate of Gigo's code is given as a hypothetical legal brief in response to litigation and what lessons may be learnt from the activities undertaken by a GI professional. A summary suggests what may transpire in the near term—that contracts and licences rather than the present model might better serve copyright protection.

3.3 Intellectual Property Rights (IPR)

Intellectual property (IP) refers to the property of the mind or intellect and is any intangible 'thing' that gives one an operational and functional advantage over others. IP can be an invention, original design or the practical application of a good idea. IP law protects the property rights in these creative and inventive endeavours of inventors and developers and gives them certain exclusive economic rights for a limited time to profit from these creative works or inventions. However, this claim will also depend on particular circumstances and relationships because employers may have prior claims, as do independent contractors. An employer is the first owner of copyright in a work created by a person in pursuance of their employment under a contract of service. Similarly, copyright in certain commissioned works vests in the commissioning party. Partners to an invention may also jointly claim ownership to the intellectual property.

Ownership of IP rights is the legal recognition and reward that one receives for creative effort. The legal position in initial ownership of IP is that it can be modified by agreement either expressly or by implication from conduct. Ownership apart, the law also enables access by the community-at-large to the products of IP. Note that the protection is for 'rights' rather than physical property and therefore IP is a form of intangible property. In the U.S., copyright, as a subset of IP, exists for a public good; namely to permit and advance 'the sciences and useful arts'. Therefore the purpose of copyright law was never intended to reward creators, rather, it was viewed as a means to an end and not an end in itself. But at the same time it also suggests that there is no closing down of access to ideas, information and creativity.

Intellectual property, thus is the generic name that encompasses a bundle of rights, which protect innovation and reputation (Figure 3.1). It includes the regimes of copyright, patents, trade marks, designs, circuit layouts, plant breeder's rights, confidential information (or trade secrets), and a miscellany of others such as moral rights and *sui generis* [one of a kind] rights. While each of these regimes applies to different aspects of innovation or reputation, specific legislation protects each of these forms of IP. In general the framework for these pieces of legislation are largely based on a country's obligations under international treaties which it may have signed or acceded to.

In the spatial information context it may be noted that IP is omnipresent because the GI professional is engaged in all kinds of inventive, creative, original endeavour. The GI professional may also be involved in the compilation of data to build databases for map products. The data may be expressed in graphic and textual form in order to communicate ideas.

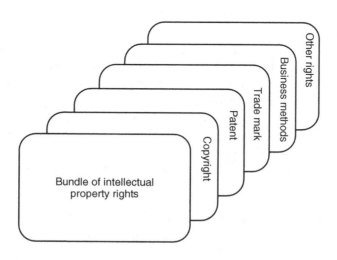

Figure 3.1 Intellectual property rights as a bundle of rights

In GI services there may be other 'clever' ways of delivering products, by way of methods or systems that are unique and special to spatial information systems. All these activities and delivery methods should be deemed valuable enough for the GI professional to seek some form of protection from loss, theft, and inadvertent slippage into the public arena. But, at the same time the GI professional needs to be mindful of avoiding any infringement of the rights of others and the consequent liability that might arise from such infringements. Especially with IPR the boundaries separating what may be in the public domain and who owns what can be 'fuzzy' and the boundaries easily transgressed. The potential for infringement is further exacerbated in the digital environment, given the ease with which information and data may be sent over the Internet with easy replication and the production of derivative works. A practical difficulty is in tracing the lineage of the data or the author of the work. The need to be able to trace lineage could arise for one of a number of reasons: to give attribution to the work; to seek permission; or to sheet home liability for damage that may have been caused from errors emanating from the data.

In GI there appears to be a preoccupation with copyright in regards to maps and recent developments in the dynamic electronic/Internet maps produced 'on-the-fly'. The copyright debate has focused on the issue of maps being 'factual' products and thereby being excluded protection since most copyright laws do not protect 'facts', only their expression. Today GI maps are stored as files on digital databases, raising the dual dilemma as to whether these factually accurate compilations in databases are given protection under copyright law. Even if the database were protected in some way, for example, the E.U. Database Directive discussed below, it is arguable that if one were to extract the facts and express these in a different graphical representation that very act may not amount to an infringement of the copyright subsisting in the database. Moreover, with databases it is often very difficult, if not impossible, to separate facts from expression.

There is no clarity as to how much protection existing law affords GI databases. In the U.S. the *Feist* case and in Australia the *Desktop Marketing* case raise serious problems for the copyright protection of maps and digitised GI systems in particular and databases in general.[3] This lack of clarity is seen very much as a live debate in academic and policy arenas

---

[3] *Feist Publications Inc. v Rural Telephone Services Company*, 111 S.Ct. 1282 (1991), 499 US 340 (1991); *Telstra Corporation Ltd v Desktop Marketing Systems Pty Ltd* [2001] FCA 612, *Desktop Marketing Systems Pty Ltd v Telstra Corporation Ltd* [2002] FCAFC 112.

as to how much and what kinds of protection is available—an issue raised by the two cases—to be discussed in depth later in this chapter. Also, further complications arise when considering the dual protection afforded by the E.U. Database Directive discussed next. The first protection of a database is via copyright for intellectual creations. The other is the database right that prevents unauthorised extraction or re-use of the contents of a database.

Similarly, 'deep linking' produces a new-age problem. Deep linking involves one website linking directly to the material buried in another website bypassing its home page with its advertisements, disclaimers, and conditions of use.[4] The IPR issue here depends on what one says about the material being shown on the second website.[5] A neat example is the case of deep linking to depict a street or tourist map from the original site using appropriate coding. The second website showing the maps from the original site is not reproducing anything and hence there is apparently no breach of copyright because it depends on the location of the source of the map. If it is a true link and the map is dynamically built by the original site, arguably this is acceptable, albeit discourteous if unacknowledged, and no copyright is breached.[6] On the other hand, if the deep linking is to images and maps in another's web page to create one's own web page, a derivative work is said to have been created. Under such circumstances, there is an apparent infringement of IPR.

The dilemmas in the electronic environment, coupled with the uncertainty in application of the law, the opacity in separating different rights, obligations and the duration of protection, and varying interpretations of the law within and between jurisdictions, provide interesting challenges for GI professionals working across international boundaries.

Various Australian government publications and reviews return to the theme that intellectual property is crucial to the promotion of innovation.[7] Innovation is integral to the competitiveness of organisations and nations,

---

[4] See http://www.selu.edu/Academics/FacultyExcellence/Pattie/DeepLinking/intro.html for a definition.

[5] See Wood, DJ 2002 'Best practices for avoiding linking and framing legal liability' at http://www.gigalaw.com/articles/2002-all/wood-2002-06-all.html.

[6] A number of cases have been litigated in the U.S. the most well-know being *Ticketmaster Corporation v Microsoft Corporation* No. 97–3055 DDP (C.D.Cal.12 April 1997). The IPR issues include copyright infringement, trade mark infringement, and unfair competition.

[7] See Copyright Convergence Group 1994 *Highways to Change*, Canberra: AGPS, Department of Industry, Science and Resources 1999 *Shaping Australia's Future. Innovation Framework Paper*, Canberra: AGPS and at http://industry.gov.au/library/content_library/shaping.pdf and Australia House of Representatives Standing Committee on Legal and Constitutional Affairs 1999 *Advisory Report on the Copyright Amendment (Digital Agenda) Bill 1999*, Canberra: Australian Parliament House.

and innovation drives economic growth. The conferring of enforceable rights on those who produce creative intellectual output gives the incentive for people to engage in these endeavours and intellectual property protection offers rewards for the investment of time and effort.

The publications also discuss the continuing debate as to how to cope with the changes brought about by information technology to the copyright regime. One view suggests that copyright law must be strengthened by extension and tightened in the form of a 'big stick' transmission right that should be protected. An opposing view argues that copyright law is dead in the on-line environment and that there must be other ways to protect rights, and to fulfil obligations and responsibilities. To date, it appears that a middle ground is gaining ascendancy, given that the law has been forced to deal with technological change, both by reference to precedent and by activism on the part of owners and users of copyright material. Despite its imperfections, the intellectual property regime appears to be working and adapting to the new environment. But, the continuing task is to create laws that can be flexible, adaptive and innovative in this new environment and to make clear whom the beneficiaries such reforms should serve.

In the emergent electronic environment two dilemmas are readily apparent: one is the concept of the 'form of expression' and the other is that of 'browsing'. The quandary in the first is that copyright owners seemingly have been disadvantaged with the advent of photocopying and the various exceptions under copyright law. Now however, it may be possible to use technology to track and charge every 'use' of the work on-line, no matter how small a part of the work that is 'used'. In the second dilemma, it used to be the case that on buying a book a purchaser had unlimited access to it for reading, reflection, review and critique, and other scholarly activities. However, in the electronic environment it may be possible that any browsing may be monitored and fees may have to be paid to a collecting agency on a pay-per-view basis.[8]

The copyright world of the 16th century printer's monopoly with its Royal Prerogatives and Charters empowered by the *Statute of Anne* of 1709 that gave copyright owners protection has been radically transformed by the digital era of 'bits and bytes' and instantaneous, perfect copies.

---

[8] In Australia the Copyright Agency Ltd (CAL) is the collecting society for both 'authors' and publishers and whose function is the collection of payments for use of copyright material as well as the distribution to members. Similarly in the U.S. the Copyright Clearance Centre Inc. (CCC) performs the same functions. (See http://www.copyright.com). The U.S. Copyright Act created the Copyright Royalty Tribunal, an independent agency appointed by the President to set royalty rates and distribute royalties for compulsory licences.

Since the first formulations of copyright protection much has been asked of the original ideals of copyright in particular and IP in general. There is persistent and intense pressure to look for new forms of protection and reward and yet safeguard the public interest in the free flow of information. Ironically, while the electronic world is borderless, IPR have attained international recognition founded on domestic laws.

3.4 Intellectual Property Rights Protection in Australia

The framework facilitating the protection of IPR in Australia is by common law, equity and statute. At common law the protection of IPR in goods and services is found in goodwill or commercial reputation attached to a name, mark or trade dress (get-up). Trade secrets and confidential information is protected by the equitable action of breach of confidence. An underlying objective of such protection is to foster innovation, the creation and design of new technology, and new methods of doing business. While the degree of protection is variable for GI products and services, the most relevant are copyright, database protection, patents, trade secrets and trade marks, and business methods. Other kinds of protection of peripheral importance to GI include moral rights, geographical indications, photographs and fonts. A general description of IPR protection is given below.[9]

Copyright

The *Copyright Act* 1968 (Cwlth) protects the rights of authors and creators in their 'works' of original creative effort. Such rights include the right to reproduce the work and to make copies and the right to present it to the public. Copyright is a type of property based on a person's creative skill and labour. The basis of this right is to prevent others copying the work without authorisation, in the original form in which the creator has expressed the idea or information. In Australia there are no formalities other than affixing the copyright notice and symbol—the letter c enclosed in a circle, thus ©, and the name of the creator and date. In the U.S., however, there are copyright registration requirements. The advantages for registration include establishing a public record of the copyright claim as well as the necessity

---

[9] IP Australia 2004 What is Intellectual Property? at http://www.ipaustralia.gov.au/ip and Attorney General's Department (2003) A Short Guide to Copyright at http://www.law.gov.au/.

for works of U.S. origin before an infringement suit is filed in court. Registration will also establish the *prima facie* evidence in court of the validity of the copyright, the assessment and award of damages, and to assist U.S. Customs Service to protect against the importation of infringing copies.[10]

In Australia copyright is free and automatically safeguards original works of art, literature, music, films, broadcasts, and computer programs from copying and certain other uses. Copyright exists as a bundle of economic rights pertaining to what one may do with the original work and other copyright material. Copyright subsists in an article, for example in a book, and whilst the creator may have ownership of the copyright, the book is owned by the person who purchased it.

Databases and Compilations

Copyright can and does subsist in compilations of factual information as are found in directories and databases. In a work that records facts whether it be an encyclopaedia, a map, a directory, a catalogue, a database or some other factual compilation there is neither copyright in the facts nor in each individual fact so presented. The author or creator however may have copyright in the form in which the facts are presented if there has been sufficient intellectual effort in the selection or arrangement of the facts. Copyright may also subsist if the author or creator has done sufficient work or incurred sufficient expense in gathering the facts. The *Copyright Amendment (Digital Agenda) Act* 2000 (Cwlth) has been passed in order to order to fulfil obligations under the Internet Treaties, namely for the protection of computer programs and databases.

Patents

The *Patents Act* 1990 (Cwlth) protects rights of inventors in their inventions, provided certain threshold requirements have been satisfied, such as novelty, inventiveness, utility and non-obviousness. A patent grants monopoly rights to inventors of new inventions. The monopoly is for a limited time of 20 years for the inventor to exploit the invention. Patent registration in Australia is usually filed with a patent office such as IPAustralia. Any disclosure of the invention before the patent application is filed will prevent registration. Where disclosure about the invention is necessary, such as to employees, business partners, or advisers, it must be

---

[10] For a general statement of the position in the U.S. see Copyright Basics at http://www.copyright.gov/circs/circ1.htm#cr.

done on a confidential basis and a written confidentiality agreement with these people is advisable. Artistic creations, mathematical models, plans, schemes or other purely mental processes cannot be patented.

Business Methods and Systems

It may be noted that patent law varies widely between countries. U.S. patent law permits the patenting of inventions that could not currently be patented in Australia or the U.K. One example is that of patents for business methods and systems. In the U.S. methods of doing business may now be patented, just like any other patent. Rather than an inventive product, the subject matter relates to a new, useful, and non-obvious way of doing business. There is no general definition of a business method patent. One example is the U.S. Patent and Trademark Office's (USPTO) grant to Amazon.com for a so-called one-click patent as a business method patent.[11] This patent is for a system and method for placing an order to purchase an item via the Internet. Information associated with a user is pre-stored by a website and the user may thereafter order articles from the website with only one click of the mouse—clicking a link associated with the item.

Designs

The *Designs Act* 2003 (Cwlth) protects rights in relation to the particular appearance of an article so long as its features are novel or original. This Act increases the threshold of distinctiveness required to obtain enforceable design registration and which may make it easier to prove infringements of registered designs. The registered design protects the features of shape, and pattern of ornamentation applied to the article. It also means that the owner may prevent others from using the design without an express agreement. To be registered, the design must be *new*—not known or previously used, and *original*—that is, the design has never been applied to the product although it may have been applied to another type of product. Registration of a design may subsist for up to 16 years. Some designs may also qualify for protection under copyright.

Trade Marks

The *Trade Marks Act* 1995 (Cwlth) protects the marks used by traders in relation to their goods and services so as to indicate the origin or trade

---

[11] U.S. Patent No. 5,960,411 issued on 28 September 1999.

connection of those goods and services. A registered trade mark is one for which protection is granted to a letter, word, phrase, sound, smell, shape, logo, picture, manner of packaging, or a combination of these. The trade mark, shown as ™, is used by traders on their goods and services to indicate their origin and to distinguish such goods and services from those of other traders. Initial registration is for 10 years with possibility of further renewals. Registration means that the owner has exclusive legal rights to use, license or sell it within the country for the goods and services for which it is registered. While registration is not compulsory for its use, there is protection against misrepresentation under trade practices or fair trading legislation and it is possible to take action under common law. Sometimes the term 'trade dress' is used to describe the way a product looks and service marks are also attached to identify particular services. Thus, Beatrix Potter's character and name *Peter Rabbit*™ is protected by trade mark law. It may be noted that trade mark law is very jurisdiction oriented, more so than either copyright and patent law. The duration of the patent and trade mark registration and protection also varies by jurisdiction.

Moral Rights

The *Copyright Amendment (Moral Rights) Act* 2000 (Cwlth) recognises the moral rights of creators and authors in their works even if they may have assigned their copyrights to someone else. Moral rights are a separate category of right apart from 'economic' rights normally associated with copyright. Here the rights protect the creator's honour and reputation and can prevent changes to a work. Unlike Australia and Europe, in the U.S. the protection of moral rights is for works of fine art only.[12]

Geographical Indications

The protection under this heading is that it identifies a good as originating in a given place. Geographical indications are the identification of a country or region where the quality, reputation or other characteristic of a product is essentially attributable to that geographic region, for example, *Champagne*, *Bordeaux* and *Cognac* readily identify particular regions in France that produce these wines and spirits. There are exceptions to protection, however, where the geographical indications become generic so that the names used refer to the 'process' or the grape variety, for example, Dijon mustard, Shiraz wine.

---

[12] *Visual Artists Rights Act* 1990 17 U.S.C. § 106A.

Protection of Other IPR

Trade secrets and *confidential information* are protected by an action in equity for the breach of confidence. This protection may be achieved by a confidentiality agreement with employees to stop them from revealing trade secrets or proprietary knowledge during and after their employment. The common law action of passing off or an action for misleading or deceptive conduct protects the business reputation and goodwill in unregistered trade marks or trade names.

Photographs are protected by copyright the moment they are taken and the right to protection is automatic and free. The length of time a photograph is protected depends on when the photograph was taken and when it was first made available to the public. Celebrities and well-known personalities may have a right to control the manner in which their name or likeness is used.

Fonts as in typography for print and other works are capable of special copyright protection. Copyright may be found to exist in fonts *per se* as artistic works and hence its unauthorised use can be an infringement of someone's rights. Thus, fonts may be licensed to others for their use. Such artistic works may also be capable of protection as a trade mark under the appropriate legislation.

Table 3.1 gives a summary of the basic features and characteristics of IPR, with a comparison of the basic features of the different mechanisms for protecting IP in Australia, the types of works protected, duration and formalities.

3.5 *Quid Pro Quo* and the International Environment for Intellectual Property Rights Protection

The rationale for protecting IPR, it is claimed, is to ensure that a proper balance between the rights and interests of copyright holders and the public is maintained. This rationale is found in the 1971 Paris revisions of the *Berne Convention* and *Universal Copyright Convention* where the incentive/dissemination, morality/fairness, and natural law arguments were debated.[13] Creators need to be given incentives to create, and economic theory suggests that most individuals are rational, profit-maximising creatures. To benefit from their creations these works need to be disseminated and shared with

---

[13] The *Berne Convention for the Protection of Literary and Artistic Works* (1886) as revised in Paris in 1971and the *Universal Copyright Convention* 1952 (UCC) as revised in Paris in 1971. Use of the copyright symbol © was significant when the U.S. was not a member of the Berne Convention and would recognise copyright only where the © was used in accordance with UCC. Other treaties that do not require any formalities have since overtaken the UCC.

Table 3.1 Summary of the basic features and characteristics of intellectual property rights protection in Australia

Right	Types of property/interests protected	GI examples	Duration	Formalities
Copyright	(a) Original literary, dramatic, musical or artistic works (b) Sound recordings, films, broadcasts, or cable programs (c) Typographical arrangement of published editions	Code and scripts for automating data processing Codes to generate databases from computer output	Generally life of the creator plus 50 years after death. *TRIPs life of author plus 50 years after death*	None. Automatic upon the work being 'created' and put into a tangible medium
Digital Agenda	Computer programs with technical protection mechanisms	Encryption, digital watermarks	As with copyright. *TRIPs 50 years*	None
Database Right	Databases	Databases that are *sui generis* [of its own kind]	*Sui generis* right runs for 15 years. *TRIPs 50 years*	None
Patents (a) Standard patent	New inventions including industrial processes	New type of printer design, computer casing, new method of making computer chips	20 years from date of patent registration. *TRIPs 20 years*	Application to IP Australia to be registered
(b) Innovation patent (previously petty patent)	All other types of invention that do not satisfy a standard patent application		8 years from date patent is sealed	Application to IP Australia to be registered
Business method and systems	Automated mapping systems, contour following devices, software mapping solutions that dynamically produce map locations and maps	Computer system to identify local resources, interactive automated mapping system	Up to 20 years from first registration	Application IP Australia to be registered

Table 3.1 (Continued)

Right	Types of property/interests protected	GI examples	Duration	Formalities
Designs				
(a) Registered Designs	'New' designs, being features of shape, configuration, pattern or ornament applied to an article and having 'eye' appeal	Aesthetically pleasing designs including typographical fonts and masthead designs for brochures and advertising material	Up to a maximum of 16 years from creation, but initially for a period of 12 months. *TRIPs 10 years*	Registration by application to the IP Australia
(b) Design Right	'Original' designs, being any aspect of shape or configuration of the whole or part of an article. Applies to functional and aesthetic designs excluding spare parts and surface decoration	CD cases, templates for drawing map symbols	Up to a maximum of 16 years from creation or 10 years from first use in public	None. Automatic as with copyright
Registered Trade Marks ™	A mark—device, brand, label, name, signature used to indicate a connection in a course of trade between the goods or service and the owner of the mark	ESRI, MapInfo, MapQuest, GeoMedia, Trimble	Initially for 10 years but renewable in 10-year periods indefinitely. *TRIPs 7 years and renewal indefinitely*	Applications for registration to the IP Australia
Moral	Any creative work of the intellect that qualifies for copyright protection	ESRI legend information design used in Arc/Info, ArcView	Life of author plus 50 years after death other than film which ceases on the death of maker	None

Type	Subject matter	Examples	Duration	How protected
Geographical Indications	Identification of a region where the quality, reputation or other characteristic of a product is essentially attributable to that geographic region	*Coonawarra, Claire Valley, Margaret River*	Indefinite as long as reputation is maintained with continued use. *TRIPs as with trade marks.*	Approved by Geographical Indications Committee on the basis of set criteria
Trade secrets and confidential information	Almost anything of a confidential nature	Idea for a new program code or invention (prior to patent registration), customer lists, business methods	Until such time that the subject matter falls into the public domain	Depends on confidentiality agreements with employees, partners and other parties
Passing off	Trade names, product 'get-up' and styles	Software names or acronyms that are unregistered as trade marks, but around which it has acquired a reputation associated with goodwill	Indefinite so long as the name, get-up or style is associated with reputation, for example, by continued use	None
Circuit Layouts	Topography pattern or arrangement of layers in IC chips in eligible layout (EL)	New design of IC	10 years initially at creation to a maximum of 20 years. *TRIPs 10 years*	None
Plant Breeder's Rights	New variety as distinct, uniform and stable and is clearly distinguishable from any other variety	Geographical indications and geographic regions	Up to 25 years for trees or vines and 20 years for other species	Registration with IP Australia

Note: The standard duration of protection under the World Trade Organization (WTO) Agreement on Trade-related Aspects of Intellectual Property Rights (TRIPs) is given in italics

everyone. The morality/fairness argument says that it is morally right and fair for individuals to be rewarded for their skill, efforts and expertise in producing something useful. Natural law dictates that there be some form of a reward for intellectual endeavour that arguably is for the common good.

In the U.S. the purpose of copyright law has been to promote the sciences and useful arts as enshrined in the U.S. Constitution. This edict authorises Congress '[t]o promote the Progress of Science and useful Arts, by securing for limited Times to Authors and Inventors the exclusive Right to their respective Writings and Discoveries'.[14] As President James Madison, one of the framers of the Constitution has noted, copyright provides an efficient means of achieving these constitutional goals because it was one of those fortuitous policies in which the ends of the individual citizen and the goals of the community could be made synonymous.

Together with these ideals and to enable openness in governance and the dissemination of information to citizens, all public records and public information in the U.S. are not copyrightable. These works are said to be in the public domain and are therefore are unprotected. It also means that such works may be freely copied or used in the creation of derivative works without permission or authorisation. These public domain works include U.S. government publications, judicial opinions, legislative enactments, unadorned ideas, blank forms, short phrases, titles and extemporaneous speeches.

For private works, however, whether a work has come into the public domain or not is governed by the relevant U.S. *Copyright Acts*. For works before 1978, the 1909 U.S. *Copyright Act* offers protection of up to 95 years, provided renewal formalities have been followed. The 1976 U.S. *Copyright Act* governs works produced after 1978 where the protection is for the life of the author plus 70 years thereafter. In 1999 the *Sony Bono Copyright Term Extension Act* (CTEA) added a further 20 years to most copyright protection terms thus protecting works for up to 95 years from first publication. Any work published prior to 1923 is now in the public domain, provided the term has not been extended.

Lawrence Lessig, the Stanford Law Professor has led a U.S. Supreme Court challenge in 1998 opposing the copyright legislation extending the term by a further 20 years.[15] The extension is the 11th in the past 40 years and the challenge argued that the Constitution restricts the scope of the power of Congress to grant copyright extension. Moreover the basic *quid pro quo* model

---

[14] U.S. Constitution Art. 1, § 8 Cl. 8. See also Branscomb, AW 1994 *Who Owns Information? From Privacy to Public Access*, New York: Basic Books, p. 8.
[15] *Eldred v Ashcroft* 537 U.S. 185, 65 USPQ2d. 1225 (2003).

is designed to provide monopoly rights to produce something for the public in return for protection for a limited time. Extending copyright over existing works, it was submitted, will not induce any new creativity, and the monopoly right is being granted for 'nothing' in return and therefore diminishes the public domain. However, the U.S. Supreme Court, by a majority of seven to two, held that Congress had the power to extend the duration of copyrights.[16]

The Australian IPR regime seems more liberal than in the U.S., but here too there is resistance to the pressures from large media copyright owners who wish to extend copyright terms. As in Australia, the E.U. is also preparing to tighten its IPR laws, as will be noted later in this chapter. A synoptic view of this balancing of interests is given in Figure 3.2.

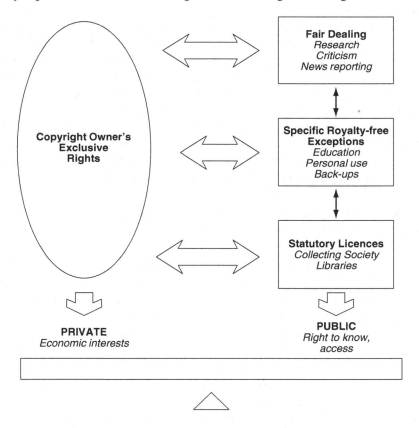

Figure 3.2 Copyright *quid pro quo*: a balancing of interests

---

[16] McCullagh, D 2003 'Supreme Court nixes copyright challenge', 15 January at http://news.com.com/2100-1023-980792.html.

3.5.1 Intellectual Property Rights Conventions, Agreements and Treaties

Modern-day rationales for protection of IPR spring from international conventions, agreements and treaties. Domestic copyright and IPR legislation in most countries is strongly influenced by a country's participation in, and signatory to, international agreements. The first and still most important is the Berne Convention 1998 (see below). Then there are those administered by the World Intellectual Property Organization (WIPO), based in Geneva, a specialised agency of the United Nations, and the requirements of the agreement on trade-related aspects of intellectual property rights (TRIPs) emanating from the World Trade Organization (WTO) Agreement of January 1995. Also there is the WIPO Copyright Treaty (WCT) and the WIPO Performances and Phonograms Treaty (WPPT) adopted by the WIPO Diplomatic Conference on 20 December 1996. Both these 'Internet treaties' update the Berne Convention, the Rome Convention and the TRIPs Agreement, and came into force in 2002. To date Australia has not signed either treaty.[17] Member countries comply with the treaties or face the prospect of losing membership and thus protection of IP materials in other countries. In cases of blatant infringement of copyright, trade sanctions may follow. Set out below is a brief overview of these international conventions, agreements and treaties relevant to IP and the GI environment.

3.5.2 Berne Convention for the Protection of Literary and Artistic Works 1998

Australia is a party to this Convention in its own right since 1928. The Berne Convention covers 'every production in the literary, scientific and artistic domain, whatever may be the mode or form of its expression'. The Convention sets out the basic categories of what can be protected under copyright (written expression, visual arts, music and films). There is also the right to reproduce, broadcast and adapt and exceptions to these rights and conditions under which their use is permitted. Finally, the Convention sets out the duration of copyright protection—basically the life of an author plus 50 years after death. In technical terms the principle of national treatment gives the same protection in member states as that provided to the country's own citizens. Copyright protection is automatic without any formal

---

[17] Details of these conventions, agreements and treaties may be found at http://www.wipo.org/about-wipo/en/index.html.

requirements such as registration. Also there is list of nine exclusive rights including the preservation of the 'moral rights' of authors within the *droit du suite* and *droit moral* principles. All of these principles embodied in the Berne Convention underpin copyright laws of most signatory countries including Australia.

3.5.3 Rome Convention for the Protection of Performers, Producers of Phonograms and Broadcasting Organizations 1961

Australia became a party to the Rome Convention on 30 September 1992. The Rome Convention builds on the Berne Convention and in particular performers are protected against unauthorised recording or broadcasting of their live performances. Producers of sound recordings (phonograms) are protected against direct or indirect reproduction, while broadcasters can prohibit unauthorised recording or rebroadcasting. Copyright protection for these types of creators and investors is also known as 'neighbouring rights' in some countries. The Rome Convention also has rules about what subject matter is to be protected, what rights copyright owners have, what exceptions may be granted and the duration of protection. In general, exceptions to these rights include private use in research and news reporting of current events.

3.5.4 Agreement on Trade-related Aspects of Intellectual Property Rights (TRIPS) 1995

Australia, as foundation member of the WTO, is a signatory to the TRIPs Agreement. This Agreement seeks to strengthen the international IP environment by setting minimum standards of protection for domestic enforcement. These reiterate and add to the legal and institutional provisions specified in earlier agreements. TRIPs require that parties comply with the chief obligations of the main conventions of WIPO; the Paris Convention for the Protection of Industrial Property (Paris Convention in relation to patents and trade marks) and the Berne Convention. Under TRIPs four main provisions of relevance to the GI industry include the following. First, all computer programs and databases are protected by copyright, with computer programs receiving a similar level of protection as that given to literary works. Second, limitations have been placed on exceptions to exclusive IPR where the interests of right holders may be prejudiced. Third, there is an acknowledgment of the possibility of anti-competitive abuse for IPR and an affirmation of the right to take action where such abusive practices are apparent. Finally, there are the provisions of the Washington Treaty on IP in respect of integrated circuits. While not yet in force nor signed by Australia, domestic Australian legislation are consistent with these provisions through the *Circuit Layouts Act* 1989 (Cwlth). The main difference between TRIPs and other IP treaties is that failure to comply with the Treaty can lead to WTO trade sanctions.

3.5.5 WIPO Copyright Treaty (WCT) and WIPO Performances and Phonograms Treaty (WPPT)— the Internet Treaties

Finalised in Geneva in December 1996 these Internet Treaties update the Berne and Rome Conventions, respectively, to take account of new technologies such as the Internet. The centrepiece of the two treaties is a new technology-neutral right of communication to the public, that is, using this right, a copyright owner may authorise the electronic distribution of their creation via any technology, regardless of whether or not it is 'wired'.

The WCT came into force in 2002 and while Australian government statements have been broadly supportive of the aims of the treaty has yet to sign it.[18] The preamble to the WCT expresses the need to 'provide adequate solutions to the questions raised by new economic, social, cultural and technological developments'. The WCT requires contracting states to comply with the substantive provisions of the 1971 Paris revision of the Berne Convention. Two provisions are particularly relevant to the GI industry. The first is with regard to computer programs in whatever form and the second relates to compilations of data or other material which by reason of their selection or arrangement of their contents constitute intellectual creations, that is, original databases.

Both treaties introduce obligations to prohibit circumvention of technological measures to protect IP. Tampering with technological protection measures (TPM), such as encryption, to protect copyright and related rights, and watermarks, that contain digital rights management information, are prohibited. Amendments to the *Copyright Act* 1968 (Cwlth) in the

---

[18] About 50 countries have either signed the Internet Treaties or simply ratified them. Ratification is a necessary step for countries to be a party to the treaty. Australia has not yet signed or ratified either treaty, although it may do so at some stage given the two amendments of the *Copyright Act* 1968 (Cwlth) in 2000.

Digital Agenda and Moral Rights legislation have ensured Australia has fulfilled the spirit and purpose of the Internet treaties.[19]

3.6 Copyright and Geographic Information

Copyright is the exclusive right given to a creator to reproduce, publish, perform, broadcast and adapt a work. In Australia the *Copyright Act* 1968 (Cwlth) and the *Copyright Amendment Act* 1984 (Cwlth) applies.[20] Copyright does not exist apart from the Act and is vested in the creator as soon as the work is created, even though it may as yet be unpublished. There is no registration requirement in Australia.[21]

A *royalty* is a monetary consideration received by the owner of a copyright in a work from a person for a licence to make copies of that work.

Copyright impinges on nearly every aspect of GI as we know it. GI systems as applied to geographical data will involve a system of hardware, software and reiterative procedures that are designed to support the capture, management, manipulation, analysis, modelling and display of spatially referenced data. The present task analyses copyright issues as they relate to GIS by examining each functional step in the system where a perceived IP right might subsist. The task is complex as there are interactions between the various steps and procedures, but also interactions between the software and hardware that produce further elements that need protection.

Copyright is primarily concerned with the prevention of unauthorised copying. It is a right to prevent the unauthorised reproduction by a third party of the form in which a person has chosen to express ideas (Ricketson 1989: 480).[22] Paradoxically, copyright law developed out of a right to copy, hence, *copy-right*. In the Middle Ages, the owner of a manuscript was understood to have possessed a right to grant permission

---

[19] *Copyright Amendment (Digital Agenda) Act* 2000 (Cwlth) and the *Copyright Amendment (Moral Rights) Act* 2000 (Cwlth). In the U.S. circumvention devices are prohibited under the *Digital Millennium Copyright Act* 1998 (DMCA).

[20] In the U.S. the equivalent is the US Code Title 17 *Copyright Act* 1976 Public Law no. 94–553, 94th Congress. In Canada the *Copyright Act* 1988 and in the U.K. the *Copyright Act* 1956.

[21] In the U.S., Taiwan, and Canada registration has ceased to be a prerequisite for the subsistence of copyright, although it can be a necessary condition of suing for infringement and/or recovery of damages. In Japan, the registration provision remains.

[22] Ricketson, S 1989 'The use of copyright works in electronic databases', 1989 *Law Inst. J.* 480–482.

to others to copy it, a right that could be exploited for profit. The closest analogy being religious monasteries that charged a fee for permission to copy one of their books (Rose 1993: 9).

Copyright protects 'works' grouped by category, such as literary, dramatic, musical, artistic, and 'subject matter other than works', such as film, sound recording, television and sound broadcasts, and published editions. Each category attracts certain exclusive rights for the copyright owner, provided three main requirements are met. The first is that the author is a 'qualified person', such as an Australian citizen or resident. The second is that the work must be 'original' and is the product of an author's skill and labour. Finally, the work must be reduced to a 'material form', such as writing.

A grant of copyright lasts the life of an author plus 50 years after the death of the author. An unpublished work enjoys perpetual protection; however, publication 'triggers' the start of the time limitation of 50 years. Affixing the copyright symbol © is not relevant, but may be valuable in proving ownership and date of creation especially when a dispute arises.

The exclusive rights in 'literary works' granted under s 31(1) of the *Copyright Act* 1968 (Cwlth) include the right to:

- reproduce or copy the whole or significant portions of the work in material form;
- publish the work;
- perform the work in public;
- cause the work to be transmitted to subscribers to a diffusion service and to transmit and broadcast the work;
- make adaptations or translations of the work from one language to another or to create a 'derivative work' where the substance of the work is taken, but is converted to another form.

Such rights can also be sold or transferred to another person.[23] An agreement of the assignment of copyright in GI must be in writing.[24]

---

[23] In the U.S. the *Copyright Act* 17 USC § 106 gives an owner of a work exclusive rights to copy, prepare derivative works including making modifications, distribute, and for special categories of works, to public performances and public displays. No one else may exercise these rights without the authorisation of the owner (see Greguras, F, Egger, MR and Wong, SJ 1995 'Multimedia content and the Super Highway: Rapid acceleration or foot on the brake?' URL: http://www.batnet.com/oikoumene/mmcopyright.html).

[24] *Copyright Act* 1968 (Cwlth) s 196(3).

A 'fair dealing' or 'fair use' clause permits users to copy limited amounts of a work for personal research or study purposes. The permitted use here is in terms of the *type of use* rather than the amount used that is determinative. Thus, while it may be a technical violation of copyright to reverse engineer (or de-compile) someone else's computer program the U.S. position is that it is permissible if it is in the public interests to do so.[25] In the U.K. reverse engineering of computer programs is permitted under certain circumstances.[26]

Section 10(1) of the *Copyright Act* 1968 (Cwlth) defines 'literary work' to include a table or compilation, expressed in words, figures or symbols whether or not in a visible form and includes a computer program or compilation of computer programs. This section recognises the importance of IT and its manifestations. More importantly, it assists in placing GIS products within a proper frame of reference within copyright law.

Note also that a single work may have separate copyright interests in terms of separate 'owners' and different duration of copyright protection, because of the different commencement dates for the various copyrights embedded in the work. Thus, a video recording will incorporate a script, music and rights to making a film in which each interest may belong to a different person. There may be different types of copyright involved, lasting for different lengths of time.

3.6.1 Maps

In Australian copyright law printed and digital maps are copyrightable because these are legally interpreted as 'literary' and 'artistic' works under s 10(1).[27] Literary works include a table or compilation, expressed in words, figures or symbols, whether or not in a visible form. Map data stored in an electronic database, even though seemingly a factual compilation is considered a literary work. A drawing is an 'artistic' work, which may include a diagram, map, chart or plan. The *Copyright Act* 1968 (Cwlth) also requires 'originality' in the sense that the author has expended a minimal degree of skill, judgement and labour to achieve that result.

---

[25] U.S. reverse engineering cases include *Sega Enterprises Ltd v Accolade Inc.* (Unrpt. 9th Cir. No. 92–15656, 20 October 1992) and *Atari Games Corp. v Nintendo of America, Inc.* 975 F.2d 832 (1992). See the relevant sections in the *U.S. Copyright Act*: 17 USC § 102(b). See also Samuelson, P 1992 'Computer programs, user interfaces and § 102(b) of the *Copyright Act* of 1976: A critique of *Lotus v Paperback*', 55(2) *Law and Contemporary Problems* 311.

[26] *Copyright (Computer Programs) Regulations* 1992 (U.K.).

[27] The position in the U.S. is similar. See 17 USC § 101 and § 102.

The concept of originality can be most problematic when one interprets maps as electronic databases, that is, those that store facts can be thought of as merely compilations. For a compilation to be 'original' an author must demonstrate a certain degree of skill, not only in selecting the factual content, but also in the expression of those facts so that the arrangement in presentation can be deemed to be a truly original product.[28]

Any map thus comes within both the 'drawing' and 'literary work' definitions of the Copyright Act. This dual categorisation raises uncertainty, particularly in relation to a number of statutory defences to infringement of copyright.

Initially, maps, as with directories and other fact-based works, enjoyed a flexible judicial approach. Prior works could be used by a later cartographer on the basis that it was in the public interest to produce more accurate maps.[29] However, by the eighteenth century, the approach had become similar to those adopted in cases involving directories and other types of works. Thus, while a later cartographer was free to make a map on the same subject as an earlier cartographer, the obligation was to do this independently and not to copy or use any of the results of work done by a predecessor. Indeed as PAGE WOOD VC indicated by way of *obiter* in *Kelly v Morris*,[30] that it would even be impermissible for the later author to refer to the original map at all!

The leading Australian map case is that of *Sands & McDougall Pty Ltd v Robinson*.[31] The High Court held that the defendant had produced the plaintiff's map of the Balkans, even though the former had changed the colour of the political divisions, corrected the Balkan boundaries, introduced some places which had acquired recent prominence and deleted some other places. ISSACS J at 52–53 said that 'the map was not a mere copy in the ordinary sense of the term, but it was clearly a reproduction of a substantial part of the respondent's map in a material form which necessarily violated the respondent's copyright if his work is protected by the Act'. If this is the correct approach involving maps then it suggests that later map makers should steer clear of all earlier publications and begin from scratch—an altogether unpalatable and impracticable prospect (Ricketson 1984: 210).[32]

---

[28] *Sands & McDougall Pty Ltd v Robinson* (1917) 23 CLR 49—a case of inventive originality.

[29] *Sayre v Moore* (1785) 1 East 316, n.5, 102 ER 139.

[30] (1866) L.R. 1 Eq. 697 at 702.

[31] (1917) 23 CLR 49.

[32] Ricketson, S 1984 *The Law of Intellectual Property*, Sydney: Law Book Company.

Maps are graphical representations of space whether in one, two, three or *n* dimensions. More conventionally, we may think of maps as depicting terrestrial space related to some feature of the Earth, from the mountains to the plains to the sea and deep ocean valleys. We may also divide such terrestrial space into discrete regions to depict both administrative boundaries and social space in terms of cultural groups. In a GI context the use of 'layers' of information superimposed on a common base map produces a map (Figure 3.3). The map composition comprised of layers may have multiple interests and rights—a fragmentation of IPR and interests—since the data may be derived from different sources and at different times. While each of the layers is self-explanatory, that of the data commons and IPR statement needs explanation. Data commons refer to public domain data that belongs to everyone and the IPR statement such as a copyright notice is one that is found on both the final map and on the map or digital database.

A further feature of maps is that they are always scaled to suit the purposes for which they will be used. Town planning maps will necessarily be on a very large scale, perhaps 1:200, whereas weather maps may be on a continent-wide scale of more than 1:1 000 000.

Some map features are stylised by colour, line width and name placement, involving a high degree of cartographic licence and some 'cartographic silences' as well. Moreover, a cartographer will have to contend

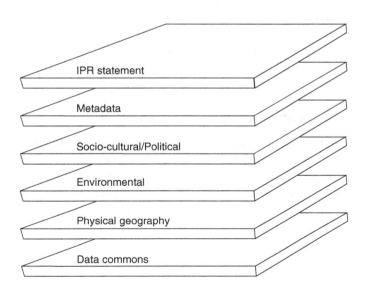

Figure 3.3 Layers of information in a map

with attempting to show what are essentially three-dimensional curvilinear features of a spheroid on a two-dimensional piece of paper. In essence this is cartography in its element, depicting, interpreting and creating an expression of reality. While some features may be both geometrically correct and positionally accurate on a map, the totality of the map form is but representative, and is one of the many possible ways of depicting the real world.

A further issue for copyright protection of maps is the rule that ideas, facts or information are not copyrightable, only the form in which those ideas, facts or situations are expressed.[33] The courts in the U.S. have characterised this issue as the 'merger doctrine' which provides that when the expression of an idea is inseparable from the idea itself, the expression and idea have merged. This therefore prohibits copyright protection.

The following U.S. court case where this map doctrine was argued is found in *Kern River Gas Transmission Co. v Coastal Corp.*[34] In this case the plaintiff produced a map of the proposed gas transmission lines based on a 1:24 000 scale U.S. Geological Survey (USGS) topographic map. The mapping was based on independent field surveys. The Court denied the plaintiff's claims of copyright protection and held that, while the mapping of transmission lines did not lack originality, the maps used the only effective way of expressing the idea of the location of the pipeline. The implication from this court's decision is that any map which tries to be 'correct', as most makers of maps will invariably claim, will be characterised as a 'pictorial presentation' which is not protected by copyright. Furthermore, to protect such types of maps may give the plaintiff a 'monopoly over the facts'. It is to be observed that this is the first case in the U.S. that classified maps as pictorial, graphic works following the 1976 *Copyright Act*.

Digitising from and Scanning of Paper Maps

The data used in GIS are digitised versions of maps. To these data have been added other information, such as various forms of indexing, cross-matching, unique identifiers and other software generated markers. These extra pieces of information make the digital versions of maps more 'functional' in a GIS. However, while these have made the digital data very

---

[33] See *Victoria Parks Racing and Recreation Grounds Co. Ltd v Taylor* (1937) 58 CLR 479 per LATHAM CJ at 498 'The law of copyright does not operate to give any person an exclusive right to state or describe particular facts.'
[34] 899 F.2d 1458 (5th Cir. 1990).

complex, the result makes the data easier to use and access. Reference to any text on GIS will show that the data may be used for depicting graphical representations, giving attribute information, for integrating with other data, and for spatial analysis.[35]

Of the four functions of GIS noted previously, graphical representation has potentially the greatest risk to copyright infringement, advertently or otherwise. Data capture and output includes digitising, scanning and display. The conventional practice is either to digitise map features from an existing map or to use data captured electronically through scanning paper maps and/or captured by remote sensing instruments. The general rule is that the use of works that are already protected by copyright remains protected, even in electronic form.[36] The database owner's permission is required (for example, through a licence agreement) for converting a hard-copy 'literary work' into electronic form.

In digitising or scanning a paper map to create a base layer, a user may be described as 'translating' a work into a form that is compatible with the requirements of an information system. Since 'translation of a work' is the sole prerogative of a copyright owner, an infringement of copyright will occur if no license or permission for digitising, scanning or display has been granted.

In addition, the scanning of a paper map is comparable to that of duplicating and the use of a 'performing right'. This is because the information system in depicting the map on a screen monitor is giving a 'performance' of the copyright data. This is an infringement if undertaken without the permission of the copyright owner. The physical print on paper or on the computer screen involves a 'republishing' of the copyright data stored in a GIS. Simply changing the mode of publication from conventional paper maps to electronic views on a computer screen may not discharge the user from observing the rights of the copyright owner. The change of format and technology does not intrinsically alter pre-existing rights in the data (Henry 1975: 71).[37]

The question whether computer screen displays constitute a reproduction in material form is not yet settled in Australia. The considered view is that screen displays are not a reproduction. This may be because screen displays are ephemeral and transitory and technically are not a

---

[35] Longley, PA, Goodchild, MF, Maguire, DF and Rhind, DW 2001 *Geographic Information Systems and Science*, Chichester: John Wiley & Sons, Ltd.

[36] See *Copyright Act* 1968 (Cwlth) s 35.

[37] See Henry, N 1975 *Copyright: Information Technology Public Policy*, New York & Basel: Marcel Dekker.

storage media so that what is shown on the screen is not tangible and in a material form.

Figure 3.4 is a suggested copyright notice for maps. While there is a provision for fair dealing/use for purposes of research and private study by individuals, caution should be exercised if any other use is contemplated. Once one begins to deal commercially with, or exploit a copyright work, fair dealing/use is negatived, and one may be technically infringing copyright in that work. It may be a simple matter to get permission and, if the copyright owner cannot be found, to refrain from using the material altogether. If one does not have permission then such material cannot be used because of the uncertainty it creates together with the liability implications.

There may be other material that may not be protected by copyright, either because the copyright has expired or because the owner(s) have wished to place the material in the public domain. One example is where copyright owners insert a statement to the effect that it is permissible to use the material, but request that due acknowledgment of the source of the material be given. For example, the author '... welcomes reproduction of any information ... in this publication, so long as credit is given ... and a copy of the reproduced material is sent to our office' (Heidemann 1992).[38]

This map is copyright. GeoScience Australia is the owner of the copyright subsisting in this map and any unauthorised copying of this map is unlawful. It is a condition of sale that copying in any form, or by any means is not permitted. Without in any way limiting the generality of the above the purchaser, or user of this map shall not:

- copy by raster scanning, or

- copy by digitising

Figure 3.4 A copyright notice with conditions

---

[38] Heidemann, MA 1992 'Copyright and Copy wrong', *Planning* (American Planning Association) v. 58(2), pp. 22–23.

Implications for Geographic Information Systems

The aim of GIS is to provide tools for decision-making. To be useful these artefacts of GIS should be accurate, contain as little minor variations in definitions and presentation of form and structure as possible, possess no discrepancies between the real world and depictions of it and have the means to standardise the presentation of facts. These requirements appear to suggest that there may be no scope for creativity and originality, whether in the presentation of geographical facts or in the selection of such facts. A map presentation cannot create an 'original' result by eliminating a kink in the river or softening the slope of a hill by rearranging the contours. Moreover, in GIS databases the geographical facts tend to dictate their arrangement and selection. The problem of applying such concepts as creativity and originality to a compilation and database was recognised by Onsrud (1993: 8) who put it this way:

> Facts, algorithms, physical truths and ideas exist for everyone's use. It is difficult to argue that the outline of a building, the bounds of a land parcel or a line of constant elevation on a map (that is, contour line) are expressions of originality. Any other person or sensor attempting to represent these facts would have little choice but to do so in much the same way. To represent the features by other than points, lines and polygons or image bits would make the representation non-standard, greatly decreasing the value to others and make the data useless or cumbersome for computer processing.[39]

Longley *et al.* (2001) are of the view that the position expressed by Onsurd is untenable because representations of 'facts' can incorporate considerable originality and creativity in whatever form, cartographic or photographic. The view is also untenable because many geographic features are fuzzy, and different people can have different spatial interpretations of a geographic object. Finally, there may be no business case for the commercial remote sensing industry if there were to be no copyright protection for such data products because they contains no originality or creativity. These differences in opinions are one of interpretation and conception.

It is believed that Onsrud was simply describing what would be the case in the digital data—the binary 0s and 1s in a data file. For example, where the coordinates of each of the points, lines and polygons, when accurately digitised, will be exactly the same for each geographic feature. To be sure there must be some originality and creativity in digitising the

---

[39] Onsrud, HJ 1993 'Law, Information Policy and Spatial Databases', *NCGIA Working Paper* (April), Orono, Maine: NCGIA, 18 pp.

features, but the digital representation of these will be the same on the digital database. It is true that there can be no clear-cut boundaries for physical, social and cultural features and at best these are both 'fuzzy' and arbitrary. Just as drawing a line on the sand will have 'fuzzy edges', the reality is that such fuzziness are a matter of scale and interpretation—a characteristic of GI work.[40] The copyrightability of remote sensing data stems not from its originality or creativity, but rather from the way the data are organised, arranged and presented. The right to copyright protection is precisely because of the expenditure of resources in capturing the data and the programs used to make such data useable and commercially saleable.

Geographic Information Material on the Internet

The Internet has already become an integral part of much of society today. This technology has revolutionised many aspects of journalism, science, and publishing, and many other fields. The two features of this technology, namely the wide distribution of content and its timeliness, make it a truly powerful medium. It seems that this medium will continue to grow, develop and become even more embedded in modern societies. Similarly GI has the potential to do the same in increasing the effectiveness and efficiency of GI systems from the way we obtain, use and share GI in all its forms: maps, graphics, text, and data. Today, many ingenious on-line GI systems have been built and developed that have been targeted at delivering the products and services over the Internet.

Distributed geographic information (DGI) refers to the entire field of widespread dissemination in any of the forms mentioned above. A cursory examination of the contents of recently published books on Web-based GIS will confirm various types of DGI applications.[41] Such applications range from raw data download, static map display, metadata search, dynamic map browsers, data pre-processors, Web-based GIS query and analysis, and 'Net-savvy' GIS software.[42] This last mentioned application is one where a desktop mapping program is able to use 'live' data from the network. Net-savvy GIS software is one of the primary goals of the Open GIS Consortium (OGC),[43] a collective of GIS vendors, developers, and other

---

[40] Longley *et al.* 2001 pp. 338–342.

[41] Plewe, B 1997 *GIS Online: Information Retrieval, Mapping, and the Internet*, Santa Fe, NM: OnWord Press; Zhong-Ren Peng and Ming-Hsiang Tsou 2003 *Internet GIS: Distributed Geographic Information Services for the Internet and Wireless Networks*, Hoboken, N.J.: John Wiley & Sons Inc.

[42] On-line map services include MapQuest (http://www.mapquest.com); MapOnUs (http://www.maponus.com); and MapBlast! (http://www.mapblast.com).

[43] http://www.opengis.org.

parties whose mission is to increase the ability of disparate GIS platforms to operate together.

Given the architecture of the Internet, one of the major concerns therefore is the protection of IPR. As a data producer, for instance, the concern is the prevention of copyright infringement. In general, most material published on the Internet is protected by copyright unless otherwise denied. The concept of fair use/fair dealing simply says that if someone obtains a copyrighted product legally then that person may legitimately use such copies for personal purposes. However, when that person redistributes the material, then copyright is infringed. This is despite the fact that the information has been value added and augmented, or even given away free by the latter user.

While it is nearly impossible to prevent illegal copying using technology, there have been attempts to ensure the origins of material are maintained. On maps and images one could use watermarks (sometimes called 'fingerprinting') inserted unobtrusively and invisible to the naked eye, to discourage users from breaking the law. Alternatively one could introduce 'copyright traps' in the form of including false, but insignificant features such as mis-spelling names, inserting (or omitting) *cul-de-sacs* in streets, encoding messages or random colours into the bits that constitute a raster image which will be preserved even if the image is altered in anyway.

If data were distributed for viewing only then the data could be degraded sufficiently to make it useful for viewing only, but not much else. Alternatively a raster map in GIF or JPEG format[44] may look nice, but would be of little use without the extensive metadata on the projection used to generate the pixel coordinates. Vector maps may be drawn with false coordinates using map centimetres instead of real world metres which would look all right on a computer screen, but will not be positionally correct in a GIS. All these technological features are meant to provide an audit trail of evidentiary material if litigation were to take place and proof of ownership were to be required by the courts.

The term digital rights management (DRM) describes a technology for identifying, trading, protecting, monitoring and tracking all forms of

---

[44] Graphics interchange format (GIF) is a common raster graphics data format to display images. Raster data in this format is compressed for simple graphics with up to 256 colours. Joint Photographic Experts Group (JPEG) is an industry committee that develops standards for digital photographs. This acronym is commonly used for raster data formats and the data files are compressed to optimise the true-colour images of photographs and remote sensing images. However, this format is 'lossy' in that a small amount of data is lost in the compressing process giving a lower quality image than the original.

rights usages over both tangible and intangible information assets, including management of rights-holders' relationships. Geospatial DRM manages all rights, not only the rights applicable to permissions over digital geographic data. The current inability to confidently control the flow of such information activities has been a barrier to broader adoption of Web-based geospatial technologies. The hope is that the adoption of an open DRM capability by all in a Web services transaction environment will enhance the flow of geographic data and information.

However, given that some users are so intent in using the information, the strategy adopted should be one that simply facilitates the use of the data and information. This may be achieved via reseller licenses, and by selling rights to alter and/or redistribute the data and maps, with appropriate restrictions, acknowledgments and price to be paid. The licensing process can be as simple as that described by Brinson *et al.* (2001).[45]

In a number of documents that are available on the Web, you will see a statement that says that it is permissible to copy the document for certain purposes. Here are three such examples:

- 'This article may be copied in its entirety for personal or educational use (the copy should include a Licence Notice at the beginning and at the end).'
- 'Permission is granted to freely copy this document in electronic form, or to print for personal use.'
- 'All the text and pictures on this Web server are copyrighted. You may use the pictures for any non-commercial purposes if you attribute the source.'

These 'limited permission grants' are placed on the documents because the owners want the documents to be shared and used for certain purposes, but do not wish to be bothered with requests for permission for such personal uses. However, do not be confused between a limited permission grant and a waiver of copyright. A limited permission grant is just a licence to use the work in ways stated in the limited grant permission—and only in those ways. If there is a need to use the work in any other way that is not covered by the document's limited permission grant, then it is advisable to contact the copyright owner and get permission; if not you will be infringing the copyright on the work.

---

[45] Brinson, JD, Dara-Abrams, B, Dara-Abrams, D, Masek, J, McDunn, R and White, B 2001 *Analyzing E-Commerce and Internet Law*, Upper Saddle River, NJ: Prentice Hall, pp. 428–429.

3.6.2 Electronic Databases

In GIS where a map is reduced to digital form, one of the first attributes that can become blurred by data processing is the control over the ownership of IPR in the digital data. In the past agencies have claimed copyright of the paper maps that they have provided to users. However, with digital data the task is not so easy. With traditional paper maps and books, the article including the copyright notice may be handed over the counter to a customer. With digital products it may be as impersonal as pressing a button to transfer the data electronically. Similarly, the copyright notice on a digital copy may easily be deleted with the press of a key. While map data may be handed over in a diskette one is unable to examine the article or to preview its contents as one could do with a map or a book.

Furthermore, when one digitises from a published map there is a strong element of ownership and possession in claiming 'my digital map' even though it is based on someone else's original. Also, when one adds other information and performs operations on it such as re-formatting and combining it with other data, the 'paternity' of the original source(s) becomes obliterated, blurred or may disappear after several iterations of the data. The 'offspring' now takes on a new form and identity indistinguishable from the source data. The product could now be claimed to be an 'original' in its own right. The issue therefore is: at what stage does the new map become an 'original' and assume a new identity in order that copyright and other IPR may be claimed on its own merits?

There may be no answer to this dilemma short of tracing 'paternity' and ownership at each step of the operation. This is where a metadata layer becomes useful, not only to describe what is being done, but also by whom, and at what stage. The end product could be one where there may be multiple copyright owners and a 'new' copyright work will conceivably be the whole of the work.

In terms of maps and electronic databases it is said that the originality requirement is 'low' under Australian copyright law with little or no scope for a cartographer to create a fully 'original' result. This may be because GIS aim for accuracy and standardisation in order to make the database more accessible (Eldred 1995: 14).[46] The information in the database makes it valuable and the arrangement, selection and formats gives the work sufficient originality. Significantly, it has been noted with

---

[46] Eldred, P 1995 'Geographic information systems and copyright: Are we on the right road?', *WALIS News* (March), pp. 13–18.

some irony that the more comprehensive a database the less scope is there to lay a claim to copyright protection on the basis of originality. This is because in a comprehensive database, there may be nothing left to 'select' (Karjala 1994) and thus satisfy the requirements of the *Copyright Act*.

Electronically stored databases may have to be considered to be no different from non-electronic databases and therefore should raise no new copyright issues. Non-original databases do not quality for copyright protection because there may be no originality in the selection and arrangement (see Lindsay 1993).[47]

Authorship of a Database or Compilation

For databases gathered by remote sensing satellites there has been expressed considerable doubt as to whether the data in its raw form is a 'work' and merits copyright protection. The decision whether raw data and databases qualify for copyright protection will depend on an assessment of the degree of originality, the arrangement of the data and the skill in constructing the database. There is also uncertainty as to who the 'author' of the resulting work is, if indeed there is a human 'author' at all. It seems likely that Australian courts would find some person as the author, even though this issue was not addressed in a recent case.[48] By analogy, in the computer program context, the author may be one of the following persons: the programmer, the person who conceived the idea which was implemented by the programmer, the operator of the computer on which the program runs, or the person who commissioned the work produced by the computer.[49] In other words, the author is a 'natural' person.

The position in the U.K. is similar in that s 9(3) of the *Copyright, Designs and Patent Act* 1988 (U.K.) states that an 'author' of a computer-generated work is 'the person for whom the arrangement necessary for the creation of the work are undertaken'. Although this is distinguished by s 178 where a 'computer-generated work' is defined as 'a work that is generated by computer in circumstances such that there is no human author of the work'.

The question of the 'authorship' of a compilation is further complicated by technological developments. Intelligent computer software are

---

[47] Lindsay, P 1993 'Copyright protection of electronic databases', *Journal of Law and Information Science*, v. 4(1) pp. 287–292.

[48] *Telstra Corporation Ltd v Desktop Marketing Systems Pty Ltd* (2001) 51 IPR 257.

[49] See Ricketson (1984: 442) *op. cit.* and the English position in *Express Newspapers Inc. Plc. v Liverpool Daily Post and Echo Plc.* (1986) 5 IPR 193 which distinguishes between 'works' generated by a computer and 'works' produced with the aid of a computer.

now available in which a compilation may be assembled by a 'robot' computer (bot) to add further information to a database. There is doubt whether the original compiler of the database still retains a claim to copyright. The original compiler logically cannot claim copyright in the information added by either a bot or another user (see Cross 1994: 115).[50]

Database Rights in Derivative Works

In constructing a database for GIS, distinctions must be made between those that are 'constructed' and those that are 'acquired' from other sources as well as a combination of both.[51] The crucial question is what rights are affected and what rights arise from the creation of GIS map data derived from a number of different databases? *Constructed databases* are less problematic because the developer has complete control over the content and form of compilation of the database. Here copyright issues may be determined in terms of 'what' is copyrightable and who owns such rights. *Acquired databases* may be purchased or obtained through a licence agreement.

Contractual agreement may need to ensure those specific rights and obligations concerning the use, distribution and ownership of the data are preserved. The greatest difficulty with these types of databases in regard to copyright is whether such protection will subsist in favour of both the developer and the data provider. A further complication arises when attempting to apportion the compensation that may be due either to one or both parties. Equity suggests that the compensation should be settled in terms of the contributions of each party to the whole. The E.U. Directive on the Legal Protection of Databases has addressed the issues of database rights and such protection at length and this is discussed later in this section.[52]

Revisions

Databases, especially electronic ones are subject to constant revisions and updates. One view is that each revised and updated version should be entitled to a separate copyright protection. The problem occurs when someone makes minor changes to an existing database other than the owner. It is here suggested that that person may not claim copyright in

---

[50] Cross, JT 1994 'Protecting computer databases under the United States copyright laws: Implications of the *Feist* decision', in Carr, I and Williams, K (eds) *Computers and Law*, London: Intellect Books, pp. 113–127.
[51] See Milrad (1994) for a further discussion of the legal issues of the respective rights of ownership, use and distribution of different kinds of databases.
[52] E.U. Directive 96/9/EC.

the amended version. Indeed, that person may have infringed copyright in the absence of permission to use and to amend the data.

Compilations and 'Sweat of the Brow'

In assessing whether a work attracts copyright protection the test of originality holds that the author has expended a minimal degree of skill, judgement and labour to achieve a result and that the work has not been copied from elsewhere. This concept of originality is most difficult to apply in relation to factual compilations in the form of a database since copyright protection may not subsist in raw factual data, as shown below. The selection of facts, its inclusion or exclusion and its arrangement and presentation may be crucial in deciding whether copyright protection is available. In IT there is doubt whether computer-produced data can claim copyright protection, whether such data are original and if so, who is the author? Also collated information in the form of an electronic database has become subject to interpretation. This is because the Australian *Copyright Act* 1968 (Cwlth) s 10(1) makes it clear that compilations of information will be protected as a 'literary work' *only* if a court is satisfied that the compilation represents a suitable degree of originality. Subject to these provisos, an electronically stored database will be entitled to the same amount of protection under the *Copyright Act* 1968 (Cwlth) as a conventional database in hardcopy format.

Facts, as such, are not protected under copyright law (Cross 1994: 113).[53] The principle that facts *per se* are not copyrightable is reflected in various statutes.[54] However, courts in various Commonwealth jurisdictions and in the U.S. have found ways to protect factual *compilations* within copyright law. Protecting such compilations serve important social policy objectives such as the dissemination of public information and promoting an open democratic government where the citizenry is constitutionally entitled to know what its government is doing.

Cross (1994) provides an interesting analysis and commentary on the central issue of copyright protection in compilations and the kinds of protection that are available to different types of electronic databases. The arguments for and against the protection of databases are centred on the

---

[53] *op. cit.* Cross, JT 1994.

[54] In the U.K. s 35 of the *Copyright Act* 1911 and s 48(1) of the *Copyright Act* 1956 include 'compilation' under the scope of protected subject matter. In the U.S. § 103 *Copyright Act* 1976 Pub. L. No. 94–553 *as amended* by the *Berne Convention Implementation Act* of 1988 Pub. L. No. 100–568 *currently codified* at 17 USC. §§ 101–810 allows copyright protection to extend to compilations of information but only to the new material contributed by the author of the compilation.

case of *Feist Publications v Rural Telephone Service Company* (see Case Note and hereafter referred to as *Feist*).[55] This analysis may be contrasted by the Australian case of *Desktop Marketing Systems Pty Ltd v Telstra Corporation Ltd*[56] which is presented as a Case Note and commentary below.

In the U.S. compilations have been give copyright protection because of a compiler's organisational efforts in the selection, arrangement or coordination of information. This protection is for the whole compilation and not just the individual pieces of information within it. Books of maps[57] and stock market indices have been granted copyright protection.[58] There are, however, two shortcomings with this approach. First, the theory protects the compilation *as a whole* and therefore arguably no cause of action will arise against someone who either took only a small part of the data or arranged the data differently.[59] Secondly, for certain types of databases, for example, phone books and sports statistics, because the effort of the compiler is in the *accumulation* of information rather than its arrangement, these kinds of factually based compilations have attracted no IP protection.

A second method by which compilations have been protected is through what is described as the 'sweat of the brow' and 'industrious collection' theory. This theory rewards a compiler by giving protection because of the time and effort spent in accumulating the data.[60]

Case Note: Feist Publications Inc. v Rural Telephone Service Company Inc. 111 SC 1282, 113 L.Ed.2d. 358 (1991)

Facts of the case

Respondent Rural Telephone Service Company Inc., is a certified public utility providing telephone service to several communities in Kansas.

Continued on page 148

---

[55] 111 SC 1282, 113 L.Ed.2d 358 (1991).

[56] [2002] *FACFC 112*.

[57] *Rockford Map Publishers Inc. v Directory Service Co.* 768 F.2d. 145 (7th Cir. 1985) cert. denied 106 SC 806 (1986).

[58] *Dow Jones & Co. v Board of Trade* 546 F.Supp. 113 (SD NY 1982).

[59] *Triangle Publications Inc. v Sports Eye Inc.* 415 F.Supp. 682 (ED Penn. 1976).

[60] *Hutchison Tel. Co. v Frontier Directory Co.* 770 F.2d 128, 131 (8th Cir. 1985); *Jeweler's Circular Publishing Co. v Keystone Publishing Co.* 281 F.83, 89, 95 (2d Cir.) cert. denied 42 SC 464 (1922); *Natural Business Lists Inc. v Dun & Bradstreet Inc.* 552 F.Supp. 89, 92–93 (ND 111. 1982).

Continued from page 147

Pursuant to state legislation, Rural publishes a typical telephone directory, consisting of white pages and yellow pages. It obtains data for the directory from subscribers, who must provide their names and addresses to obtain telephone service. Petitioner Feist Publications, Inc. is a publishing company that specialises in area-wide telephone directories covering a much larger geographic range than directories such as Rural's. When Rural refused to license its white pages listings to Feist for a directory covering 11 different telephones service areas, Feist extracted the listings it needed from Rural's directory without Rural's consent. Although Feist altered many of Rural's listings, several were identical to listings in Rural's white pages. The District Court granted summary judgement to Rural in its copyright infringement suit, holding that telephone directories are copyrightable. The Court of Appeals affirmed this ruling.

Supreme Court ruling

Rural's white pages are not entitled to copyright, and therefore Feist's use of them does not constitute infringement. pp. 1287–1297.

Article I, s 8, cl. 8 of the U.S. Constitution mandates originality as a prerequisite for copyright protection. The constitutional requirement necessitates independent creation plus a modicum of creativity. Since facts do not owe their origin to an act of authorship, they are not original and, thus, are not copyrightable. A compilation of facts may possess the requisite originality because the author typically chooses which facts to include, in what order to place them, and how to arrange the data so that readers may use them effectively. Copyright protection extends only to those components of the work that are original to the author, not to the facts themselves. This fact/expression dichotomy severely limits the scope of protection in fact-based works. pp. 1287–1290.

The Copyright Act of 1976 leaves no doubt that originality is the touchstone of copyright protection in directories and other fact-based works. 17 U.S.C. §§ 101, 102, 103. pp. 1290–1295.

Rural's white pages do not meet the constitutional or statutory requirements for copyright protection. While Rural has a valid copyright in the directory as a whole because it contains some foreword text and some original material in the yellow pages, there is nothing original in Rural's white pages. The raw data are uncopyrightable facts,

Continued on page 149

Continued from page 148

> and the way in which Rural selected, coordinated, and arranged those facts is not original in any way. Rural's selection of listings—subscriber's names, towns, and telephone numbers—could not be more obvious and lacks the modicum of creativity necessary to transform mere selection into copyrightable expression. pp. 1295–1297.
>
> Court of Appeals, 10th Cir. 916 F.2d 718 (CA 10th 1990) reversed.

In the case of *Feist* the U.S. Supreme Court ruled that the white pages of a telephone directory was not copyrightable. The decision by Justice SANDRA DAY O'CONNOR re-emphasised the need for at least a minimal degree of 'creativity' in having a work qualify for protection. In so ruling, the U.S. Supreme Court rejected the 'sweat of the brow' or 'industrious collection theory' as a basis for protecting 'compilations' under copyright laws (see Westermeier 1994).[61]

The 'sweat of the brow theory' holds that it would be inequitable for a third party to appropriate information without having to reimburse a compiler of information for the time and effort in gathering it. This theory suggests that the compiler should be able to prevent others from copying any of the pieces of information that comprise the compilation. The theory, however, was previously rejected by the Second Circuit Court of Appeal in *Financial Information Inc. v Moody's Investor Service Inc.*[62] There it was held to be inconsistent with the Constitutional edict that facts cannot be given copyright protection for policy reasons.

The *Feist* case now casts serious doubts on whether copyright can protect factual compilations in the U.S. The constitutional requirement of originality[63] as reiterated by the U.S. Supreme Court in two previous decisions[64] precludes an author from claiming copyright protection in certain compilations of fact. Also, there was clear Congressional intent to reject the 'sweat of the brow' theory in the debates leading up to the 1976 amendments to the U.S. *Copyright Act*. Originality, not effort, is the litmus test for copyright protection. Thus, while the *Feist* opinion on its facts is limited to telephone directories, the reasoning used by the Justices there may have serious repercussions on other types of compilations such as spatial databases.

---

[61] Westermeier, JT 1994 ' "Sweat of the brow" protection survives', *GIS Law*, v. 2(1), pp. 1–3.

[62] 808 F.2d (2nd Cir. 1986).

[63] U.S. *Constitution* Art. 1 § 8 Cl. 8.

[64] *Trade Mark Cases* 100 US 82 (1879) and *Burrow-Giles Lithographic v Sarony* 111 US 53 (1874).

Facts are not copyrightable because the author has not created them.[65] However, a compiler of a database may obtain copyright protection for the selection and arrangement of the facts in an original and creative way. It is easy to identify the point at which the database takes on sufficient structure, coherence and order to qualify the database as an original work. Much of the effort in a database is expended in collecting and transforming the data into a computer. Its real value depends on the amount and quality of information it contains.

Significant differences in approaches have been taken to give protection to compilations and database around the world. Even within the U.S. there is continual debate on the impact of the *Feist* decision. On the one hand, there are fears of an unbridled 'free-rider' problem where there may be those who will take the fruits of a compiler's efforts without having to pay for it (Ginsburg 1992).[66] On the other hand, there are those who take the view that the decision in *Feist* is making a return to sound copyright principles because the reduction of copyright protection lessens the possibilities of putting monopoly power over information in select hands (Litman 1992).[67]

In the U.S. confusion reigns post-*Feist* especially the copyright protection of maps digitised in a GIS. As argued by Karjala (1994) the lower courts are in a quandary.[68] On the one hand, they are unable to directly reject *Feist* by recognising that there is validity in the 'sweat of the brow' doctrine and on the other lower courts are hamstrung to offer greater protection so that valuable GIS products become vulnerable to misappropriation by 'free riders'. Such courts are duty bound to uphold the long-established principles of copyright, which does not protect ideas, systems, procedures and concepts as legislated in the U.S. *Copyright Act*[69] while protecting the expression of ideas in creative and original ways. The tensions between these differing roles has become too great and something may have to give. The *Feist* decision leaves much room for thought. 'It underlies the inadequacies of the judicial system as a vehicle for crafting new legal tools to cover changing economic circumstances

---

[65] See *Feist* 111 S.Ct.; 113 L.Ed.2d. at 370 per O'CONNOR J.

[66] Ginsburg, JC 1992 'No "sweat"?' Copyright and other protection of works of information after *Feist v Rural Telephone*, 92 *Columbia L. Rev.* 338 (1992).

[67] Litman, J 1992 'Copyright and information policy', 55 *Law and Contemporary Problem* 185 (1992).

[68] Karjala, DS 1994 'Copyright in electronic maps', *Proceedings of the Conference on Law and Information Policy for Spatial Databases*, Tempe AZ, October, NCGIA and Center for the Study of Law, Science and Technology, Arizona State University College of Law, *URL*: http://www.spatial.maine.edu.tempe/karjala.html. Also in 35 *Jurimetrics J.* 395–415 (1995).

[69] 17 USC § 102 (b).

brought about by new information technologies' (Branscomb 1994: 39).[70] There is also another view that asks whether market considerations can be introduced directly into copyright law so that the nature of the market can be considered on a case-by-case basis in determining the scope of protection. On that view, it is said that the *Feist* decision is not, and should not be the law in Canada (Siebrasse 1994).[71]

The *Feist* should now be contrasted to the following case that has recently been decided in the Federal Court in Australia. The contrasts in approach could not have been more different.

Case Note: Desktop Marketing Systems Pty Ltd v Telstra Corporation Ltd [2002] FACFC 112

Facts of the case

Telstra publishes the White Pages and Yellow Pages telephone directories. The headings used in these directories were taken from unpublished 'headings' books which Telstra had developed over many years.

Desktop Marketing Systems used data from Telstra's directories to build searchable CD-ROM based directories called 'CD Phone Directory', 'Australian Phone Disc', and 'Marketing Pro' which it sold to marketers. The primary data used to produce these three CD-ROMs was taken directly from Telstra's White Pages and Yellow Pages.

Telstra objected to Desktop's conduct and commenced proceedings alleging infringement of copyright in its White Pages and Yellow Pages directories and the heading books for the years in which Desktop sold its CD-ROMs.

Decision at first instance

FINKELSTEIN J in the Federal Court found in favour of Telstra and held that Desktop had infringed Telstra's copyright by producing the CD-ROM products.

Continued on page 152

---

[70] Branscomb 1994 *op. cit.*

[71] Siebrasse, N 1994 'Copyright in facts and information: *Feist* is not, and should not be the law in Canada', 1 *Canadian Intellectual Property Review*, 191.

Continued from page 151

Decision on appeal

Desktop appealed to the Full Court of the Federal Court. It submitted that FINKELSTEIN J had erred in holding that copyright could subsist in a mere industrious collection of facts. Desktop did not dispute that the necessary conditions for copyright in Telstra's telephone directories and headings books had been satisfied, except for the requirement that the directories and headings books be 'original'. Further, it was common ground between the parties that if Telstra was correct that industrious collection was enough for copyright to subsist in compilations of facts, then Telstra's investment of labour and expense was sufficient to exceed any quantitative threshold requirement.

On appeal the matter came before BLACK, CJ, LINDGREN and SACKVILLE JJ, who again found in favour of Telstra. Based on previous English authority, their Honours held that 'original' in Australian copyright law does not require any 'creative spark'. Instead, according to their Honours, a low standard of originality is required, namely that the material was 'not copied, but originated from the putative author'. This was on the grounds that a database can have originality due to the labour and expense involved in researching and culling the information contained in the database.

On special leave application to the High Court

Desktop sought special leave to the High Court and the matter came before HAYNE and CALLINAN JJ on 20 June 2003. Desktop argued that the principles of copyright law had been misapplied in the lower courts in this case. This error arose from the application of the industrious collection test as opposed to some sort of intellectual effort. Their Honours were of the view that an appeal by Desktop would have insufficient prospects of success to warrant a grant of special leave.[72]

The Case Note on *Desktop Marketing Systems Pty Ltd v Telstra Corporation Ltd*[73] (hereafter *Telstra*) suggests that under Australian law only a low threshold of originality is required and copyright protection is available for the effort of gathering and listing of data. This decision appears out of

---

[72] *Desktop Marketing Systems Pty Ltd v Telstra Corp Ltd* (Unreported M85/2002, HCA, 20 June 2003).
[73] *[2002] FACFC 112.*

step with international approaches for the protection of databases. As well certain assumptions regarding authorship have been made which may open the way for future challenges. As previously observed in *Feist*, the decision emphasised originality in the sense of requiring a minimum degree of creativity in selection and arrangement of the data in a database.

Practice Notes: Checklist for Copyright Protection for Compilations and Databases

In Australia, following the *Telstra Corporation Limited v Desktop Marketing Systems Pty Ltd* case, each answer to the following questions must be in the affirmative if copyright protection is to be available.

■ Is the compilation a collection of 'intelligible information'?

For copyright purposes the compilation must have useable information such as contact details in a phone book.

■ Is the compilation 'original'?

Originality must be applied to the compilation as a whole and not to individual parts of the compilation.

■ Is the original creator claiming copyright?

The compilation must originate from the person(s) claiming copyright.

■ Is the creation one that is expressed in one of a number of ways of a particular arrangement?

■ Has the compiler exercised skill, judgement or knowledge in selecting and arranging the material for inclusion in the compilation?

■ Has the compiler undertaken substantial labour or incurred substantial expense in collecting information recorded in the compilation?

The labour or expense must meet a minimum threshold. It is sufficient that labour or expense was outlaid for the purpose of producing the compilation.

In Canada the Federal Court of Appeal in *Tele-Direct Publications v American Business Information*[74] held that copyright would not subsist

---

[74] (1997) 154 DLR (4th) 328.

in a yellow pages directory if insufficient skill or judgement was evident in the overall arrangement of the work, regardless of the industrious collection of the information. This requirement is also contained in the North American Free Trade Agreement (NAFTA) where protection is provided to compilations which constituted intellectual creativity by reason of the selection or arrangement of the contents.[75] The TRIPs Agreement also contains similar requirements, as does the WIPO Copyright Treaty 1996.

The Australian approach in *Telstra* is enigmatic for several reasons. First, it ignores the basic principles of copyright law because it protects the information and facts in a work rather than the form of expression of that work. Second, the decision prevents 'second comers' from building on the ideas and the creation of value added or derivative products. Finally, the decision may have the effect of potentially conflicting with Australian competition laws where the database owner could choose to refuse to licence the copyright in the compilation or charges prohibitive licence fees.[76] While s 46 of the *Trade Practices Act* 1974 (Cwlth) deals with the misuse of market power and anti-competitive practices s 31(1)(a) the *Copyright Act* 1968 (Cwlth) bestows exclusive rights to reproduce and adapt a work on the copyright owner.

Several alternative solutions to the impasse have been suggested such as contract law in addition to the use of technological protection measures (TPM). The tort of unfair competition was proposed, but rejected by the High Court in *Moorgate Tobacco Co. Ltd v Philip Morris Ltd (No. 2)*.[77] Compulsory licensing and *sui generis* rights have also been given as alternatives. In regard to compulsory licensing, also known as statutory licensing, use of the data is permissible, but users have to pay a royalty to the copyright owner. This is thought to be administratively cumbersome and there is no agreement as to what the 'correct rate' of charging might be. While this solution may partially solve the 'free-rider' problem, access to socially significant data is available to the general public for data collected with taxpayer funds.[78] The idea of a *sui generis* right has been implemented in the E.U. Directive on the Legal Protection of Databases (discussed in the next section). This is 'one of a kind' right prevents unfair extraction and provides for a 15-year protection period

---

[75] *North American Free Trade Agreement Act* 1993 (Canada).

[76] See Lee, C 2003 'High Court hangs up on issue of copyright and compilations', *Australian Intellectual Property Law Bulletin*, v. 16(4) August, pp. 45–60.

[77] (1984) 156 CLR 414; 3 IPR 545.

[78] See Givoni, S 2003 'Pushing the boundaries of copyright: protection of databases', *Australian Intellectual Property Bulletin*, v. 15(8) January, pp. 113–118.

from the creation of a database. In addition the database owner is given protection over the raw data contained in the database.

In Australia, compared with other common law jurisdictions, the time may have arrived for legislative intervention on a review of copyright protection of databases and compilations. The present path seems to show that the copyright laws are going in the direction of virtually granting a monopoly in mere information, patently out of step with the rest of the world.

Practice Notes: Checklist for the Creation of a Database

- Identify the database, size, format, content and method of creation. When was it created and when was it substantially updated?
- For what purpose was the database created and what is it used for now?
- Identify the author and the maker. Who owns the database and who actually owns the rights to exploit the use of the data?
- Identify how much data are required to be extracted and assess whether the amount may be quantified in a quantitative and qualitative manner as 'substantial' or not.
- Is the database publicly available?
- Does a public body own it?
- Is the database subject to copyright or to an extraction right? Is copyright claimed for the database as a whole?
- What licence terms, formal or informal, are associated with this database and are there any fees that apply for its use?

Maps as Factual Compilations and Maps as Mere Representations

Wolf (1992, 1993) has described why courts, erroneously, have treated maps like 'directories' just as in the white pages listing in *Feist*.[79] Arguably, the

---

[79] Wolf, DB 1992 'Is there copyright protection for maps after *Feist*?' 39 *J. Copyright Society* 224; and Wolf, DB 1993 'New landscape in the copyright protection for maps: *Mason v Montgomery Data Inc.*', *GIS Law* v. 1(4), pp. 14–17.

source of such errors in interpretation may be traced to an over-dependence on the 'sweat of the brow' theory as a substitute for originality as in *Telstra*. More importantly, it reflects the confusion between maps as factual compilations and maps as a pictorial representation of reality.

In a recent case of *Mason v Montgomery Data Inc.*[80] the Court of Appeals for the Fifth Circuit interpreted *Feist* to support copyright protection for maps. The plaintiff in this case undertook legal and survey research in order to draw parcel maps on USGS maps. Mason sued Montgomery Data claiming that the defendant had infringed Mason's copyright on 233 real estate ownership maps of Montgomery County, Texas.

At first instance, in *Mason v Montgomery Data Inc.*[81] the District Court held that Mason could not recover statutory damages or attorney's fees for any infringement of 232 [sic] of the ownership maps. The court in a later judgement held that Mason's maps were not copyrightable because of the idea–expression merger doctrine and granted summary judgement for the defendants. The idea embodied in the maps was thought to be inseparable from the map's expression of that idea.[82] The court dismissed Mason's claims with prejudice and awarded the defendant costs and attorney's fees.

On appeal, the court agreed with Mason that the maps were copyrightable and so reversed the decision of the District Court's judgement. The court acknowledged the creativity and hence, originality, in the plaintiff's selection and arrangement of information including the reconciliation of conflicting information. The Appeal Court noted that historically most courts would have treated maps solely as compilations of facts. Moreover, the U.S. *Copyright Act* amendments in 1976 have categorised maps, not as factual compilations, but as 'pictorial, graphic and sculptural works', a category that includes photographs and architectural plans.[83] The court also observed that since it is the pictorial or graphical form of expression in maps that are protected and not the ideas underlying them, maps should be distinguishable from other non-pictorial fact compilations for which the merger doctrine may appropriately apply.[84]

---

[80] 976 F.2d. 135 (5th Cir. 1992).

[81] 741 F.Supp. 1282 (SD Tex. 1990).

[82] *Mason v Montgomery Data Inc.* 765 F.Supp. 353 (SD Tex. 1991) at 356.

[83] 17 U.S.C. § 101 (West Supp. 1992).

[84] *Mason v Montgomery Data Inc.* 741 F.Supp. 1282 (SD Tex. 1990) at 142.

'Thin' Copyright Protection

The idea–expression merger problem encountered in copyright law is an especially difficult one when applied to computer databases. This is because the solution may depend on how one defines an 'idea' in a database. One interpretation could be the configuration of the database itself, a 'file' of information that is arranged in a certain way. A broader interpretation is one where the idea underlying the database is the *purpose* rather than the parameters of that database (Cross 1994: 124).[85] Whichever is preferable, users of information systems need the assurance that the databases they build would be protected in some way.

As an example, in GIS maps may be stored as a matrix of digits in a database. A database developer will have an incentive to produce maps in a creative way if there is certainty of protection. But the protection will only be extended to how the 'facts' are expressed and not the facts *per se*. This policy has a 'social' advantage since there will be no need to produce a unique map without the advantage of 'borrowing' facts from previous maps. In this borrowing of facts there is a fine policy judgement for balancing the competing needs of social policy and social costs. The relevant considerations in this balancing act includes whether the information taken is used to make a non-competing product or whether it is used to improve an existing product, and whether a large amount of information has been taken. The term *thin copyright* has been coined especially for map products (Karjala 1994).[86] Thin copyright provides protection against the 'grosser forms' of copying and encourages the adding of value either through verification or revision or through improvements in presentation (see Gorman 1992).[87]

In other words, a veneer of copyright protection is afforded to, say an electronic database which is made up of various elements, including compilations that are 'normally' not copyrightable and those 'creative' elements which are copyrightable. When assembled together the new work may attract copyright in its own right because it is ostensibly a newly created original work. Thus, while there may appear to be infringements of copyright works, the new work has value added to the original by way of revision, updating and improvements in presentation. The infringement of copyright is thinly veiled and explains the term 'thin' copyright.

---

[85] *op. cit.* Cross 1994.

[86] *op. cit.* Karjala 1994.

[87] Gorman, RA 1992 'The *Feist* case: Reflections on a path-breaking copyright decision', 18 *Rutgers Computer and Tech LJ* 731 (1992).

More generally, it appears that many national copyright laws provide protection for databases irrespective of whether they contain protected works or not. However, most also require, that the databases be characterised by some minimum level of originality or intellectual creation by reason of its selection and arrangement. Different levels of originality and creativity between jurisdictions introduce uncertainty for database producers as to whether the database will qualify for copyright protection. Thus, some countries provide special protection to those works that lack the minimum level of originality to safeguard the investment in producing the database. Nordic countries, for example, have adopted a common provision which states that 'Catalogues, tables and similar productions in which a great number of items of information have been compiled, as well as programs, may not be reproduced without the consent of the producer until ten years have elapsed from the year in which the production was published.'[88]

In Japan, similar provisions apply when an amendment to the *Copyright Act* in 1985 makes specific reference to 'data base works'.[89] Data base is defined as 'an aggregate of information such as articles, numerals or diagrams, which is systematically constructed so that such information can be searched for with the aid of a computer'. These are protected as independent works 'where, by reason of the selection or systematic construction of information contained therein, [they] constitute intellectual creations'.[90]

From the above discussion, the emergent unresolved issues seem to be the special protection accorded to database producers to authorise exclusive reproduction, in whatever manner or form; limitations to this right in respect of literary and artistic works including the database; and the minimum protection the database. These unresolved issues remain contentious as they may fall outside the scope of international copyright conventions and the principle of national treatment. Also the special protection might seem to dilute copyright protection where it might otherwise be applicable. It would therefore be appropriate to examine the case of database protection in the E.U. next as an exercise in observing future

---

[88] *Copyright in Literature and Artistic Works*, Act No. 158 of 31 May 1961 (Denmark), Art. 49; *Copyright in Literature and Artistic Works*, Law No. 404 of 8 July 1961 (Finland), Art. 49; *Property Rights in Literary, Scientific or Artistic Works*, No. 2 of 12 May 1961 (Norway), Art. 43; and *Copyright in Literary and Artistic Works*, Law No. 729 of 30 December 1961 (Sweden), Art. 49.
[89] *Copyright Act* 1970 (Japan) Law No. 48 of 1970, Art. 12.
[90] Law for Partial Amendments to the Copyright Law of 23 May 1986 (Japan), Art. 2(1) (xter).

trends in the making, but also reading from it implications for the GI industry and GIS databases.

3.6.3 European Union Database Directive

As everywhere else, the pressing issues concerning databases include duration of protection, authorship, ownership, and protection of non-original databases and rights to unfair extraction. The European Union Directive on the Legal Protection of Databases[91] applies to 'a collection of independent works, data or other material arranged in a systematic or methodical way and individually accessible by electronic or other means'. The Directive does not apply to computer programs as this is given protection in a separate E.U. Software Directive of 1991.[92] The Database Directive introduces a unique two-tier protection scheme for electronic and non-electronic databases. First, Member States will provide for the protection of databases by copyright as intellectual creations. Secondly, the Database Directive introduces a right *sui generis* to prevent the unauthorised extraction or re-utilisation of contents of a database—the database right. In this section the focus is on the database right as this right differs fundamentally from copyright protection.

Sui Generis—Database Right

The database right protects the 'sweat of the brow' of a database producer who has made a substantial investment in either obtaining, verifying or presenting the contents of a database (Art. 7 s 1). The investment must be substantial assessed either qualitatively or quantitatively. However, there is no definition of what is 'substantial'. The owner of the database right is the 'maker of the database'—a person who takes the initiative and risks of investments and hence excludes subcontractors and employees to a claim of ownership.

The scope of the database right extends to 'preventing the extraction and/or re-utilisation of the whole or a substantial part of the database evaluated qualitatively and/or quantitatively of the contents of a database'. Extraction is taken to mean the permanent or temporary transfer of all or a substantial part of the contents of a database to another medium by any

---

[91] 96/9/EC (OJ L7, 27 March 1996).
[92] European Union Council Directive 1991 Legal Protection of Computer Programs 91/250/EEC OJ L 122/42 of 17 May 1991.

means or form, while re-utilisation is any form of making available to the public all or a substantial part of the database. It is to be noted that extraction and re-utilisation of insubstantial parts are permitted unless the acts are committed in a 'repeated or systematic' manner and conflict with the normal exploitation of that database and unreasonably prejudice the legitimate interests of the maker of the database.

There are limited statutory exceptions (Art. 9) such that there are no special privileges for journalists, quotation rights, library privileges or the reuse of government information. In the debates leading up to the Directive it was considered that the extraction and re-utilisation of insubstantial parts of the database right were considered sufficient for such purposes. Exceptions to *sui generis* rights include data that are extracted for private purposes, the contents of non-electronic databases, the use of the data for illustration in teaching and scientific research, and for purposes of public security and the proper performance of an administrative or judicial procedure.

Art. 10 s 1 provides a 15-year duration of the database right from the date of the completion of the making of a database or later when it is first made available to the public. Any substantial change, including substantial new investment triggers new protection of the database. Only citizens of a Member State qualify for protection under the database right (Art. 11) or companies formed in accordance with Member State or registration offices within the Community. However, the Council of the E.U. may extend protection to nationals of third countries on the basis of special agreements.

The database right granted is independent of the eligibility of the database for copyright protection. In other words a database right may exist where the database and contents are not copyright works and it is possible for both rights to exist in the same database and for some or all of its contents. The right does not extend to the contents of the database and the right is granted without prejudice to any copyright or other right subsisting in the content. *Sui generis* rights do not extend to non-substantial parts of a database. Thus, legitimate users may extract and reuse non-substantial parts of the database.

The *sui generis* character of the new right has been tested in court in several jurisdictions within the E.U. and these have repeatedly come up against the lack of a clear-cut definition of several terms which have been used in the legislation. What 'substantiality of investment' means is undefined, and there is vagueness in the use of the terminology such as 'repeated and systematic' extraction, 'insubstantial', database and maker. This has led one observer to state that the contours of the new database right remains obscure after five years of implementation.[93] While the Directive may have given the database industry a one-time boost by extending IP protection, some observers have said that these have come at a high cost.[94] Recent court rulings, some of which are summarised in Table 3.2, show that the Directive has eroded the public domain, overprotected databases of doubtful worth, and raised new barriers to data aggregation.

In the U.K. the *Copyright and Rights in Databases Regulations* 1997 came into force on 1 January 1998, implementing the E.U. Database Directive of March 1996. A test for copyright protection under this Regulation is the intellectual creativity test which thus excludes 'sweat of the brow' compilations. The Regulations reiterate the form and structure of the E.U. Directive and the Database Right as the following case will demonstrate.

British Horseracing Board (BHB) v William Hill Organisation Ltd[95]

This is the first U.K. case where a court has provided detailed guidance on the E.U. Database Directive and its interpretation. The BHB claimed William Hill was making unlicensed use of its database by publishing lists of runners in forthcoming horseraces on its Internet site. The BHB maintains an extensive database of information in respect of horseraces in the U.K. According to the BHB the cost of establishing the database was considerable, as was the cost of continuing to obtain, verify and present its contents. This expenditure has been estimated to be about £4 million each year. Essential elements of the database are made available to various parties within the horseracing industry on a daily basis through various Internet sites, newspapers, and to bookmakers.[96]

William Hill provides betting services over the telephone and the Internet. Its website publishes a list of runners in forthcoming races, and details of race meetings. BHB argued that this information on horses in forthcoming races was obtained from its database. Furthermore, this

---

[93] See Hugenholtz, PB 2001 'The New Database Right: Early Case Law from Europe', paper 9th Annual Conference on Intellectual Property Law and Policy, Fordham University School of Law, New York, 19–20 April 2001. Also at http://www.ivir/publications/hugenholtz/fordham2001.html.

[94] See Maurer, SM, Hugenholtz, PB and Onsurd, HJ 2001 'Europe's Database Experiment' *Science*, v. 294 (26 October) pp. 789–790. Also available at http://www.sciencemag.org.

[95] High Court of Justice Ch. Div. 9 February 2001, Case No. HC 2000 1335 available at http://www.courtservice.gov.uk/judgements/judg_home.htm.

[96] See Taylor, G 2001 'Protect your online database', *International Internet Law Review, Issue 13* (April) pp. 33–36. Also at http://www.netlawreview.com.

Table 3.2 E.U. Database Directive 1996: summary of some early case law as at 2001

Database right	Case law	Jurisdiction	Subject matter/decision
Notion of a database	*MIDI Files* Müchen 30.03.00	Germany	Midi files on-line not a db.
	C-Net Berlin 09.06.2000	Germany	Entertainment event list a db. Extraction of single item not an infringement of the db. right
	KPN v XSO Hague 14.01.2000	Netherland	On-line phone directory. Search engine infringes db. right
	NVM v De Telegraaf Hague 12.09.2000	Netherland	Real estate agent unauthorised extraction and re-use even of small amount taken of great value to end users
	NVM v De Telegraaf Hague 21.12.2000	Netherland	Collection of real estate objects not a db. but a spin-off of other works.
	Algemeen Dagblad a.o. v Eureka Rotterdam 22.08.2000	Netherland	Automatic hyperlinks to newspaper articles on-line is unauthorised use of headlines
	Editorial Aranzadi Elda 02.07.99	Spain	Listing of case law and legislation a db. Unauthorised reproduction as db. right infringed
	British Horseracing Board v William Hill London 09.02.2001	U.K.	Betting website lists protected by db. right
Copyright	*Dictionnaire Perment des Conventions Collectives* Lyon 28.12.98	France	Compilation of collective bargaining agreements. No cprt. infringement
	Groupe Moniteur v Observatoire des Marchés Publics Paris 18.06.99	France	Public procurement tenders. No cprt. in work Unfair competition
	Medizinisches Lexicon Hamburg 12.07.2000	Germany	Medical lexicon on CD, Web. Linking, reproduction unauthorised
	NOS v De Telegraaf Hague 30.01.2001	Netherland	Cprt. in non-original writing. Refusal to licence is anti-competitive behaviour
Subject matter Sweat of the brow	*France Télécom v MA Editions* Paris 18.06.99	France	Phone directory a db. Piracy held

		Country	
Investment			
Quantitative	*Berlin On-line* Berlin 08.10.98	Germany	Classified ads db. Search engine unreasonably damaged interests of db. owner
Qualitative	*Süddentsche Zeitung* Köln 02.12.98	Germany	On-line real estate db. Search engine infringe db. right
Maker of the database	*TeleInfo CD* Bundesgerichtshof 06.05.99	Germany	Piracy from protected db. and unfair competition
	Baumarkt.de Dusseldorf 29.06.99	Germany	Website for DIY products not a db. because no substantial investments made
	Kidnet/Babynet Köln 26.08.99	Germany	Alpha listing of parenting catalogue. Substantial investment gives db. right. Copying not authorised
Scope of the database right Extraction and reuse	*UNMS v Belpharma Communication* Brussels 16.03.99	Belgium	Unauthorised extraction of pamphlet information unauthorised
Insubstantial use	*Électre v TI Communication and Maxotex* Paris 07.05.99	France	Unlicensed use on Website of CD ROM of bibliographic information
Exemption limitations	*De Telegraaf v NOS & HMG* Hague 10.09.98	Netherland	Newspaper publisher refused a licence. Abuse of dominant position. No cprt. or db. legal argument made
	Mars v Teknowledge London 11.06.99	U.K.	Semi-conductor chip decipher encryption. Spare part exception denied
Duration			E.U. Directive Art. 10 gives 15 years from date of completion of making of the database
Beneficiaries of protection			E.U. Directive Art. 11 only nations of a Member State or Community citizens

Abbreviations: cprt. copyright, db. database

Source: Hugenholtz 2001 'The New Database Right: Early Case Law from Europe' at http://www.ivir.nl/publications/hugenholtz/fordham2001.html.

taking of raw data is deemed to be an extraction and a re-utilisation of a substantial part of the BHB database. In the alternate, William Hill's activities could be construed as the systematic extraction and re-utilisation of insubstantial parts of BHB's database.

The court held that database rights subsisted in BHB's database and that William Hill had infringed such rights by extracting and re-utilising substantial parts of its database. In assessing substantiality, the court said that a comparison must be made as to what had been taken or used with what was in the claimant's database. The importance of the information to the defendant was relevant as it could shed light on whether the information was an important or significant part of the database. In all of these it was important to keep in view the purpose of the E.U. Database Directive, which was to protect the investment in obtaining, verifying and presenting the contents of databases. Further the court held that the Directive was concerned with the unlicensed use of data without permission from the owner and the taking conflicted with the normal exploitation of the database and unreasonably prejudiced the legitimate interests of the copyright owner.

This case provides useful guidance from the courts in interpreting and applying the E.U. Database Directive. It demonstrates the vigour with which U.K. courts are prepared to go in protecting investment expended in the creation and maintenance of databases. It seems that other European decisions are also consistent with the U.K. approach.[97] A summary of some early case law as at 2001 under the E.U. Database Directive is given in Table 3.2.

Concluding Note on Database Protection

Databases represent a great investment of time, skill and capital. If the prevailing view were not to offer copyright protection for databases, then economics would dictate the size, scope and scale of the production of such databases, by either public or private enterprise. On the other hand, if there were copyright protection for databases, whether by the special or usual kind, private enterprise may perceive this to be a lucrative new market niche ripe for exploitation. The issue of copyright in databases thus distils to questions of access and monopoly rights, economics and policy. The ascendancy of one over the other may delineate the boundaries of protection for creators of electronic databases. Little or no protection will

---

[97] See Hughenholtz 2001 *op. cit.* See also Hugenholtz, PB 2003 'The Database Right File' which maintains up-to-date case law at http://www.ivir.nl/files/database/index.html.

give insufficient incentives to develop and maintain databases. Too much protection may stifle research, development and learning and at the same time encourage the negative aspects of monopoly power as well as civil disobedience.

An important implication for GI industry is that data providers should be alert to the opportunities for licensing arising from database rights while data users should be careful in avoiding violating either the copyright or database right of owners of databases. Such owners should also be vigilant to unauthorised use, especially if the database is on-line and to label databases with clear notices to show that it is protected by a database right. Developers of databases should also keep a record of work they undertake in building the resource, including time and money spent on its maintenance and the dates when databases are significantly updated. This latter is important if a new database right is to apply and a new period of protection to begin.

Practice Notes: Practical Tips for Maximising Legal Protection of Databases

- Ensure that any contracts dealing with the creation or licensing of databases adequately cover the new rights where other intellectual property rights may have overlooked them.
- Ensure that you know what rights subsist (or will subsist) in your databases and those that are being created. Better still get expert help to audit existing databases and contracts governing their creation.
- In granting licences for use of databases, ensure that the licence agreement adequately defines and deals with all rights, which exist in the database.
- Avoid situations where the ownership of any copyright and database right is held by different people.
- Regularly update any new databases in order to maximise the term of protection available, but keep good records of the work which is undertaken, any financial or other investment in the database and the date(s) on which it is carried out.

Source: adapted from Westell, S 1999a 'Legal Protection of Databases' http://www.geoplace.com/ ma/1999/0899/899law.asp.

3.6.4 Moral Rights and *Sui Generis* Regimes

Moral Rights

Copyright protects only a limited class of acts such as copying and publishing for a limited period of time. However, another set of rights, recognised by Australia, Europe and the Berne Convention—*droit moral*—protects the integrity of a creator's work in terms of reputation and honour.[98] Moral rights are not transferable and remain with the author and heirs for 50 years from the year of the author's death. Moral rights include:

- a right to attribution, that is, the right to claim authorship of a work;
- a right not to have authorship of a work falsely attributed;
- a right to integrity, that is, the right to object to any distortion, mutilation or other modifications which might be prejudicial to the artist's honour or reputation.

Under traditional copyright protection, the right to control publication has economic implications because an author may be permitted to take a benefit on the price every time the work is sold subsequent to its initial disposal. *Droit moral*, however, introduces a non-economic component to copyright protection and may be exercised by individuals and co-authors. Moral rights subsist in literary, dramatic, musical and artistic works and films, but not sound recordings, broadcasts or published editions.

The moral rights regime was introduced in Rome (1928) and was contained in the Paris Revision (1971) of the Berne Convention. The revision gives protection of moral rights such as the right to paternity and right to integrity and it would be up to member countries to impose either civil or criminal sanctions. Australia, as a Convention signatory, is bound by the principle of national treatment, which means it recognises moral rights and is now part of the Copyright Act as amended in 2000.[99] National treatment means that works are protected in countries of the Convention on a reciprocal basis, foreign nationals are to be treated the same as citizens.

---

[98] Members of the E.U. and eastern European countries recognise a *droit de suite* right that allows creators to benefit from resale of their work. This is an optional right which member countries of the Berne Convention may recognise. It operates only in those countries that choose to recognise such rights (Ricketson 1984: 380, 432–434 *op. cit.*). It is doubtful if such a right is being contemplated either in Commonwealth jurisdictions or in the U.S. See generally, Groves, P 1991 *Copyright and Designs Law. A Question of Balance. The Copyright, Designs and Patents Act, 1988*, London: Graham & Tratman, Ch. 6 Moral Rights, pp. 114–135.

[99] *Copyright Amendment (Moral Rights) Act* 2000 (Cwlth).

In 1988 the U.K. introduced specific legislation for the protection of non-economic rights of creators for the first time.[100] The U.S. *Copyright Act* does not recognise moral rights as such and few court decisions can be said to have helped establish any recognised moral rights. IP lawyers have argued that U.S. legal practice has provided many attributes of moral rights, despite not having amended the legislation to make such rights, explicit. Case law is said to adequately protect such rights under the rubric of related legal regimes such as defamation, privacy and unfair competition. Thus, while the U.S. signed the Berne Convention in 1991, and in so doing recognised the moral rights that protect a creation, such rights have not been explicitly established by legislation in the U.S. (Branscomb 1994: 86).[101] The U.S. has never adhered to recognising moral rights mainly because of a lack of such a tradition.[102]

In the GI and IT industries generally there does not appear to be a moral rights regime *per se*, but rather one that is interpreted as a national treatment issue. Thus, the ownership of foreign copyright by virtue of international conventions, agreements, and treaties may bring with it moral rights if the work is to be published in a jurisdiction which recognises such rights. The author's rights to attribution and acknowledgment, the right to integrity of the work to be published unaltered and the inalienable rights of the author, may be preserved under such circumstances.

In so far as moral rights in maps are concerned, it appears that all such rights may subsist, except the right of integrity. A digital map produced by an 'author' using GIS may object to having the subtle colours of the work reduced to 8 bits rather than the millions of colours that are available on modern computers. An author can rightly complain if the aspect ratio of the digital map is changed or if the work is truncated, clipped and cropped to fit into a particular format design. How authors of such GIS products can ensure 'integrity' is an issue that will rear its head more frequently in future. In practice, cautionary notes attached to margin of the map as well as technical specifications placed in a box annotation may declare the details under which the original map was produced. The use of such information in the metadata layer of a database will help ensure proper acknowledgment of authorship and integrity of the data. This may also serve to assert the moral rights of the 'author'. Employees, subcontractors and others who

---

[100] *Copyright Designs and Patents Act* 1988 (U.K.) ss 77–89.

[101] Branscomb 1994 *op. cit.*

[102] In a limited way one may read moral rights as having appeared in a statute. The U.S. *Visual Artists Rights Act* 17 USC §106A provides protection for moral rights for works of fine art only—prints, drawings and some types of photographs.

Figure 3.5 A moral rights notice

have no right of claim to copyright likewise will be unable to assert their moral rights to the works.

Defences to infringement include those that have been undertaken in good faith or where the author had given genuine consent or where an exception applies. For example, in the building industry in Australia an architect may have a right of integrity to the building and plans. In such cases where extensive renovations are contemplated, the practice is for the owner to give the architect or author of the work three weeks notice of the impending changes or relocations. A further three weeks must also be given for the architect or author to access the work or site to make a record of the work before it is changed.[103] Remedies include an injunction, damages, or a declaration that any false attribution or derogatory treatment of the work be removed or reversed. An assertion of moral rights could take the form shown in Figure 3.5.

Sui Generis Rights

In the light of the limited protection provided to databases and compilations of fact under copyright law, there have been suggestions that a law should be created for there to be a *sui generis* right in databases, independent of copyright law and other common and statutory law. In the U.S. two basic models have been proposed for such legislation: the IP model and the unfair competition model.

---

[103] Sexton, J 2001 'Its his gallery and he'll sigh if he wants to', *The Australian* 29 June, p. 10; See Doherty, M 2002 'Gallery reveals new entrance. Controversial plans dumped after protest', *The Canberra Times* 24 February, p. 3 for reports concerning the Australian National Gallery Building and the architect Colin Madigan.

Under the IP model, legislation would create a new exclusive property right in databases. This right would be granted for a limited period time, alienable by contract and subject to various statutory exceptions, defences and compulsory licences. This model was proposed in the *Collections of Information Antipiracy Act* (CIAA) 1999.[104] Under the unfair competition model, proposed under the *Consumer and Investor Access to Information Act* (CIAIA) 1999 legislation would prohibit particular methods of competition that undermine competitive markets for databases.[105] This model would impose liability for conduct that unfairly appropriates commercial value of a database created by another.

Although both these bills were not enacted their future introduction is a possibility. One reason the CIAA has not been enacted is that organisations of scientists and a coalition of Internet-based firms recognised the serious threats that the legislation posed to the digital public domain and mobilised against this legislation. The CIAIA forbids duplicating another firm's database and then engaging in direct competition with it. This bill will also affect the public domain, and in a much narrower and more targeted way than CIAA. There does not seem to be persuasive evidence that market failures were occurring or imminent in the database industry and as a result the bill has not found greater support for its passage.[106]

Pamela Samuelson's (1995) observations suggest reasons why we should be using *sui generis* rights to protect IP in information that currently attract only uncertain protection in IP law.[107] Some of her propositions follow. *Sui generis* legal regimes may be used to protect certain classes of products that have been vulnerable to market-destructive appropriations which existing regimes are unable to correct. *Sui generis* regimes borrow some concepts from either copyright or patent law and have been described as 'legal hybrid regimes' (Reichman 1994).[108] In the main, *sui generis* regimes go against the grain of patent, copyright and trade secrecy laws by protecting those subjects which ordinarily are unprotected. The adoption of *sui generis* laws may have salutary effects on industries whose

---

[104] H.R. 354, 106th Cong. (1999).

[105] H.R. 1858, 106th Cong. (1999).

[106] See discussion by Samuelson, P 2003 'Mapping the digital public domain: Threats and opportunities', 66 *Law and Contemporary Problems* 147(Winter/Spring).

[107] Samuelson, P 1995 'A manifesto concerning the legal protection of computer programs: Why existing laws fail to provide adequate protection' in Brunnstein, K and Sint, PP (eds) *Intellectual Property Rights and New Technology*, Proceedings KnowRight 95 Conference, Wein & München: *Österreichische Computer Gesellschaft* and Oldenbourg, pp. 105–115.

[108] Reichman, JH 1994 'Legal hybrids between the patent and copyright paradigms', 94 *Columbia L. Rev.* 2432 (1994).

products they govern. Without such laws, the legal landscape appears to be a quilted patchwork of *ad hoc* solutions, hamstrung by legislation created for a different era.

The electronic age has brought about the ready appropriability of information *borne on or near the face* of widely distributed products.[109] In manufacturing economies, trade secrecy laws provide significant lead times for firms to recoup investments. In information economies trade secrecy protection offers little because of the greater amount of know-how that is borne on the face of an information product. Thus, in order to address the needs of emergent information economies, there is an urgency for a new legal regime that will provide, artificially at least, a lead time that is functionally equivalent to that under trade secret law.

Sui generis rights may provide a middle position for information products that cannot fit into any of the established IP regimes or where the negative aspects under current IP law provide little or no protection. The difficulties in applying copyright principles to electronic databases need not be rehearsed, suffice it to say that the weight of legal opinion suggests that it is time to try and find novel solutions for new products that accompany information systems.

Concluding Note—Moral Rights and *sui generis* Rights

The theme of moral rights has already been discussed, with the conclusion that, even if an author has sold or given away proprietary rights in the work, there remain the other rights that might subsist in the works. While slow in attaining universal recognition and status, many countries are gradually coming to recognise that such moral rights exist. Given this, there is also the view that perhaps IT is pushing the law in new directions. There is considerable support that, like moral rights, there is merit in making efforts to establish a *sui generis* regime at least for IT. The E.U.'s push for a *sui generis* right since before the Database Directive has been taken up by lawyers in the U.S. who have issued a manifesto for such a right to protect IPR in IT. This is especially so if one accepts that information is considered as facts. Then such facts may be subject to misappropriation by those who have not worked to discover these. As a legal hybrid, *sui generis* rights may have to be written in a way that takes advantage of the best that the IPR regime has to offer. Moreover, some subjects, that have

---

[109] Know-how borne 'on the face' of software products refers to the information and knowledge revealed through the use of the program or in the user manuals. 'Near the face' refers to the underlying program that may be revealed through de-compilation of publicly distributed object code.

either uncertain protection or no protection at all under other IPR regimes may arguably find protection most certainly under a *sui generis* regime.

3.6.5 Business Methods and Geographic Information Patents

While business systems have become patentable in Australia a report has noted that such patents may need to be closely monitored. The concerns were whether business system patents actually encourage innovation and whether patents have been granted to truly novel ideas. Furthermore, much of the controversy surrounding such patents is a consequence of 'the relative inexperience' of patent offices around the world in this new field of patenting.[110]

A business method patent is like any other patent, except that the subject matter happens to relate in some way to a method of doing business. There is generally no definition of a business pattern, save by way of an example. The Amazon.com 'One-Click' patent, often cited as a classic example, is a system and method for placing an order to purchase an item via the Internet.[111] The methodology involved is one where information associated with one user is pre-stored by a website. When that user orders items from the website on another occasion, with only one click of a mouse (clicking a link associated with the item) the user is able to order that item without having to do much else.

Amazon.com brought an infringement action against Barnesandnoble. com (BN) alleging that the latter's 'Express Lane' checkout system infringed its 'One-Click' business method patent.[112] To show infringement of a patent a plaintiff must demonstrate that the defendant's conduct matches up exactly to *all* of the elements of the plaintiff's patent claims. In addition, a court might find infringement under the 'doctrine of equivalents', even if no literal infringement has occurred. The doctrine prevents would-be infringers from escaping liability by making trivial changes, but copying the essence of the invention.

A district court rejected BN's arguments that the Amazon.com patent was invalid as obvious and anticipated by the relevant prior art. Further, Amazon.com presented other evidence of non-obviousness that the court found convincing. The district court granted Amazon.com's request

---

[110] See Australian Advisory Council on Intellectual Property (ACIP) 2004 '*Report on a Review of the Patenting of Business Systems*' at http://www.acip.gov.au/library/bsreport.pdf.

[111] US Patent No. 5,960,411 issued 28 September 1999.

[112] *Amazon.com Inc. v Barnesandnoble.com, Inc.* 73 F.Supp.2d 1228 (W.D. Wash. 1999).

for a preliminary injunction and requiring BN to remove the 'Express Lane' feature. This injunction however was overturned on appeal.[113] The appeal court found that BN had mounted a substantial challenge to the validity of the patent in suit. As Amazon.com is not entitled to preliminary injunctive relief under the circumstances, the court ordered a vacation of the district court's findings and preliminary injunction and remanded the case for further proceedings.

A search of the U.S. Patent Office website for business method patents will show that many different types of such patents have been issued, some of undue scope.[114] Rappa (2002) has summarised the various kinds of business models and the variety of ways these types of patents that may be categorised and implemented.[115] Most other countries are taking a more cautious stance compared with the expansiveness of the U.S. approach. Many are working towards the international harmonisation of IP law and the globalisation of commerce. For example, the Patent Offices of U.S., Japan and E.U. have met on several occasions and report a 'consensus' opinion consisting of two propositions: (1) 'A technical aspect is necessary for a computer-implemented business method to be eligible for patenting'; and (2) 'to merely automate a known human transaction process using well-known automation techniques is not patentable'.[116]

The position of the European Patent Office (EPO) is given in the Convention on the Grant of European Patents, known as the European Patent Convention (EPC). The EPO will grant a European patent for any new invention that involves an inventive step and is susceptible to industrial application (EPC Art. 52(1)). However, EPC Art. 52(2)(c) states that 'methods for doing business, and programs for computers' are excluded from the definition of inventions capable of patenting although this is qualified by Art. 52(3). The position therefore is that business methods

---

[113] *Amazon.com Inc. v Barnesandnoble.com, Inc.* 239 F.3d 1343 (Fed. Cir. 2001).

[114] http://www.uspto.gov/patft/index/html for access to the U.S. Patent and Trademark Offices' searchable databases.

[115] These include various models: *brokerage* with brokers as market makers, *advertising* as an extension to the traditional media broadcast model, *infomediary* or information intermediaries, *merchant* for wholesalers and retailers of goods and services, *manufacturer* where buyers are contacted directly and the distribution channel compressed, *affiliate* in which purchase opportunities arise wherever people may be surfing on the net, *community* based on user loyalty, *subscription* where users are charged a fee to subscribe to a service and *utility* a metered service pay-as-you-go approach. See Rappa, M 2002 'Managing the Digital Enterprise' at http://www.digital-enterprise.org/models/models.html.

[116] See Report on Comparative Study Carried Out Under Trilateral Project B3B, 14–16 June 2002 at http://www.european-patent-office.org/tws/front_page.pdf. E.U. 1999d *Directive on the harmonisation of certain aspects of copyright and related rights in the Information Society* (COM (1999) 250 Final (Directive)).

and computer program inventions that are of a 'technical character' may be patentable. But this position is now under debate with Member States proposing a change to the *status quo*.[117]

Geographic Information Patents

While patenting has traditionally been for the protection of new manufactured goods, increasingly patents have been sought and obtained for use in the IT and GI industries. A search of the U.S. Patent and Trademark Office's on-line database for 'Geographic Information System' AND 'GIS' reveals that there were over 105 'hits' that satisfied the Boolean search parameters. The abstracts of selected GI type patents reveals an interesting mix. Some patents are from the related field of computing such as database design. Most are in the areas of image processing, GPS navigation and routing, vector- and raster-based geographic data, map-based directory systems, three-dimensional interactive image and terrain modelling systems, and methods for mapping and conveying product location. There is one patent in particular which is currently of great interest and dismay to the GI industry at large.

U.S. Patent No. 6,240,360 was assigned to the inventor Sean Phelan of London trading as Multimap.com on 29 May 2001 for a 'Computer system for identifying local resources'.[118] The patent has also been granted throughout the E.U. and patent pending in various other countries worldwide. While the title is unassuming the abstract to the patent reveals how the system transfers spatial data from a server to a remote computer based on location information requested by the remote device. The patent abstract reads as follows:

> A map of the area of a client computer (10) is requested from a map server (11). Information relating to a place of interest is requested from an information server (12) by the client computer (10). The information is superimposed or overlaid on a map image of a position on the map image corresponding to the location of the place of interest on the map. The Information (or 'overlay') server (12) may contain details of, for example, hotels, restaurants, shops or the like, associated with the geographical coordinates of each location. The map server (11) contains map data, including coordinate data representing the spatial coordinates of at least one point on the area represented by the map. [Numbers in brackets refer to a drawing provided with the patent application. See Figure 3.6 for an illustration.]

---

[117] See McCoy, MD and Spence, AT 2001 'Lessons from the United States and Europe on Computer-related Patents' at http://www.gigalaw.com/articles/2001-all/mccoy-2001-08-all.html.

[118] See http://patft.uspto.gov and search under patent number 6,240,360.

The legal scope of the monopoly granted by this patent is set out in Claim 1 and is made up of seven steps. Broadly, these are: (1) storing on a map server computer, map data; (2) storing on the map server computer coordinate data; (3) storing on an information server computer information data of at least one place of interest; (4) transmitting a map request to the map server computer from the client computer; (5) using map data to display an image of the map on the client's computer monitor; (6) transmitting an information request to the information server computer from the client computer; and (7) displaying the information data relating to at least one place of interest on the client's computer monitor. The test to assess an infringement of this patent is where each and every requirement of the relevant claim is satisfied.

This Multimap.com patent appears to have 'cornered the market' in the art of Internet mapping, and potentially has a fundamental effect on the whole of the GI mapping industry. While seemingly similar to a number of such products, the very existence of the patent till at least the year 2016 would prevent the future product development in this direction unless developers are prepared to pay premium licences fees for its use. Other than ambiguities in the terms used, for example, 'place of interest', the patent is apparently not 'new', novel and non-obvious. Radcliffe (2003) is of the view that the patent is essentially the automation of a manual process that of itself does not have any technical effect, and prior art has been identified pre-dating the patent which anticipates the entirety of

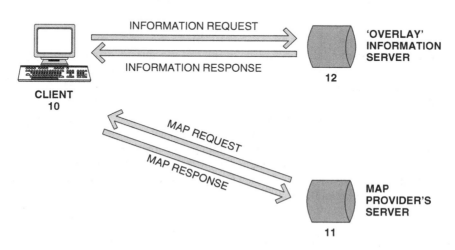

Figure 3.6 US Patent for a computer system for identifying local resources (US Patent No. 6,240,360)

Claim 1 to render the patent invalid. The other subsidiary claims are either its obviousness and/or anticipated by prior art with the result that the remainder of the patent is invalid as well.[119]

The implications of the validity of the Multimap patent can be very wide-ranging. The patent remains valid and the owner can rightfully claim royalties from users until the monopoly runs out or is revoked by a court of law. This also means that GI product developers working in a similar area would have to ensure that the Multi-map patent is not infringed in any way by undertaking due diligence audits and to write appropriate warranties when licensing products to indemnify third parties. The complexity, costs and difficulties involved in such legal compliance requirements can be considerable. Already there have been various groups discussing the mounting of a court challenge. A crucial part of the challenge will rest on gathering evidence that the work is obvious and that a prior art has been in place before to the submission date of the patent in August 1996 in the U.S.[120]

It appears that the day of the patent in the GI sector has now unavoidably arrived, and that every sensible organisation in the sector should be patenting their inventions and improvements (Radcliffe 2003). But patent law is territorial and as such there is a need to file patents in many jurisdictions to ensure its global enforcement. While the lodgment of patents can be an expensive exercise it would have to be weighed up against the potential revenue streams it might generate. Forum shopping suggests that the U.S. market would be a lucrative one, not only because of its size, but also because of its 'expansive' patent laws that also permit the patenting of business methods and systems. The marriage of GI technologies and business methods would offer organisations that create and patent significant breakthroughs a market niche and dominance of undreamt of proportions.

3.6.6 The Digital Agenda

The 'Digital Agenda' is a broad term that is used to describe the implementation of the 1996 WIPO Copyright Treaty (WCT Art. 11) and WIPO Performances and Phonograms Treaty (WPPT Art. 18) that focuses

---

[119] Radcliffe, J 2003 'Death of Copyright—Long Live Patents', 2003 Cambridge Conference, Ordnance Survey, Southampton, U.K. Paper 4D.2B. Rose, N and Radcliffe, J 2003 'Death of copyright – Long live patents and database rights', 2003 Cambridge Conference, Ordnance Survey, Southampton, U.K. Paper 4D.2A.

[120] The electronic mail discussion on this patent challenge may be found at http://mapserver.gis. umn.edu/wilma/mapserver-users/0203/msg00495.html and Multimap is at http://www.multimap.com.

predominantly on the protection of copyright content in the digital networked environment. The *Copyright Amendment (Digital Agenda) Act* 2000 (Cwlth) which came into operation on 4 March 2001 and the U.S. *Digital Millennium Copyright Act* 1998 (DMCA) are of a similar genre. Both enactments and similar ones in other jurisdictions provide legal protection and legal remedies against circumventing technological protection measures (TPM) used by copyright owners to protect their work from piracy.[121] These enactments have implications for the electronic industry and those involved in the development and implementation of on-line technologies or those who work with other digital media. Users and developers in the GI industry in general should thus become familiar with some of the provisions and IPR protective measures that have been put in place.

The definition of a 'work' under Australia's Copyright Act is much enlarged, including computer programs and code, digital movies and pictures, e-mails, compilations and a wide variety of different works incorporated in multimedia presentation. Web-based GI products—interactive maps, dynamic map queries, databases—would thus be caught under this Act. The amendments to the Act extend the traditional protection of copyright laws to original electronic and digital works. As with the DMCA in the U.S. to ensure a balance between the increased rights of copyright owners and the public interest in access to information, a series of exceptions to these new owner rights have been created. A majority of these are an extension to existing rights in the digital environment. The relevant exceptions are:

- A user may copy 10% of an electronic text work without permission, but only for research or study purposes. However, this exception excludes computer programs.
- Making a reproduction of a website is permitted if it is for 'fair dealing' (fair use) purposes. Fair dealing includes research or study, criticism or review, news reporting, and legal advice.
- Making a temporary reproduction of a work that may be required, as part of a technical process of making or receiving an electronic communication, is not an infringement of copyright, for example, in browsing and certain types of caching on computer systems.
- A computer program may be reproduced or adapted to obtain information necessary to enable the development of an interoperable product. Copyright protection extends to the source and object code of a computer program and any expression of

---

[121] s 116A (Australia) and s 1201 (U.S.) respectively. See also *Copyright, Design and Patents* Act 1988 (U.K.) s 297A, and *Copyright Act* 1994 (NZ) s 226(2)(a).

systems or methods, but not the functionality of the program (which is covered by the patent system).

- A copy of the program may be reproduced or adapted for security testing purposes and to correct errors.
- The owner of a copy of a computer program may make back-up copies.

New provisions in the U.S. Copyright Act make the infringement of copyright a criminal offence and authorise the award of additional civil damages. There have been a number of cases in the U.S. alleging infringement under the DMCA the most well known being *U.S. v Elcom*.[122] In this case a Russian scientist was charged with 'wilful trafficking' in or providing to the public, software that could circumvent technological protection on copyrighted material under the DMCA's anti-circumvention provisions. The charges were later to be withdrawn. The other cases are *Universal City Studios, Inc. v Reimerdes* and *Universal City Studios, Inc. v Corey* concerning the decryption program known as 'DeCSS' that circumvents a protection system and enables motion pictures to be copied and played on devices that lack the licensed decryption technology.[123]

In Australia, under the *Copyright Act* 1968 (Cwlth) the severity of the penalty depends on the type of infringement and whether the offender is an individual or a company. A new technology-neutral right of communication is provided to enable communications over the Internet. The legislation also introduces prohibitions against the development and sale of copyright-circumvention tools which could be physical (a hardware key or dongle) or non-physical (such as software tools). It is also an infringement of copyright to remove digital rights management (DRM) information or to deal with material that do not have DRM information.

Implications for Geographic Information

Various forms of copy protection have been used in GI software such as 'dongles' (a hardware key), password protected software, and the matching of software registration with hardware serial numbers of computers. These devices have worked well in the past and future developments would depend on the business model adopted by GI software vendors. Increasingly

---

[122] Documents relating to this case can be found at http://www.eff.org/Cases/US_Elcom/. The trial brief is Case No. CR 01-20138 RMW U.S. District Court Northern District of California, San Jose Division, 21 October 2002.
[123] 111 F.Supp.2d 294 (S.D.N.Y. 2000), *aff'd* 273 F.3d 429 (2d Cir. 2001); and 273 F.3d 429 (2d Cir. 2001) respectively.

the business model adopted would be one where copyright protection and security measures are embedded in digital content to enable use and access to the content. However, even with the adoption of the DRM model, while it may not stop all forms of piracy, such as 'insider' copying of master files, the system can still prevent the illegal and unauthorised uses of digital content. In addition, it may be possible to track every use of the content, such as how many times it is used, when it is used and so on which may bring with it grave privacy implications.

3.7 Atypical Developments and Other Legal Issues

3.7.1 The 'Copyleft' Movement and No Rights Reserved

Given the difficulties of policing and enforcing IPR in the IT industry, there has grown up a movement under the rubric of free and open source software (FOSS). FOSS makes available its source code free for public use and encourages users to debug, change the source code and redistribute the derivative software. 'Free' in this sense is the freedom to use the software program for any purpose, to study how it works and adapt it to one's own needs, to redistribute copies and to improve and share it with the community. It does not mean that the price is zero, since FOSS can be traded in the market. Users are encouraged to think 'free speech' not 'free lunch'.[124] The idea of 'free' software is not new given that during the 1960s and 1970s when most people were using mainframe computers, source code was freely accessible and taken for granted.

In the U.S., AT&T's Bell Labs led the way in developing the UNIX operating system and associated language 'C' that could run on different and varied hardware. Under terms of a regulated monopoly deal with the U.S. Department of Justice, AT&T could not sell UNIX for profit and so it distributed source code to universities and others to improve the software and fix bugs. The thinking of the day was that software was given away free as a hook to encourage people to buy hardware.

The logic of free software broke down in the late 1960s when the U.S. Department of Justice filed a massive anti-trust suit against IBM, forcing it to 'unbundle' its systems and to sell its software separately. IBM

---

[124] Free Software Foundation (FSF) 1996 'The free software definition' at http://www.fsf.org/philosophy/free-sw.html.

sold mainframes with operating systems that did not distribute source code. In fact buyers had to sign non-disclosure agreements simply to get executable copies.[125] These decisions marked the birth of the modern commercial software industry and by 1975 the start of the widespread use of Microsoft software. The arrival of the personal computer in the early 1980s also made software extraordinarily valuable in its own right. This is where FOSS came into its own and reinvented itself in the mid-1980s.

To replace traditional copyright, the Free Software Foundation (FSF) developed a standard copyright agreement, the GNU Public Licence (GPL) often called 'copyleft'.[126] The central idea of the GPL is to prevent cooperatively developed open/free software source code from being turned into proprietary and restrictively copyrighted software. The GPL states that users are permitted to run the program, copy the program, modify the program through its source code, and distribute modified versions to others. What programmers may not do is add restrictions to the program. This is the 'viral' clause of GPL since source code that has been 'copylefted' is to be released using GPL as well. The rapid spread of the Internet in the early 1990s accelerated FOSS activity with the development of the GNU/Linux PC operating system. The official GNU/Linux version 1.0 operating system was released in 1994 and became a credible competitor to Microsoft in the PC operating system market. A GNU GPL is illustrated in Figure 3.7.

In the 1990s there was also an alternative institution for 'free' software – the Open Source Initiative (OSI) as a response to the decision by Netscape to publicise the source code for its World Wide Web (WWW) browser. OSI requires entities distributing FOSS to satisfy the Open Source Definition (OSD) in its copyright statement.[127] While copyleft GPL *requires*

GNU General Public License
GPL
Version 1.2, November 2002
Everyone is permitted to copy and distribute verbatim copies
of this license document, but changing it is not allowed.
Copyright © 2000, 2001, 2003 Free Software Foundation,
Inc.
59 Temple Place – Suite 330, Boston, MA 02111-1307, USA.

Figure 3.7 GNU General Public License (GPL)

---

[125] DeLamarter RT 1986 *Big Blue: IBM's Use and Abuse of Power*, New York: Dood, Mead.

[126] See GNU GPL see http://www.gnu.org/licenses/gpl.html.

[127] See http://www.opensource.org/osd.html and http://www.opensource.org/docs/definition.php for more details.

any redistribution of GPL software to be released only under GPL, the OSD *allows* redistribution under the same terms, but does not require it. OSD source code must be distributed with the software or otherwise made available for no more than the cost of distribution. Anyone may redistribute the software for free, without owing royalties or licensing fees to the author. Anyone may modify the software or derive other software from it and then distribute the modified software under the same terms. OSI removes the viral impact of the GPL. Towards the end of the 1990s the FOSS process had proved its viability, not only in building complex software packages that could compete with proprietary products, but also in a number of other market segments.

In the new millennium the Creative Commons (CC) project was given worldwide impetus with the release of Lessig's book *The Future of Ideas*.[128] The CC is an organisation based at and sponsored by Stanford University, offers a number of licenses designed to allow the broader use of copyright material. The licence allows the licensee to use copyright subject matter on the basis of one or all of the following conditions: attribution, non-commercial distribution, no derivative works (verbatim copies only), and share and share alike. The CC is also engaging in an international commons (iCommons) project seeking to post and translate their licence to other legal jurisdictions.[129]

The CC has since released its new computer-readable copyright licence designed for artists, writers and programmers who want to give the public access to their works.[130] The Founders Copyright licence, in the form of code that can be added to a digital work, allows authors to specify exactly how other people can use the work and lets those details show up in a search engine. The CC is dedicated to expanding the number of works of any nature in the public domain. Under the terms of the Founders Copyright between the CC and a contributor, a contract will guarantee that the relevant creative work will enter into the public domain after 14 years. During the first 14 years the creator will enjoy all the rights that conventional copyright protection permits. In version 1.0 of the Licensing Project CC will help build licences that tell others about works that are free for copying and other

---

[128] Lessig, L 2001 *The Future of Ideas. The Fate of the Commons in a Connected World*, New York: Random House.

[129] For an example of an organisation using the CC licence see the Public Library of Science site at http://www.plos.org.

[130] See http://www.creativecommons.org/projects/founderscopyright.

If you publish your educational materials on-line Creative Commons provides an interface so that the world knows how they can distribute or re-use your educational materials. We can also help you find material to share and build upon.

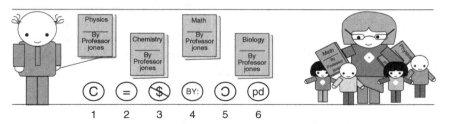

Notes to 'rights' icons:
1 refers to the copyright notice, 2 = no derivative works permitted, 3 = may copy but only if non-commercial, 4 = attribution permits others to copy, distribute etc., 5 = share alike, permits others to distribute derivative works only under a licence, 6 = a public domain document.

Figure 3.8 Creative Commons 'rights' notice for authors

uses, but on certain conditions.[131] Figure 3.8 gives a CC 'rights' notice for authors.

Implications of FOSS for Geographic Information and Intellectual Property Rights

Examples of free and open source software (FOSS) implementation abound, for example, Sendmail—a program for routing approximately 40% of the e-mail transmitted over the Internet. Another application of the use of FOSS is in the Beowulf Project. The Beowulf Project is a scheme for connecting computers to form a high-performance computer (Beowulf cluster) that approaches the speed of a 'super-computer'.[132] Since a Beowulf cluster can be developed from common, off-the-shelf computers using FOSS, a Beowulf cluster 'super-computer' can be built and implemented at a fraction of the cost of other systems with similar computing capacity. The use of this architecture in research projects in developing countries has the

---

[131] There is also a variant to the CC licence developed by Open Content. The Open Publication Licence lets anyone freely see, modify, and redistribute the content. Changes or additions must be marked as such, and anyone who releases a modified version must describe where to find the original work. The author retains copyright and can prohibit redistribution. See http://www.opencontent.org/openpub/.
[132] Information about the Beowulf Project and other details may be found at http://www.beowulf.org and at http://www.canonical.org/~kragen/beowulf-faq.txt.

potential to bring great benefits as well as the simplicity of putting together a super computer while at the same time saving valuable financial resources.

There is no doubt that access to IT is a means of enabling and hastening economic development of poor countries. In doing so, it will enable developing countries to leapfrog developmental stages of IT implementation. FOSS will enable such countries from depending on a few software suppliers who are often located in other countries. In addition, with the sharing of applications across the public sector in these countries will reduce duplication of work and licence expenses on proprietary software. The introduction of a diversity of software reduces the possibility of failures caused by viruses that tend to attack a software monoculture. Moreover, FOSS may help avoid the use of expensive proprietary software and closed data formats.

FOSS is not an alternative to the IPR regime. Rather FOSS requests users to respect the IP of the software's author(s) as outlined in the GPL or OSD licence. Governments are expected to provide the legal protection and remedies when this is necessary and deserved. Moreover, under the restrictive IP licensing regime there may be no encouragement for ICT development in developing countries. Ironically the increasingly stringent enforcement of IPR demanded by proprietary software producers may further encourage and hasten the switch to FOSS. With free access to source code, not only is there empowerment, but also a simple price advantage and an ever-growing demand. One example is the FOSSFA initiative for Africa where the Action Plan for 2003–2005 proposes three distinct approaches a government might take in formulating its FOSS policy.[133] A neutral approach ensures a choice is supported and discrimination against FOSS is eliminated. The enabling approach sees policies geared towards creation of the capacity to use FOSS. The aggressive approach is one where governments actively encourage the development of FOSS through both legislation and policy.[134]

3.7.2 Other Intellectual Property

Geographical Indications

Geographical indications are place names and in some countries words associated with a place that are used to identify the origin and quality,

---

[133] The text for the FOSSFA statement is at http://www.prepcom.net/wsis/1046170300.
[134] See FOSSFA Action Plan 2003–2005 at http://www.fossfa.org/resources.html.

reputation or other characteristics of products for example, 'Tequila', 'Champagne', 'Basmati' rice, or 'Parmagiano Reggiano' cheese. Art. 22 of the TRIPs Agreement defines a standard level of protection, that is, geographical indications have to be protected in order to avoid misleading the public and to prevent unfair competition. Art. 23 provides a higher or enhanced level of protection for geographical indications for wines and spirits. There are a number of countries that want to extend this list to a wider range of products, including food and handicrafts. Among the allowable exceptions to the agreement are when a name has become a common or generic term for example 'cheddar' now refers to a particular type of cheese, not necessarily made in Cheddar in the U.K. Also, when a term has already been registered as a trade mark, for example in Italy, 'Parma' is a type of ham from the region of the city of Parma, but in Canada it is a registered trade mark for ham made by a Canadian company. To date since the Doha mandate of 1997 and the Cancún Ministerial Conference of 2003 member countries of the WTO have been working towards creating a multilateral register for wines and spirits; and extending the higher level of protection beyond wines and spirits.[135]

In Australia the *Australian Wine and Brandy Corporation Act* (AWBC) 1980 (Cwlth) was amended in 1992 to comply with Art. 23 of the TRIPs Agreement and Art. 6 of the E.U. on Trade in Wine Agreement.[136] There appears sufficient protection conferred by domestic law to ensure that the public would not be misled or for there to be unfair market competition.[137]

The GI industry can and does play an important role in determining geographical indications. Prior to 1992 and the TRIPs Agreement there was no legal mechanism for defining Australian wine regions. States were carved into zones and then regions within zones with possible sub-regions on arbitrary and populist grounds. With amendments to the AWBC criteria were put in place to clearly define regions and zones. Other than minimum requirements of growers, hectarages and output per annum there were other geographical factors. The geographical elements include the geology, soil types, temperature, harvest dates, use of water, State and

---

[135] See http://www.wto.org/english/tratop_e/trips_e/gi_e.htm.

[136] See Cheung, C 2004 'Feta cheese—geographic indication or generic term?' *Intellectual Property Law Bulletin*, v. 16(9), pp. 133–135. See also Blakeney, M and Kretschmer, M 2004 *Intellectual Property and Geographical Indications: A Legal and Economic Analysis*, London: Edward Elgar Publishing.

[137] For example, the *Trade Marks Act* 1955 (Cwlth); *Trade Practices Act* 1974 (Cwlth); common law tort of passing off, labelling legislation in *Commerce in Trade Description Act* 1905 (Cwlth); and State and Territorial food legislation.

Local government planning Acts, history of grape growing, and documentary proof that the region is known and recognised to the outside world.

Geologists have determined that many of the best wines in Australia come from vines grown in soils on top of limestone. Evidence from radiometric maps show how geophysical tools can be used to locate better-drained soils. For example, air-borne remote sensing radiometric tools measure natural radioactivity of potassium, uranium and thorium. In addition soil types and soil profiles including soil chemistry all play important parts in determining the quality of the grape harvest.[138]

Photographs

While the topic of photographs might appear to be peripheral to the interests of GI professionals, the unique, dynamic and multimedia possibilities of the Internet to store, present and deliver information have created a whole new methodology of both fair use/dealing and what constitutes copyright infringement. Photographs in general are protected under relevant copyright legislation worldwide, whether in print form or on the Internet. The difficulty is to decide what constitutes fair use.

In the U.S., for example, the case of *Kelly v Arriba Soft Corporation*[139] revolved around the use of 'thumbnail' pictures. Kelly is a professional photographer well known for his pictures of the American West. Arriba Soft Corp. (now Ditto.com) is a search engine company that locates and indexes images and presents these as small pictures called 'thumbnails' as opposed to traditional text listings of search results. The issue before the court was one of what constituted fair use. The court held that where the thumbnail could be enlarged to a higher resolution, then the copying is not excused as fair use, but an infringement of the exclusive right to display the pictures for which there is copyright protection. Also the case held that 'in-line linking' or 'framing' constituted a copyright infringement where market harm is caused. In-line linking refers to the importation of an image from another site and to display this as if it were part of the importing website. Framing refers to one where a website displaces another website in a captured window or frame.

---

[138] Pain, C and Nightingale, J 2001 'Geology proves fruitful for viticulture', *AusGeo News* 63 October/November, p. 3. See also Farmer, D 2002 'The red soils of Coonawarra, part of a unique *terroir*', *The Australian and New Zealand Wine Industry Journal*, v. 7(6), December. Also at http://www.winetitles.com.au.
[139] 280 F.3d 934 (9th Cir. 2002).

There are several implications for the GI industry. First, aerial photographs and Landsat images are now obtainable from various Internet sites where an index may be shown to customers to assist in the ordering process. Such re-sellers must ensure that users are unable to download the photographs and images to avoid purchasing these products. Second, photographs and images may be no different from maps displayed on Internet sites. In the case of maps it is important that all the marginal IPR references are preserved on the map as a whole. This is because it may be possible to capture such maps in a frame and a screen-shot taken of the display that may then be printed. To do so is to violate fair use and the rights of the copyright holder. Also, copyright owners have an exclusive right to display and publish works and a licence or permission is required before anyone else may do the same with the material. Furthermore in-line linking and framing may infringe IPR. A Web designer may be able to divide a Web page into distinct frames, permitting some regions to remain constant on a page while other frames contain content that is reloaded as necessary as the user navigates through the site. A Web design may incorporate contents of a third party website into one frame and surround this with content contained in the local server. The Scottish Court of Session case of *Shetland Times Ltd v Wills*[140] illustrates how easily copyright infringement can occur on the net.[141]

Fonts

In Australia, a newspaper masthead comprising Chinese calligraphic characters is an 'artistic work' capable of copyright protection as well as being a trade mark capable of protection under the *Trade Marks Act* 1995 (Cwlth). In *Australian Chinese Newspapers Pty Ltd v Melbourne Chinese Press*[142] the court was satisfied that the requisite level of thinking had gone into the creation of the whole work, including the masthead and therefore was an original artwork capable of copyright protection. The art and the creation of the characters were sufficiently expressive and the artist had exhibited a sufficient degree of thinking in producing the work.[143]

---

[140] (1996) 37 IPR 71; and [1997] SCLR 160.

[141] See earlier reference to deep linking and the *Ticketmaster Corporation v Microsoft Corporation* No. 97-3055 DDP (C.D.Cal.12 April 1997) case cited earlier in this chapter.

[142] (2003) 58 IPR 1.

[143] Pappas, C 2004 'Can copyright subsist in Chinese characters?' *Intellectual Property Law Bulletin*, v. 16(9) February/March, pp. 150.

A claim of copyright in computer-aided typography fonts in a brochure has also been argued. *Lott v JBW & Friends Pty Ltd*[144] is an appeal to the Supreme Court of South Australia from a decision in the Magistrates Court where copyright was found to exist and to have been infringed. The respondents had stated that they were commissioned to produce a brochure *Opera in the Outback*. There was a dispute over unpaid fees, and the respondents had sought damages for 20 occasions of infringement of copyright. The respondents claimed that they designed and crafted the graphic bar, using words supplied by choosing the typefaces and font that were used as well as all the graphic design in the letters, the angle of curves above and below the words and setting the words in colour, using different hues. It was claimed that the graphic bar was only licensed for use on the brochure.

The Supreme Court of South Australia found that the respondents had designed the graphic bar; that they owned the copyright in it; but that the Director of the Company who had commissioned the design and authorised the publication of the graphic bar in advertisements was personally liable for the breach of copyright. Additional damages were awarded for the flagrancy of the infringements, even though it was argued that there was an implied licence and a customary arrangement to use commissioned designs for additional advertisements. The Court found that the agreement was clear and the licence was limited only to the use of the graphic bar on the brochure.

The decision is consistent with a long line of authorities. Even very simple works, especially graphic works, enjoyed copyright protection. The decision highlights the need to write agreements carefully so that the use of commissioned works is expressly agreed to between the parties, and the need for a written assignment in place where relevant and not to rely on custom and practice in an industry.

For the GI industry, fonts used on maps have been carefully chosen to provide effect and contrast. The size, shape, colour and weight of the font used can illuminate and bring out important features. Many of the fonts used in the industry are invariably those which are either in the public domain or those used under licence from major corporations such as Monotype Corporation, International Typeface, and Agfa Monotype. The design and artwork that accompany the fonts on maps may have a degree of creativity, originality and artistic merit

---

[144] [2000] SASC 3.

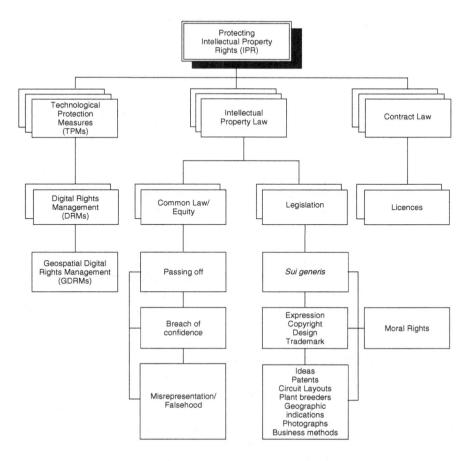

Figure 3.9 Protecting intellectual property rights: a summary

that together are capable of protection under copyright laws. More importantly, the person commissioning the work may claim copyright ownership, unless there is an assignment of those rights. This is similar to the case of bespoke computer software discussed elsewhere in which the programmer retains copyright.[145] A summary of the various means of protecting IPR is given diagrammatically in Figure 3.9.

---

[145] Cho, G 1998 *Geographic Information Systems and the Law*, Chichester: John Wiley & Sons, pp. 67–68.

3.8 Infringements, Defences and Remedies

Infringements

In Australia, the *Copyright Act* 1968 (Cwlth) provides that it is an infringement of copyright to do any of the acts exclusively reserved for the copyright owner without the copyright owner's permission. The Act also covers the importation of works into Australia for sale where the making of such works would have constituted an infringement. Moreover, to 'authorise' infringements is also proscribed. In practice such statutory provisions have been difficult to enforce, not only because of the way the law is framed, but also because such difficulties are directly attributable to the technology involved. With the convergence of IT there is a gradual a move away from physical media towards electronic digital media. Nationally and internationally copyright law protects the bulk of IPR with little need for registration. Trade mark and patent protection requires country-by-country registration. However in the case of trans-border flow of physical as well as electronic data and materials the enforcement of IP law has become problematic.

Fair dealing and fair use of copyright protected materials are limited to private, non-commercial or educational uses. Once one begins to deal commercially with the work, fair dealing or fair use may no longer apply, and copyright may be infringed. However, it is not an infringement of the reproduction right to use the work in a way that does not produce copies of that work. This means that the work must be used as a 'whole' and unchanged; the user must acknowledge the original source of the work and the user may read or orally convey the contents of a work to another.

Territorial copyright and parallel importation. Sections 37 and 102 of the *Copyright Act* 1968 (Cwlth) entitle owners of copyright to prevent others importing into Australia articles such as books, records, computer software which contain copyright material. Sometimes referred to as a '*territorial copyright*', it applies to all articles whether or not made with the consent of the owner of the copyright in the country of manufacture. Where the imported article is manufactured with the consent of the copyright owner it is sometimes referred to as '*parallel importation*' or 'grey marketing'. Sections 38 and 103 of the *Copyright Act* 1968 (Cwlth) make it an infringement of copyright by selling unauthorised imports or dealing commercially with them. One reason for these sections of the Act is that the articles would be 'devalued' if such work were imported. But, consumers of books, music and computer software have criticised this reasoning because it artificially supports a higher price for the product.

Also some articles are unavailable in Australia because the copyright owners have failed to import these articles.[146] New provisions for the importation of books have come into force since 1991.[147]

Section 37 of the *Copyright Act* 1968 (Cwlth) prevents parallel importing. This permits publishers to carve up national monopolies and manipulate prices, making the market uncompetitive and with undesirable consequences. There is a view that parallel importing should not be maintained in the face of the irrelevance of national boundaries for works distributed through the Internet (Appel 1995: 7).[148]

In the U.K. and Hong Kong, copies made by licensees of copyright owners are restricted by the terms of their licences to particular territories outside the country into which the product is being imported.[149] There are no such parallel importation restrictions either in the U.S. or in Japan (Knight and FitzSimons 1990: 48).[150]

Substantiality requirements. Section 14(1) of the *Copyright Act* 1968 (Cwlth) provides that an infringement will be made out where a copy or reproduction relates to the whole work or a substantial part of it. However, the Act gives no guidance as to how the word 'substantial' is to be tested or applied. Case law suggest that it need not be a large part, but it must be an important or vital part in relation to the whole work. The test therefore is a question of fact and degree having regard to the particular circumstances of each case. It is the quality of what is taken in relation to the work as a whole rather than the quantity that is abstracted (Leonard and Spender 1989: 43).[151]

Early English common law suggests that the 'substantiality' requirement may be interpreted as a 'safety net' for the free circulation of ideas and open debate. In *Hawkes & Son (London) Ltd v Paramount Film Service Ltd*[152] SLESSER LJ considered that the term pertained to quality of reproduction not quantity. However, in *Nichols v Universal Pictures*[153] LEARNED HAND J considered that it was necessary to compare the pattern

---

[146] See *Interstate Parcel Express Co. Ltd v Time-Life International (Nederlands) BV and Anor* (1977) 15 ALR 353; and *Bayley & Co. v Boccacio Pty Ltd and Ors.* (1986) 8 IPR 297.

[147] *Copyright Act* 1968 (Cwlth) ss 44A and 112A.

[148] Appel, R 1995 'Copyright in a digital age. Chaos in the debate', *ANU Reporter*, 13 December, p. 7.

[149] *Copyright, Designs and Patent Act* 1988 (U.K.).

[150] Knight, P and FitzSimons, J 1990 *The Legal Environment of Computing*, Sydney: Addison-Wesley Pub. Coy.

[151] Leonard, PG and Spender, PA 1989 'Intellectual property protection of databases', 9 *Information Services and Use* 33–43.

[152] [1934] 1 Ch. 593.

[153] (1930) 45 F.2d. 119.

of the two works in question in order to establish whether there had been a substantial reproduction. MAUGHAM J endorsed this approach in *Kelly v Cinema House Ltd.*[154]

Reproducing facts without copying the expression or arrangement of facts will not be considered a reproduction of a substantial part of the work. In the context of databases there is often great difficulty in separating facts from expression of facts as demonstrated with the E.U. Database Directive arguments presented in the previous section. This is because in such cases it is not simply the arrangement or the expression of the facts, but rather the computer program, the so-called database management system (DBMS), which organises and searches the database that is important. When the material form in which the copyright exists is different from the form the infringing work is embodied in, it may not be possible simply to look for visual or literal similarities in the arrangement to test whether the two works are substantially the same.[155]

Databases also pose other types of copyright problems. One problem relates to whether products of databases have to respect the copyright existing in works that are selectively being put into a computer or its memory. The legal position is that if the data are not protected by copyright there is no legal obstacle to their being inserted into a database. But where the work is subject to copyright, the possibility of breach of copyright will depend on the size and scope of the use made of the data as a whole or each datum of the database. To copy the whole database would obviously infringe copyright. Infringement will not arise if only key words or abstracts are drawn from the previous work since this merely amounts to the taking of ideas that are not protected.

Reproduction in material form. The phrase 'reproduction in material form' is particularly difficult to apply to electronic databases. This is because the work may be stored 'internally' in the computer as well as stored 'externally' on some storage media or device. The problem arises where a work is stored on a database without having been previously embodied in some other material form. The issue is whether a hardcopy of this database is a 'reproduction of the work' and is in a similar form prior to its storage on the database. Alternatively, there is considerable doubt at the time the work is keyed directly into the computer database whether the printed copy or any other form of electronic storage or display device

---

[154] (1928–35) MacG. Cop. Cases 362.
[155] See Gaze, B 1989 *Copyright Protection of Computer Programs*, Sydney: Federation Press, pp. 142–165 and 181–194.

is an objectively similar 'reproduction' of the original work (see Ricketson 1989: 480 *op. cit.* 1990).[156] Thus, documents originating (or created) in digital form present considerable difficulties because it is more problematic in determining what constitutes a 'work', and indeed a 'page' of that work.

Defences to Infringement

There may be three types of defences to copyright infringement. In the first place, public interest may allow certain kinds of infringement of copyright. This view holds that the public interest may outweigh private economic gains.

In the second instance, there may be a defence if a creator can demonstrate independent creation. Copyright protection, unlike other IPR, is not absolute and is concerned with preventing someone taking improper advantage of someone else's work. An improper advantage is the copying of someone else's work. Unauthorised copying is one important component of 'improper conduct'. In order to evidence copying the question of access to pre-existing works may be determinative of the issue. Where there has been access to pre-existing works a case of substantial reproduction may be made out in contrast to a claim of no access, in which case a court could infer the existence of a defence of independent creation.[157]

A final defence to infringement is where copyright laws allow copyright to be overridden in particular cases of fair dealing/fair use. Here the balance between monopoly rights and a public's unfettered access is tipped in favour of the latter to allow for scholarly research, study, critique and news reporting.[158] The concept of 'fair dealing' as such is not defined, and the difficulty for doing so has been recognised. It is a matter of degree so that the user of the work does not derive an unfair competitive advantage.[159] One qualification to this defence is the requirement of due acknowledgment of the work that has been extracted. Giving the name of the author and title of the work may be deemed to be sufficient acknowledgment.

In the U.K., in terms of news reporting *Beloff v Pressdram Ltd*[160] established a limitation on the use of this defence on public interest

---

[156] Ricketson, S 1990 'Copyrights and databases' in Hughes, G (ed.) *Essays on Computer Law*, Melbourne: Longman Cheshire, pp. 67–98.

[157] Compare *Bradbury v CBS* 287 F.2d. 478 (1961) with *Sutton Vane v Famous Players Film Ltd* (1923–28) MacG. Cop. Cases 6.

[158] See sections 40, 41, 42 *Copyright Act* 1968 (Cwlth).

[159] LORD DENNING M.R. at 1027 in *Hubbard v Vosper* [1972] 1 All ER 1023.

[160] [1973] 1 All ER 241.

grounds in so far as the publishing of a leaked private confidential memorandum was concerned. However, in the case of *Commonwealth of Australia v John Fairfax & Sons Ltd*[161] the Australian High Court was more lenient in accepting the availability of the defence to copyright infringement where some public interest in the publication of the confidential Department of Defence document could be established. MASON J at 495 distinguished between private and government documents.

Remedies

In Australia and elsewhere, the remedies available to an infringement of copyright are formidable including restraints by injunction, the award of damages, and an account of profits. Criminal offences attach to illegal importation of copyright works and there are seizure and forfeiture provisions as well. In some countries including Australia, the 'infringing copies' are deemed to belong to the copyright owner. Interlocutory injunctions forbid the continuance of an allegedly infringing activity pending a full hearing of the defence. Such injunctions may be obtained on short notice so long as the plaintiff acts quickly, that a court is satisfied that a *prima facie* case exists and that on the balance of convenience, a defendant is not seriously and disproportionately prejudiced by the injunction. The plaintiff also undertakes to provide compensation if the defence ultimately succeeds. All injunctions are discretionary and may not be awarded if damages will adequately compensate the plaintiff.[162] A plaintiff is entitled to damages sufficient to compensate for the loss suffered. Exemplary damages are available for flagrant infringements of copyright, as are conversion damages where the infringing copies are deemed to be the property of the copyright owner.

Alternatively, a plaintiff may elect to claim an order equivalent to profits made through infringement. This is a discretionary equitable remedy and will not be awarded unless damages are inadequate compensation. Under the *Copyright Act* 1968 (Cwlth) criminal offences attach to the importation of infringing copies of works for commercial purposes. The penalties are severe and can be very effective since a person with a criminal record would have difficulty in successfully tendering for government contracts or obtaining entry visas to other countries. The *Customs and Excise Legislation Amendment Act* 1988 (Cwlth) has also amended seizure and forfeiture provisions. The powers of the Australian Customs Service (ACS) have been increased in relation to investigation of the importation

---

[161] [1981] 32 ALR 485.

[162] *Copyright Amendment Act 1992* (Cwlth) ss 115 and 116.

of goods. Infringing items may be seized and forfeited to the Commonwealth. The remedies awarded are always at the court's discretion.

3.9 Intellectual Property Rights: Employees and International Research

3.9.1 Intellectual Property Rights in the Work of Employees

The general rule under copyright law in most jurisdictions is that the person who creates a work is the author of that work, with the exception to this principle that for 'works made for hire', an employer not the employee, is considered the author.[163] However, in the U.S. there is also the 'academic' or 'teacher' exception, where such people are considered special kinds of employees and in order to preserve academic freedom the law cannot allow universities to control the circulation of their work. This means that academic teachers can retain copyright of their work.[164]

Thus, where an independent contractor who is not an employee creates the work—a commissioned work—the contractor is the owner of the copyright. However, this must be expressly agreed in writing signed by the parties that the work shall be considered a work made for hire. Examples of works created in employment relationships include software programs or work within the scope of duties of a staff programmer; newspaper articles written by staff journalists; and musical arrangements and sound recordings written and recorded for a company by a salaried arranger. The rule therefore is that the closer an employment relationship is to regular, salaried employment, the more likely that the work is created within the scope of employment.

The discussion here is particularly pertinent in ascertaining the IPR of academic researchers, students and teachers, and independent contractors and researchers throughout the common law world. In the case of an academic researcher and teacher who is, say, offered $500000 by a publisher in exchange for assigning or transferring 'the exclusive right of producing, publishing, and selling the work in any language throughout the world', will have to pass on this 'windfall' profit to his/her university. Should the employer be entitled to the whole or part of this sum of

---

[163] *Copyright Act* 1968 (Cwlth) s 35; and the statutory definition for 'work made for hire' in 17 U.S.C. § 101 (*Copyright Act*).

[164] McSherry, C 2001 *Who Owns Academic Work: Battling for Control of Intellectual Property*, Boston, MA: Harvard University Press.

money? Invariably the employer can rightfully claim the entire windfall because of the basic rule. The employer owns creations by employees because the works are made during the course of employment. The rationale being that employers should get a reasonable return for their risks in monetary investments in their employees creations, and that the employers provide opportunities for creative works to be made that might otherwise not exist. However, this rule is subject to any agreement to the contrary between an employer and employee. A further decisive factor is whether the work is made pursuant to terms of employment. This is because an academic might be permitted time to undertake consultancies over and above teaching.

In the U.K. judicial authority on the issue of ownership of academic creations, the work of employees and contracts is found in the salutary case of *Cyprotex Discovery Ltd v The University of Sheffield*.[165] This case was an appeal by Cyprotex Discovery from a decision concerning the ownership of copyright in a set of computer programs arising out of research carried out at the University of Sheffield and subsequently developed into a potentially commercially exploitable form by an employee of Cyprotex. While the background is complicated the appeal turns upon a question of contractual construction. The three appeal justices upheld the decision at first instance.

It is to be observed that the terms and conditions of service of academics do not deal with copyright. Academics are expected to teach and carry out research work as part of their employment. However more recently, universities in Australia and the U.K. have begun to discuss the issue of ownership of IP. For example, the Australian Vice Chancellors Committee (AVCC) have broached this topic in *Ownership of Intellectual Property in Universities: Policy and Good Practice Guide* (2002).[166] The advice on this matter is that for non-patentable IP such as books, CD ROMs and the like, universities and staff should strike an agreement to vest copyright in the authors.[167] Where third party interests are involved the terms of the

---

[165] [2004] EWCA Civ. 380. Also at http://www.courtservice.gov.uk/judgementsfiles/j2444/cyprotex-v-sheffield.htm.

[166] In the U.K. the Association of University Teachers (AUT) has prepared its own code of conduct in regard to IPR and authorship. See http://www.aut.org.uk.

[167] Melbourne University's Council agreed to change its IP statute to vest IPR in academic staff members 'provided the university has access, for purposes of education, to works produced by its own academic staff without charge, and provided that the university can recover any substantial direct costs contributed to the production of the publication, it will approve assignment of licensing of copyright by members of academic staff without reservation'. See Powell, S 1995 'Melbourne changes tack on copyright', *The Australian*, 18 July.

agreement for all parties should be spelt out as clearly as possible and one that also includes works undertaken previously, and works being undertaken and to be accomplished during the life of the contract.

The problems are especially more difficult where an academic has developed special ideas and programs prior to the current employment with the university, but which might be put to use in the new university employment. The copyright may vest in the academic, but not in a former employer. Under this circumstance, on equity grounds, an academic should not be made to enter contracts of employment, which would severely limit the ability to profit from their expertise in future. That academic is said to have brought a 'stock in trade' to enrich the new position when taking up employment.

For those who are likely to make inventions in the course of their duties, the staff member and university may acquire IPR in those inventions in accordance with patent law. While the university may not apply for a patent in its own right, it is suggested that a flexible approach be adopted so that, if an invention is successful, the university receives an appropriate share of the royalty income. As an employer, the university is expected to underwrite the costs of patenting and other expenses. In summary, in the absence of specific contractual specifications of duties, courts would likely engage in a balancing act to work out whether the particular task was within the terms of employment. Even with an express stipulation, courts may be willing to imply a term vesting copyright in academics in given circumstances.

In the case of postgraduate students, the ownership of IPR is unequivocal and resides with the student who created and authored the work. Research students are also considered independent workers in a patent application by a university or staff supervisor/researcher and hence can claim authorship as inventor in their own right or as a member of a team.

Practice Notes: Checklist for Employer–Employee IPR Agreements

- State clearly in a contract of employment who will own the work created by an employee during the course of employment whether created during working hours or not.
- Specify what work created by the employee prior to employment the employee wishes to retain and what is to be licensed.

Continued on page 196

Continued from page 195

- If the employee will be using work he or she has previously developed (and therefore owns) to licence to the employer that work to be used. If the licence includes a right to further develop that work, then specify who will own those modifications to that work.
- If the employee wishes to work on a separate project during the course of employment there must be a documentation of who will own the IPR at the outset.
- Ensure that there are security measures and policies in place, which are known and understood by employees and to separately protect the work created in the work place.

3.9.2 Intellectual Property Rights and International Research

It is widely accepted that in the era of globalisation and the new information economy, developing countries would require greater assistance in undertaking research targeted at economic growth and development. To achieve the UN's Millennium Development Goals, particularly in the area of poverty alleviation, there is general agreement that internationally funded, non-profit, scientific research is required. Such research could be aimed at improving the productivity, profitability, and sustainability of agriculture system of farmers in developing countries. One example of an international organisation that espouses such objectives is the Consultative Group on International Agricultural Research (CGIAR). Members of the CGIAR include nearly 60 countries, private foundations, and regional and international organisations. Research partnerships with farmers, scientists, consultants, and policy makers in poor countries help to alleviate poverty and increase food security while protecting the environment.

It is highly likely that in a multi-national multi-participant research project for a developing country, a research team could be made up of scientists from various countries with different skills and traditions from the hi-tech remote sensing expert to the village-level social anthropologist expert on extended families. There may also be geographers and geologists who have expertise in GIS, cartography as well as soil science, land cover and vegetation, catchment management, and information systems. The local government may be involved in providing the local expertise and infrastructure as well as whatever base data it can contribute to the project.

Invariably some of the research scientists from the local university as well as from overseas would be involved including private consultants.

In a research project in developing countries such as this a GIS conceptual framework could be adopted to provide both the database management system (DBMS) and the information system to manage the attribute data, and analyse the data for the report writing and implementation stages. Some of the key IPR issues that need to be considered and questions asked include the IPR culture among the various participants, the ownership of data and information, its protection from an economic and security point of view, resolving disputes, jurisdictional questions, and access. It behoves Chief Investigators of research projects to ask questions about each of these IPR issues, taking the necessary steps to minimise potential exposure to legal liability and IPR infringements. It is important to establish policy guidelines at the inception of a research project and to implement these as quality control procedures during and after the study. Especially with multi-participant GIS projects it is equally important to define the respective rights and responsibilities of each member in the consortia, however difficult it may prove to be.

Intellectual Property Rights Tradition

At the outset, in recognising that various participants contributing to the consortia come from different backgrounds and traditions there must be an agreement as to whether there is a 'public good' involved in this research initiative or whether it should be driven by a strong commercially oriented objective. It seems that where the motivation is for the public good, then an open access, cost of reproduction policy could be adopted to ensure that the research project is undertaken as smoothly as possible at a low cost. This agreement among the participants would set the parameters of the development of the IPR policy that will also resolve other questions that arise.

Ownership of Data

Issues of ownership of data must be addressed, before, during and after the project. The data audit at the *before* stage would need to set out where the data are coming from and who owns the IPR in the data and what use rights might be available either as public domain data or as licensed data. One example is the use of remotely sensed images from SPOT (a commercial organisation) to be integrated with U.S. public domain Landsat 7 (ENVISAT in future) imagery is illustrative of emerging complexities. At the *during* stage of using and amalgamating the data, invariably 'new' derivative data sets emerge and the management of the IPR for these have to be considered so that consortia members are clear as to ownership, use and later dissemination of the data. At the final *after* stage, when the project is completed, the question of who owns the information resulting from the investigation will arise. The data so developed can prove economically valuable to farmers, the government as well as to the researchers and consultants. Having a clear policy as to who has ownership of the data is thus very important. The policy would also spell out who has rights of release, on what conditions, and how costs are either to be recovered or shared.

Economic Protection of Intellectual Property Rights

Given that the project could conceivably be a foreign aid project it might well be that the information is to be shared as liberally as possible among all other developing countries with similar needs. The methodology, the frameworks and the working models could make a valuable template for implementation in a different developing country context. To facilitate widespread dissemination, it could be decided that the data be shared with all comers on the basis of disclaimers. If this decision were to be taken then there must be mechanisms put in place both in the data itself, such as in a metadata layer, as well as in the 'release of data' agreements. These disclaimers and release agreements could be designed to protect the use of the data and also to retain copyright credit for its creators. Such a release of the data would need the consent of the members of the project team as well as the governments involved and aid providers.

Security of Intellectual Property Rights

Some management structure for conserving the data and providing security of IPR will need to be established. This is because, once the research team has completed its tasks, it will be disbanded and members will disperse. The data could be kept in escrow—a trusted third party who would release the data to future would-be users who have fulfilled certain conditions of use. Alternatively the developing country itself could be the custodian of the data. Some aid agencies and research grant organisations prescribe the lodgement of research data with a clearinghouse so that later researchers may gain access to the data.[168]

---

[168] For example, the Australian Research Council (ARC) encourages social science research data to be lodged with a 'clearing house' such as ACSPRI, the Australian Consortium for Social and Political

Resolving Disputes

IPR arises at different stages of a research project and unless carefully defined for each participant or group of participants there may be disputes, some minor, but always contentious and at times intractable. To avoid these, dispute resolution mechanisms will have to be put in place in the research plan. Disputes may arise concerning ownership, allowed use and dissemination and rights and responsibilities of members. It is recommended that alternative dispute resolution mechanisms be set up such as mediation, arbitration and conciliation before seeking legal remedies via litigation in the courts. In the event of litigation 'where' the case was lodged would decide the jurisdiction and the law to be used. A master agreement for this research project must state the law and the jurisdiction to be used for legal purposes.

Jurisdiction

The question of legal jurisdiction is one to be decided by the research management team. It is normal to use the law of the land, that is, the law of the developing country in which the research is being undertaken. Where there are no laws regarding IPR within the jurisdiction, it is recommended that international IPR as contained in TRIPs be used because these rules are readily recognisable and reflect best practice. This may be despite the fact that the developing country may not be a member of the WTO or may have different cultural views regarding IPR. Using TRIPs as a benchmark would be a useful starting point.

Access to the Information

IPR have an intrinsically economic component and such rights can either be sold or given away. Some projects are based on the fact that in order to pay for the research some cost-recovery mechanism is put in place so that the revenue stream will offset some of the costs of the research. The research team may establish a cost-recovery policy if it is decided that the results were sufficiently valuable that there was a market for such data. It is more probable that the scenario would be one where there is a user-pay

---

Research Inc. to facilitate access to Australian social science data. See http://acspri.anu.edu.au/. ACSPRI is also a member of the Inter-university Consortium for Political and Social Research (ICPSR) that maintains a worldwide archive of social science data (see http://www.icpsr.umich.edu/org/index.html). Similarly the European Science Foundation (ESF) has encouraged assigning IPR for the public good. See ESF 2000 *European Science Foundation Policy Briefing: Good Scientific Practice in Research and Scholarship*, December at http://www.esf.org/publications/93/ESPB10.pdf.

regime of minimal costs for assembly and delivery of the data. A more difficult problem would be that of trying to locate data and negotiations thereafter to acquire the data on a *gratis* or licensed basis. Here the IPR of the data owner and provider would need to be respected and processes put in place to ensure that these are not infringed by the research team. Some countries have very strong database rights (for example, the E.U.) and the acquisition of data, its use and transmission from countries within the Community can be very difficult and would have to be negotiated carefully. On the other hand, where public domain data exist and are freely available, such as the use of the Digital Chart of the World (DCW) data managed previously by the U.S. Defense Mapping Agency (DMA) now renamed National Geospatial-Intelligence Agency (NGA), use must still be carefully acknowledged.[169] Even if no IPR may be involved there might be moral and ethical reasons to so acknowledge.

Practice Notes: Due Diligence Checklist for Intellectual Property

- What is the IP?
- When was the IP created, designed, invented?
- How did the creation of the IP come about?
- Who is the inventor/author?
- Was the inventor/author an employee of the organisation?
- Was the IP purchased?
- Do records exist which establish unambiguously that the organisation is the proprietor of the IP solely?
- Was the creation, design or invention subject to a licence?
- What are the terms and conditions of the licence?
- Are current uses of the IP within the terms and conditions of the licence?
- Has IP been revealed to 'others'?

Continued on page 201

---

[169] In terms of IPR the DCW is a significant exception since the project is a cooperative effort of Australia, Canada, the U.K. and the U.S. governments. Participating countries have agreed to certain waivers of enforcement of statutory copyright to promote the public distribution and use of DCW products.

Continued from page 200

- Are those 'others' subject to a licensing/confidentiality agreement with your organisation?
- Is the IP a modification of previous IP?
- Does the IP need to be registered?
- What is the status of IP registration? Is there a register of IP?

3.10 Lessons, Litigation and the Fate of Gigo's Code

The storyline given in the introductory paragraphs brings up many issues with dire legal consequences. The seemingly mundane and run-of-the-mill customary practices become very significant in the face of litigation and need careful analysis.

- Gigo apparently owns the class of computer code called Gigo's App Builder as author and creator.
- The work with RESI Pty Ltd, where proprietary software was used and developed using SML and GML, evidently belongs to RESI and the IPR retained by that firm as Gigo was an employee.
- On joining MAS and Gigo's use of the library code in App Builder as well as that taken from RESI would need a written agreement regarding the use of the codes. Apparently Gigo has rights to the previously developed codes, but not that of RESI. Depending of the substantiality and extent of the unauthorised 'borrowing', there may be an issue of conversion or one where RESI's authorisation may be required. The additional extra curricula work at home and off-work hours may need to be negotiated between Gigo and MAS as to the extent of ownership of IPR, rewards for the work and what else may be done to the subsequently developed work. There may be some ambiguity in all these and will need resolution immediately via a conference between the parties and an agreement entered into.
- The overseas project p-ArcMap for the Jhai PC and the Simputer raises two major issues. First, MAS and the NGO partners must enter into an agreement and agree on several points of knowledge management. This would include the IPR brought into the project by MAS and the other partners, the IPR arising from the cooperative project and the basis of sharing the previous as well

as new knowledge. The second major issue is the use by Gigo of the p-MapArc program code in the hobby butterfly database for the group Ornithoptera. The meaning of 'open' source needs to be clarified as it may mean 'shareware', 'freeware' or simply 'open' source such as GPL.[170] Where the software is used on a shared basis there may be a breach of the copyright of MAS as well as all NGO project members. On the other hand, if the software is in the public domain this fact must be equivocal and its assignment made explicit through statements in the condition of use or a declaration on the face of the program itself.

- The work of Gigo as a teacher of SIS at the local TAFE Institute includes material previously developed as well as new material developed since joining the Institute. The ownership of IP is a matter between Gigo and the Institute since a 'teacher's exception' might be claimed so that the IPR is retained by Gigo. On the other hand, the Institute's entry into flexible delivery modes of teaching now means that Gigo might have to enter into a separate agreement to divest IPR to the Institute for the development of such materials.

- On Gigo's resignation, MAS may either request the code to be delivered or, failing that, to serve an injunction to recover such code pending litigation to sort out issues of ownership and rights. Prudent employers would require their programming staff to make and keep copies and require program code to be kept in 'escrow' precisely for situations that have now arisen. Escrow is a legal device where a third party holds property or other documents and returned (or revealed) to the parties when certain conditions are performed or fulfilled, such as in this case, a resignation.

- The issue of whether Gigo may assign interests to Map LeSerup hinges on ownership rights to App Builder, and software developed from p-ArcMap. Assignment is always for valuable consideration as in a contract, hence the token dollar.

- Gigo's further class of code in Paradime will suffer the same ambiguities of ownership and rights. Failure to assert these will leave further trails of disaster, a legal quagmire and no solutions to enduring problems.

---

[170] 'Shareware' is software made available with permission for people to redistribute copies, but under terms that require anyone who continues to use a copy to pay a licence fee. 'Freeware' is used for software that permit use, copying and redistribution but not modification either gratis or for a fee. 'Open' source refers to the source code of computer software and the code can be accessed by anyone. The material is available at no, or very low, cost.

This case is a reminder to GI developers and proprietors that IPR is a serious matter. It also demonstrates how complicated the application of a general rule in a creative field of GI can get.[171] If an employer is to benefit from an employee's inventions then this fact must be stated in the relevant contract of employment. If there are no express terms in the contract of employment it is unlikely that the term would be implied, especially in regard to inventions and other creative works.[172] The courts are ready to imply an obligation of good faith and fair dealing in the recognition of an employee's IPR so long as the latter's conduct is equivocal with regard to the other rights. IPR need to be considered separately from a contract of employment.[173] The following practice notes give a suggested IP protection plan for consideration.

Practice Notes: Intellectual Property Protection Plan

- Identify all IP associated with GI projects in your business.
- Check whether you own the IP or that you have a right to use it.
- List registered IP and put an estimated monetary value on identified assets.
- List unregistered IP and give it an estimated monetary value.
- List other valuable assets such as software code developed in-house, client lists and corporate knowledge.
- Identify key staff involved in developing, maintaining and protecting your IP and get them to sign agreements relating to confidentiality and competition.
- Educate staff on the nature of IP, how to protect it and their responsibilities.
- Consider insuring your IP against infringement and against you infringing the IP belonging to others.

Continued on page 204

---

[171] *Redrock Holdings Pty Ltd & Ors v Adam Hinkley* [2000] VSC 91 (4th April); [2001] VSC 277 (2nd August).
[172] See *Spencer Industries Pty Ltd v Anthony Collins & Anor* [2002] APO 4 (18th January 2002); [2003] FCA 542 (4th June 2003).
[173] Barton, C and Liberman, A 2004 'Who owns employee produced inventions?' *FindLaw Australia* February at http://www.findlaw.com.au/articles/.

Continued from page 203

- Be pre-emptive rather than reactive.
- Use professional help to get it right and ensure the best possible outcomes from your IP.
- Do your homework with respect to IP.
- Develop an IP strategy to accommodate past, new and future projects.

Source: adapted from IP Australia (2004).

Summary

Arguably, in the electronic age, it appears that copyright offers the widest and most appropriate means of protecting the expression of ideas contained in maps, compilations, electronic databases and GIS in general. It is very much a live debate in legal and academic circles as to what and how much protection copyright can offer GI and all its accoutrements in a globalised, electronic world where infringement is not only difficult to detect and protection may be nearly impossible. What laws and therefore what jurisdiction become important questions.

There is a view that if we are to have laws to govern IPR we should strive to articulate a generic version of rights, obligations, and prohibitions. This is opposed to passing laws confined to particular fact situations, that is, to pass laws to cure a specific fact situation that is the cause of a problem.[174] However, in some instances this ideal might not be possible. GI, for instance, does not fit neatly into any of the specific categories of 'works' whether literary, dramatic, musical or artistic, but yet is the product of human creativity. Unlike other protected works GI products and the databases accompanying these are open to copying and unauthorised use the loss of authorship in the data in perhaps a single iteration.

The law is still not settled in protecting computer programs much less GI systems of any kind. As observed previously it is impossible to use an information system without making a copy of it first. This is radically different from traditional protected works where such copying is unnecessary. The display of data on a computer screen may constitute a reproduction in

---

[174] Broekhuyse, P 2000 'Copyrights and wrongs: Lawyer's view of the net', *The Australian IT, The Australian,* 25 July, p. 55.

material form, as well as a 'public performance', albeit ephemeral since the display may be switched off. A print of the data stored in the computer may infringe copyright and if this were part of an electronic network, the network manager may be in breach for authorising an 'infringement'. The dynamic nature of a database with frequent updates may make it impossible to determine when protection for the data begins.

There is a clever and succinct adage about copyright that says 'what is worth copying is *prima facie* worth protecting'.[175] The information age will make this rule of thumb and copyright an anachronism. Computer-based information storage and retrieval systems now challenge copyright law by offering new and creative ways of disseminating ideas. Whereas in a previous age copyright was concerned with copying—books, music on audiotapes, paper products by photocopiers, film by videotape—IT is threatening to make copyright law obsolete. The new technology is part of a long line of copying methods and there is a fundamental change in that what is being copied—information in its purest form. Information has now been liberated from a physical object, and floats, as it were, in cyberspace. IT is now posing a greater challenge to a legal system that is expected to protect it. Laws take time to change since rapid changes *per se* may be counter productive. Here practitioners and users of GI are expected to contribute to changes in the law rather than to leave it in the hands of manufacturers and software developers who may look to protect their own self-interest.

At present, IP in general, and copyright in particular, provides flimsy protection to information systems, computer programs and databases. It may take some time for the law to adapt to the rapid changes brought about by IT. How quickly these changes will come about will depend on how much impact it has on society at large and the seriousness of such impacts on the socio-political economy. Previously there has been a balancing of interests between an author's rights and the dissemination of knowledge. Today, it seems that the balance is in restricting IT and the dissemination of knowledge for society as a whole. But in striving to achieve such a balance 'care must be taken not to impede the development and use of data processing tools by unduly restrictive or complex legislation' (see Risset 1979).[176]

---

[175] per PETERSEN, J *University of London Press Ltd v University Tutorial Press Ltd* [1916] 2 Ch. 601.
[176] Risset, JC 1979 *Problems Arising from the use of Electronic Computers for the Creation of Works*, UNESCO/WIPO doc. GTO/3 1979, Paris and Geneva: UNESCO, WIPO.

It may be that licences and agreements may become the definer of boundaries of permitted uses instead of copyright law or an industry market standard. The importance of contract law cannot be underestimated since under licensing conditions both parties may benefit overall. On the one hand, the copyright holder may be able to achieve revenue maximisation through the use of a variety of licences for time limitations on use, limited distributions to licences at lower prices for particular groups of users (say non-commercial) and unlimited uses for a higher price. The enforcement thus focuses on pursuing infringements of the licence terms, a breach of contract. Users of the technology could benefit because of lower prices, easier access to the data, improved variety of content, and more certainty with the law. The issues that remains then are whether users will be willing to adopt such new distribution methods and how law makers can ensure that user rights are preserved and kept in balance with those of the vendors of GI products. '... [T]he role of copyright law may in time decline in both the protection of IP at hand and in providing fair access to that IP' (Segkar 2003: 121).[177]

In the information age Walter Wriston has observed that the world economy operates on a 24×7 'information standard time'.[178] Commercial, financial and other data are disseminated continuously to millions of computer terminals worldwide. Such data are the grist of commercial, political and social decision-making and such decisions in turn become part of the data stream that informs the world. For the world to keep up with this information standard time there has to be a degree of certainty of protection of IP of data and information in terms of what is being protected, how it is being protected and for how long protection subsists. In addition, such protection must exist within an international legal system that is not only understood by all, but is also one that has international recognition and therefore mutually enforceable across jurisdictions by way of national treatment. To achieve this ideal would mean that all the 'leakages' to copyright have been staunched and civil disobedience by way of pirating, copying and other infringement would have been eradicated and the creators of original products justly rewarded for their efforts.

---

[177] Segkar, A 2003 'The declining relevance of copyright law—how contract may become the key', *Internet Law Bulletin*, v. 6(10), pp. 121–125.
[178] See Wriston, WB 1992 *Twilight of Sovereignty: How the Information Revolution is Transforming our World*, New York: Scribner; and also Branscomb *op. cit.* 1994: 151.

Chapter 4
Geographic Information and Privacy

Learning Objectives

After reading this chapter you will:

- Be able to analyse the economic and social impact of geographic information technology on informational and personal privacy.

- Recognise the impact of privacy on geographic information science.

- Know and appreciate the difference between informational privacy and personal privacy.

- Master the issues as to why geographic information systems can be an inherent threat to personal privacy.

- Understand the legal regulatory and policy framework governing a right to privacy within an Australian common law and regulatory regime, including that of industry self-regulation practices.

- Be able to compare and highlight similarities and differences between Australian and U.S. privacy regimes.

- Appreciate the nature of the evolving fair information privacy principles and practices and its influence on geographic information systems.

Geographic Information Science: Mastering the Legal Issues   George Cho
© 2005 John Wiley & Sons, Ltd   ISBNs: 0-470-85009-4 (HB); 0-470-85010-8 (PB)

- Be adept at developing techniques for identifying geospatial technologies that promote intrusiveness, enhances personal privacy protection or are sympathetic to privacy protection.

- Be able to compare the development of policy and practice in privacy protection and suggest methods for coping with the E.U. Data Protection Directive and the use of the U.S. Safe Harbour framework.

4.1 Introduction

The theme of this chapter is the relationship between privacy and geographic information (GI) science in terms of the role of regulation, self-regulation, and best practice. This discussion may yield policy guidelines for the protection of information privacy as well as the privacy of individuals.[1] One widely accepted definition of privacy is 'the claim of individuals, groups, or institutions to determine for themselves when, how, and to what extent information about them is communicated to others'.[2] This claim to privacy is a relatively recent development in the common law world. A salient characteristic is that its protection in the past has been *ad hoc*. The Australian Constitution has not been vested with powers over privacy protection and the common law protects privacy rights indirectly. The law of defamation, negligence and passing off give a semblance of an overarching protection, as does the shield provided by contract including the duty of confidence. Likewise, in the U.S. the origins of the privacy right may be traced to a law review article published in 1890 by Justices WARREN and BRANDEIS.[3] In his famous dissenting opinion, Judge Louis Brandeis reiterated the right to be let alone as 'the most comprehensive of rights and the right most cherished by civilised men'.[4] Based on the principle of the right to be left alone, U.S. law has developed along the lines of a common law right and those rights found under amendments to the U.S. Constitution. Today statutory protection

---

[1] There has been considerable writing on the topic of privacy in recent years. A good example is the comparative analysis found in Bennett, CJ and Raab, C 2003 *The Governance of Privacy: Policy Instruments in Global Perspective*, London: Ashgate Press.

[2] Westin, AF 1967 *Privacy and Freedom*, New York: Atheneum, p. 7.

[3] Warren, S and Brandeis, L 1890 'The right to privacy', 4 *Harvard L. Rev.* 193.

[4] *Olmstead v U.S.*, 277 U.S. 438 (1928).

for a diverse set of interests from interference by governmental entities has been developed.

The tools of GI science, in the main, deal with geographic data in any information system in which location or spatial relationships are the fundamental data in the collection, organisation, identification and processing activities. But, GI science may also handle a diverse range of personal information from the truly 'personal' ones through to those of a more general nature—age, gender, height, home address, social security number, tax file number, marital status, religion, reading habits, musical tastes, membership of sporting clubs, etc. By its very nature the tools of GI science allow these kinds of information to be collected, manipulated, displayed, and transmitted cheaply, easily and speedily. Some types of information transfers are heavily regulated while others are seemingly left unregulated and consigned to the marketplace to resolve. Predictably, the capabilities of geospatial tools in information analysis and transfers have raised a multitude of novel and interesting information and personal privacy issues. As well, since GI science is fundamentally one that deals with geography, 'special' issues are raised—hence, the 'spatial is special' theme permeates the treatment of privacy protection in this chapter.

While the economic and social impacts of advances in geoinformation technology have been overwhelmingly positive, concerns have been raised by many about what information is being collected, how it is being used, and who has access to it. These concerns, in turn, have led to calls for policy and regulation. Some key privacy issues are addressed in the four sections that follow. Philosophical and doctrinal issues provide the basis of discussion for grappling with the nature and structure of the problem of privacy in the first section. Since privacy is a 'right' in many circumstances, the legal, regulatory and policy framework that underlies the source of this right is analysed in the second section. As GI science is heavily steeped in technology, apart from spatial considerations, an evaluation is undertaken of the different geospatial technologies that promote intrusiveness, enhances privacy protection or are sympathetic to privacy protection. An evaluation of these technologies is found in the third section. The emergent policy and practice in privacy protection is the topic of discussion in the final section. The European Union Data Protection Directive is analysed here, given its wide-ranging impact on many jurisdictions in terms of data protection and data transfer principles. Here the 'safe harbour' framework is discussed, together with the spectrum of alternatives used by Australia, Canada, and U.K. to respond to the Data Protection Directive.

4.2 Philosophical Issues: Nature and Structure of the Problem

There are three interrelated matters that, in combination, promote the 'spatial is special' theme of this chapter. The first is whether geographic information systems (GIS) are a threat to our personal privacy. One view is that it is not, and that most GIS are not personal data intensive, even though such systems use a lot of data from diverse databases, some of which include personal information. Whether this view is defensible or not is debatable, but it does not detract from the issue of privacy. The second matter is that there seems to be a lack of understanding of privacy issues, just as there is, for instance, some fuzzy thinking about whether we are attempting data protection on the one hand or protecting the privacy of that information. Finally, there are inevitably, some ethical questions in the use of geospatial technologies especially where there are privacy issues involved and what the 'right' thing to do might be.

Each of these matters canvassed here underpin both philosophical and doctrinal imperatives because these may not only inform the legal and regulatory framework, but the discussions may produce emergent, and possibly harmonised, international policy and practical guidelines in the light of technology itself, bringing about an expansion of privacy rights. It is therefore important that these matters be raised here to provide the backcloth for latter discussions since there may also be the need to describe the nature and structure of the problem associated with privacy.

4.2.1 Geographic Information Systems are not Personal Data Intensive

Flaherty (1994) in relating his experiences with privacy protection in GIS in a national and provincial context has said that the technology is 'not personal data intensive' and that information privacy issues may be resolved by applying fair information practices.[5] The view may suggest that GIS is not a privacy threat in the least. Interpreted differently, the opinion is one that weighs up how each side to the debate values both the social benefits to be gained from GIS and the monetary costs and other sacrifices of giving up some privacy.

---

[5] Flaherty, DH 1994 'Privacy protection in Geographic Information Systems: Alternative protection scenarios' at http://www.spatial.maine.edu/tempe/flaherty.html.

GIS has the power to integrate diverse information from multiple sources. Some of the data are of a personal nature where individuals may be identified or identifiable, while others are of a spatial nature that may be used to locate individuals through geo-coded data. Contrary to conventional wisdom, the privacy threat is not from the collection of new data items obtained by a GIS, but from the new inferences that may be obtained by correlating geographic information with personal information. A credit card transaction associated with public record information from the phone book, title deeds, court documents and aggregate census data profiles of an area, other geographic and personal information, conspire to paint a 'informational picture' of a person.

Hence the conclusion by Dobson (1998), that geo-information, in combination with personal information, clearly poses a privacy threat.[6] Moreover, whilst there are trade-offs, and a tension between the two fundamental values of privacy on the one hand and the public's right to know on the other, it is felt that there may be greater threats that abound. The worry about satellites monitoring a farmer's use of water contrasts sharply with a public's acceptance of video cameras in shopping malls, buses and taxis, city streets and in dorms and apartments. There is no debate about the social implications of these surveillance cameras and technologies and seemingly the public have accepted these for the social good. 'Social surveillance isn't particular to GIS. In fact, GIS pales by comparison to many direct surveillance forms that are embraced by society'.[7]

Equally, it may be argued that privacy is diminished as the use of geoinformation technology expands and personal data are combined, cross-matched and disseminated to a greater degree that hitherto thought possible. Furthermore, it will be difficult for individuals to find out what other people know about them and in what detail. In one sense it may be said that geoinformation technology may have compromised privacy. But in another sense, the fear of a loss of privacy may appear to be based, at least to some extent, on a lack of understanding about the way in which their personal data are collected and used for whatever purposes.

The term 'geo-slavery' describes one such system that allows an individual to 'coercively or surreptitiously monitor and exert control over the physical location of another individual'.[8] While this does not discount

---

[6] Dobson, J 1998 'Is GIS a privacy threat?', *GIS World*, v. 11(7) July, pp. 34–35. Also at http://www.geoplace.com/gw/1998/0798/798onln.asp.

[7] Dobson 1998 *ibid.*, p. 35.

[8] Ball, M 2003 'Concerning ourselves with privacy', *GIS World*, February and at http://www.geoplace.com/gw/2003/0302/0302ed.asp.

the positive use of such technologies for monitoring prison inmates and other persons such as those on parole, for instance, there are other potential abuses of geotechnology. These negative impacts on privacy need discussion, debate and balanced against the many social and personal benefits of geotechnology.

Data on geographical location, when combined with other data, transforms GI science into powerful tools for tracking, storing and analysing personal information. In the business world, detailed local spatial information used in GI applications is the fastest growing and potentially the most lucrative segment of the GIS industry (see section 4.4). The inescapable conclusion is that tracking, data integration, and analysis capabilities give GI technologies the potential to be more invasive of personal privacy than many other technologies. However, the technology must not be rejected on this basis alone because there are benefits to be gained through the responsible use of databases. Such advantages may outweigh the intangible, largely subjective and non-measurable rights to personal privacy. While the abuses of geotechnology should be curbed the collection of the geospatial data itself should not be. It would be far more beneficial for society to deal with privacy abuses on a case-by-case basis than simply to prevent data aggregation and the building of databases and have these dismissed *en masse*.[9]

4.2.2 Lack of Understanding of Privacy Issues

There is a view that many in the GI industry do not really understand and appreciate the privacy problems arising from the use of geoinformation technology. 'The more that the GIS community can gain an understanding of the privacy issues and how they may be resolved, the more likely it is to be able to contribute constructively to their resolution, whether through laws, self-regulation or some combination of strategies' (Raab 1994).[10] But, what are these issues, one might ask. Are data protection, privacy of information, and personal privacy one and the same thing? If not, what are the differences?

Data protection, whether in legislation or in a code of practice, relates to the protection of data rather than the people themselves. The

---

[9] Onsrud, HJ, Johnson, JP and Lopez, X 1994 'Protecting personal privacy in using geographic information systems', *Photogrammetric Engineering and Remote Sensing*, v. 60(9), September, pp. 1083–1095. Also at http://www.spatial.maine.edu/tempe/onsrud.html.
[10] Raab, CD 1994 'European perspectives on the protection of privacy' at http://www.spatial.maine.edu/tempe/raab.html.

protection is as much of concern for both the data subject (a natural person and in certain instances, corporations and legal persons) as well as the organisation or agency that collects and processes the data in order to comply with certain principles. Such principles may include fulfilling preconditions laid out by law or the industry body itself. Better practice mandates that the consent of the data subject is paramount in the collection, use and dissemination of the data.

The privacy of information relates to undertakings to keep the information private and the various interests individuals may have in controlling and significantly influencing the handling of this data. This may seem tautological at first blush, but in reality it is not. The privacy of information is often taken for granted, and we willingly provide personal information to others who request the information because we believe that the receiver has undertaken to do the right thing. Here we are equating private with confidential. We divulge personal details in a number of ways and in different contexts; for example, at the medical clinic, at the bank and even over the phone, providing our credit card numbers, filling in mail-order forms either on paper or on the Internet. It seems that we are constantly revealing 'personal property' of a private nature to other people. The result is that the facts of our most private information, such as personal income, family information, including health and employment details, become captured in some database somewhere in our day-to-day activities. So, the enduring issues raised are: how much privacy is afforded to the individual given that the data are already in someone else's possession? Who draws the line as to what remains private and what may go on the public domain? What is the balance between the right to keep private and the right to know for the public good? When are facts kept confidential and should they be?

The rhetorical questions point towards informational privacy, which relates to knowledge about some person that is kept private and confidential. This is quite different from personal privacy, where an individual may have chosen to be left alone and facts about that person kept from view. Are these one and the same thing? The answer is that these may be two totally different approaches to the issue since the first is about the protection of personal information kept confidential and the latter is about a person acting and taking steps to keep personal facts private. Personal informational privacy becomes a problem where the datum may be traced to the individual. Another is that while one may have divulged personal information for one reason or another, should one's consent be given again? This is especially so when that information is either disguised in aggregate form or re-purposed in some other form so that new, synthesised data are vastly different from what was given by an individual initially? The threat of personal privacy invasion is therefore either in the release of data and information that contains personal information, or the release of disparate pieces of information that contain no personal information, but which may be inferred from locational information (geography), transactions, and electronic footprints and other such trails.[11]

4.2.3 Ethical Use of Geospatial Technologies

A final issue is the ethical use of geospatial technologies in which users and proponents of this potentially invasive technology need to be reminded of their heavy social responsibilities, and to use the technology ethically and in an ethical manner in all ways and at all times.[12] Geospatial technologies include GIS as a mapping tool for decision-making through to those technologies that amass data by spatial attributes including global positioning systems (GPS), and transponders and other intelligent computer chips embedded in some devices that can report location as well as an identity marker. The latter are in a class of intelligent spatial technologies that can declare both personal information as well as locational and device-specific information in response to a poll by another device either in a pre-established relationship or to a new soon-to-be-established relationship.

The word 'ethics' is derived from the Greek word *ethikos* [character] and from the Latin word *mores* [customs]. Together, the words combine to define how individuals choose to interact with one another. In philosophy, ethics defines what is good for the individual and for society and establishes the nature of duties that people owe themselves and one another. While law often embodies ethical principles, law and ethics are far from co-extensive. The law does not prohibit many acts that would be widely condemned as unethical, for example, lying or betraying the confidence of a friend. On the other hand, much of the law is far from simply codifying ethical norms.[13]

Geospatial technologies are in daily use and have heightened locational information privacy concerns of citizens. However, while it has

---

[11] See an interesting discussion in Monmonier, M 2002 *Spying with Maps. Surveillance Technologies and the Future of Privacy*, Chicago and London: The University of Chicago Press, especially Ch. 2 Overhead Assets, pp. 1–37.

[12] Ball 2003 *op. cit.*

[13] Legal Information Institute http://lii.law.cornell.edu. *Ethics*: relating to morals, treating of moral questions; morally correct, honourable, *The Concise Oxford Dictionary*, Oxford: Clarendon Press.

been argued that there have been many beneficent uses of such technology, the case that there is a potential for the misuse of locational information has also been put forward. This has not been helped by other arguments that make the resolution of the issues even more difficult. For instance, in using geospatial technologies, an apparent legal fallacy may have arisen, if only by accident. The idea is that if there is a legal right to do something then it follows that it must be the right thing to do. So, if it is permissible to undertake data aggregation activities using a number of databases, then it is lawful to do so. But really, the issue is that the legal right must only be the starting point rather than the end point for justifying one's actions. The fact that something is legal does not mean it is either right or a wise thing to do. Thus, data taken out of context—acontextual data—and used in that sense may produce results that may be highly unjust and totally incorrect in particular cases. This is where ethical questions are raised and such questions should be foremost in the thinking and practice of GI scientists.[14] Some may wish to think of ethics as a continuum in which there is both a duality of a right and a wrong way of undertaking activities as well as ethics as a way of dealing with a right way and a better way of doing things. In equitable jurisdictions and civil cases there may be a claim to a right, but it is the one who has the better claim the often is awarded the claim.

The right to be left alone is being vigorously defended and there is resistance to increased surveillance in parts of our private lives. This challenge has arisen from the enhanced knowledge, increased public awareness, and sophistication of the general public to both the data collection and the aggregation of private information in databases and information systems. In addition, corporate and governmental behaviour has come under closer scrutiny than before, coupled with new laws and enlightened policy, which encourage greater public challenge to the norms. But it seems that the real threat is the creeping acquiescence to all sorts of intrusions without the accompanying public debate, information and education. It used to be that when a video camera was installed, say in the computer lab, the spectre of George Orwell's *1984* and 'Big Brother' watching was raised.[15]

Today, it appears that we have grown accustomed to all sorts of cameras watching us in all kinds of circumstances. The reality TV genre

---

[14] Fair information practices, in terms of collection, use and dissemination of information and enforceable legal rights are discussed later in this chapter.

[15] Orwell, G 1990 (first published in 1949) *Nineteen-Eighty-Four*, New York: The New American Library Inc.

of the *Big Brother* type, where the public is invited to watch on TV the antics of four or five couples displaying their private sexual habits raises no public outrage. On the other hand, where employers keep watch on the Internet use of their employees and monitoring their e-mail usage, is a subject of much concern and public media comment. This is particularly serious when these monitoring records are used as grounds for dismissal because of abuse, and it has no doubt raised both moral and ethical outrage. The really big question is: which is the greater sin—to invade privacy or to maintain surveillance for security purposes?

The surrender to surveillance is now happening and it is taking place 'one step at a time, and each step is attractive and relatively benign' (Dobson 2000).[16] Clarke (2001) has asked how so advanced, well-educated, well-informed free societies have been so myopic as to permit what he calls 'dataveillance' to have taken place. The answer to him seems to be that these societies have been conditioned by Orwell's *1984*. In the past 50 years or so, technology has developed and delivered far superior surveillance tools than Orwell had imagined and to cap it all 'we didn't even notice'.[17] The metaphor of the lobster has been used to explain this phenomenon. A lobster is placed into tepid water and the temperature progressively warmed to lull the victim into falling asleep and in so doing fails to detect the trap and to escape the cooking pot.[18]

As Justice WILLIAM O. DOUGLAS has argued, the protection of our basic values such as privacy is not self-executing. 'As nightfall does not come at once, neither does oppression. In both instances, there is a twilight when everything remains seemingly unchanged. And it is in such twilight that we all must be most aware of change in the air, however slight lest we become unwittingly victims of the darkness' (Marx 1994).[19] So the message is that the public has to be ever-vigilant, requiring particular sensitivity in both the design and use of inherently privacy invasive technologies such as those embedded in GIS.

---

[16] Dobson, J 2000 'What are the ethical limits of GIS?', *GeoWorld*, May p. 24 and at http://www.geoplace.com/gw/2000/0500/0500g.asp.

[17] Clarke, R 2001 'The end of privacy: While you were sleeping...surveillance technologies arrived', *AQ: Journal of Contemporary Analysis*, v. 73(1), January–February, pp. 9–14.

[18] Clarke, R 2000b 'Person-location and person-tracking: Technologies, risks and policy implications', *Information Technology and People*, v. 14(2), Summer, pp. 206–231 and at http://www.anu.edu.au/people/Roger.Clarke/DV/PLT.html.

[19] Marx, GT 1994 'Some information-age techno-fallacies and some principles for protecting privacy' at http://www.spatial.maine.edu/tempe/marx.html.

4.3 Privacy: The Legal and Regulatory Framework

The legal and regulatory framework concerning privacy is characterised by both its recency and the *ad hoc* manner of its regulation and control. The complexity and multiple dimensions have led to conflicting views as to how to structure such a framework. For example, the different types and consequences of privacy and autonomy have produced varying approaches, from the purely economic efficiency market-based suggestions to self-regulation and legislation as solutions for the protection of personal and informational privacy.

Constitutional Matters

To begin with the Australian Constitution does not empower the Commonwealth Parliament to enact a general law for the protection of privacy throughout Australia.[20] However, the Australian government has been obligated to protect privacy stemming from various international covenants, agreements and treaties to which Australia is a signatory. One example of these international obligations is Article 12 of the Universal Declaration of Human Rights (UDHR) adopted in 1948.[21] The article states that 'no one shall be subjected to arbitrary interference with his [sic] privacy, family, home or correspondence, nor to attacks upon his honour and reputation. Everyone has the right to the protection of the law against such interference or attacks'. Furthermore, Article 17 of the International Covenant on Civil and Political Rights (ICCPR) 1966 is almost identical to Article 12 of the UDHR.[22] Both these covenants impose binding obligations on its parties, to which Australia is a signatory.

Common Law

Apart from international obligations and corollary domestic legislation, in Australia the main body of law controlling the protection and/or disclosure of personal information by both public and private sectors is found in law reports, that is, the common law. The *Victoria Park* case is commonly

---

[20] Australian Law Reform Commission (ALRC) 1983 *Report No. 23 Privacy*, Canberra: AGPS, p. xliv.
[21] UDHR 1948 *Universal Declaration of Human Rights*, 10 December 1948 at http://www3.itu.int/udhr/
[22] ICCPR 1976 *International Covenant on Civil and Political Rights, New York*: United Nations and at http://www.privacy.org/pi/intl_orgs/un/international_covenant_civil_political_rights.txt

cited as authority for the proposition that Australian (and also English) law does not recognise a general right of privacy.[23] The issue of a right to privacy was revisited in Australia recently in a High Court case.[24] However, the judgement in *Grosse v Purvis*[25] recognised a common law right to privacy for the first time in the particular circumstances of stalking. In awarding damages, SKOIEN, J found that the essential elements of the emerging tort of invasion of privacy. These were that the willed act of the defendant intruded upon the plaintiff's privacy in a manner which would be highly offensive to a reasonable person of ordinary sensibilities; and which caused the plaintiff detriment or distress.

In the U.K. the right of privacy of a corporation has been held to exist.[26] Also more recently privacy rights have also been extended to individuals drawn from the fundamental value of personal autonomy.[27] Courts in several other jurisdictions have also addressed the availability under common law of an actionable wrong of invasion of privacy—Canada, India, and New Zealand.[28] One Canadian court has recognised a general right to privacy and to protect privacy interests under the rubric of nuisance law.[29] In New Zealand the tort of invasion of privacy has been recognised and s 14 of *New Zealand Bill of Rights Act* 1990 (NZ), while it does not confer a right to privacy, ensures the freedom of expression.[30]

While there is no general remedy for the invasion of privacy interests, for example, in 'improper disclosure', the protection of privacy is incidental either in common law or equity. These avenues of relief in relation to misuse of information and data apply equally to both the private and public sectors.

Given these caveats, the following section examines the *Privacy Act* 1988 (Cwlth) in the first instance and the other cognate federal and state legislation supportive of this legislation. After this there is a discussion

---

[23] *Victoria Park Racing and Recreation Grounds Co. Ltd v Taylor* (1937) 58 CLR 479; 43 ALR 597 (HCA). In the U.K. see *R v Khan* [1997] AC 558 at 582.583.

[24] *Australian Broadcasting Corporation v Lenah Game Meats Pty Ltd* [2001] HCA 63, 15 November 2001.

[25] [2003] QDC 151, 16 June 2003.

[26] *R v Broadcasting Standards Commission; ex parte British Broadcasting Corporation*; [2000] 3 WLR 1327; [2000] 3 All ER 989.

[27] *Douglas v Hello! Ltd* [2001] 2 WLR 992; [2001] 2 All ER 289 per SEDLEY LJ at 120. See also the discussion of *Campbell v Mirror Group Newspapers* [2002] All ER (D) 177 (October) in Ch. 1 *supra*.

[28] *Aubrey v ... ditions Vice-Versa Inc.* [1998] 1 SCR 591; *Govind v State of Madhya Pradesh* (1975) 62 AIR (SC) 1378; *P v D* [2000] 2 NZLR 591.

[29] *Canadian Tort Law*, 6th edn (1997) at 56; *Aubry v Duclos* (1996) 141 DLR (4th) 683.

[30] See Tobin, R 'Invasion of Privacy' (2000) *New Zealand Law Journal* 216.

of the common law controlling the disclosure of personal information and the protection of privacy, followed by a discussion of self-regulation by way of industry codes of conduct. For comparative purposes the U.S. legislative and common law regime is examined in similar terms before concluding with a discussion of an evolution of fair information privacy practices.

4.3.1 The *Privacy Act* 1988 (Cwlth)

The Commonwealth's *Privacy Act* 1988 was passed in order to comply with Australia's obligations under both the UDHR as well as the ICCPR. Australia has also been influenced by regional and international developments in the protection of privacy of its citizens. These include the Organisation for Economic Cooperation and Development's (OECD) Privacy Guidelines of 1980.[31] The guidelines promulgate basic privacy principles that protect personal data and encourage industry self-regulation as a means of achieving the goals embodied in the principles. Also influential in the passage of Australian legislation were the various developments in the E.U. such as the E.U. Data Directive,[32] as well as the United Nations (UN) 1990 Guidelines for the Regulation of Computerized Personal Data Files.[33]

The first observation to be made is that the Australian *Privacy Act* 1988 does not create any general rights to privacy. The Privacy Act applies to personal information rather than to commercial data other than credit information, as discussed below. The protection provided under the Privacy Act is in addition to any intellectual property rights and other protection that may exist in that information. The Privacy Act has been amended twice to include guidelines for data matching by government

---

[31] OECD 1980 *Guidelines on the Protection of Privacy and Transborder Flows of Pesonal Data*, Recommendation by the OECD Council of 23 September 1980. See http://www.oecd.org/e/droit/doneperso//ocdeprive/priv-en.htm and at http://www.oecd.org/dsti/sti/it/secur/prod/PRIV-en.htm and at http://www.oecd.org/documentprint/0,2744,en_2649_201185_15589524_1_1_1_1,00.html. See also EEC Council of Europe 1981 *Convention for the Protection of Individuals with regard to Automatic Processing of Personal Data*, Brussels: EEC and at http://www.privacy.org/pi/intl_orgs/coe/dp_convention_108.txt.

[32] For example the E.U. Directive on the Protection of Individuals with regard to the Processing of Personal Data and on the Free Movement of such Data. Directive 95/46/CE of the European Parliament and Council of 24 October 1995, *Official Journal of the European Commission* L 281. A comprehensive discussion of this Directive is given in the final section to this chapter *infra*.

[33] United Nations 1990 *Guidelines for the Regulation of Computerized Personal Data Files* at http://www.datenschutz-berlin.de/gesetze/internat/aen.htm.

agencies and to incorporate tax file number (TFN) guidelines.[34] It is to be noted that *data matching* is the process of comparing information about a large number of individuals on different databases to identify cases of interest usually by isolating discrepancies of apparent significance. *Data linkage* on the other hand is the connection of information held on different databases to create a composite database.[35] The focus is plainly on infringements and breaches by Commonwealth Agencies of privacy principles, TFN guidelines and data-matching principles. The Privacy Act has been further amended in 2000 to include the private sector.[36]

The Privacy Act prohibits 'interferences with privacy' by a Commonwealth agency when it breaches one or more of the information privacy principles (IPPs), the TFN guidelines or the data-matching guidelines. The IPPs apply to all Commonwealth Agencies and the principles are set out in s 14 of the Privacy Act.

Practice Notes: Information Privacy Principles as Applying to Australian Commonwealth Agencies

Information Privacy Principles

IPP 1 Manner and purpose of collection of personal information
IPP 2 Solicitation of personal information from individual concerned
IPP 3 Solicitation of personal information generally
IPP 4 Storage and security of personal information
IPP 5 Information relating to records kept by record keeper
IPP 6 Access to records containing personal information
IPP 7 Alteration of records containing personal information
IPP 8 Record keeper to check accuracy, etc. of personal information before use
IPP 9 Personal information to be used only for relevant purposes
IPP 10 Limits on use of personal information
IPP 11 Limits on disclosure of personal information.

---

[34] *Privacy Amendment Act* 1990 (Cwlth); and *Data-Matching Program (Assistance and Taxation) Act* 1991 (Cwlth).
[35] See Office of the Federal Privacy Commissioner Australia 1994 *Fifth Annual Report, 1992–1993*, Canberra: Human Rights Australia, p. 121.
[36] *Privacy Amendment (Private Sector) Act* 2000 (Cwlth).

Under the Privacy Act the Privacy Commissioner is given powers of investigation, evaluation and supervision of compliance with the Act in relation to the IPPs, TFN guidelines and data-matching guidelines. Under s 27(p) the Privacy Commissioner issued data-matching guidelines which came into effect from October 1992. In the main the Privacy Commissioner may examine proposals for data matching or data linkage concerned with an interference with the privacy of individuals.[37]

4.3.2 The *Privacy Amendment Act* 1990 (Cwlth)

The Privacy Amendment Act extends privacy protection to consumers in relation to their credit records. This amendment seeks to establish a check against the risk that credit databases may be used for non-credit-related purposes and the possibility that the data may become available to unauthorised users that would breach the privacy principles established under the Privacy Act.

The Privacy Amendment Act gives consumers a statutory right to: access their credit file; correct or update information; and make a complaint to the Privacy Commissioner seeking compensation for damage suffered because of the infringement of a debtor's privacy. In addition, credit providers are restricted in their use and dissemination of information. Criminal penalties apply to credit-reporting agencies or credit providers who furnish false or misleading credit reports; or to any person who gains unauthorised access to credit information.[38]

The National Privacy Principles (NPPs) are central to the new environment for the protection of privacy in Australia.[39] These principles establish minimum standards for the handling of personal information, but are different from the IPPs because the latter apply only to Commonwealth Agencies. The NPPs that apply to the public sector agencies and particularly to the private sector are given in summary form in the practice notes below.

---

[37] *Privacy Act* 1988 (Cwlth) s 27(1)(k) and Office of the Federal Privacy Commissioner Australia 1999 *National Principles for the Fair Handling of Personal Information*, Office of the Privacy Commissioner, Australia, February 1999 and at http://www.hreoc.gov.au/hreoc/privacy/natprinc htm. Office of the Federal Privacy Commissioner Australia 2000 *National Privacy Principles extracted from the Privacy Amendment (Private Sector) Act 2000* at http://privacy.gov.au/publications/npps01_print.html [30 June 2004].

[38] *Privacy Amendment Act* 1990 (Cwlth) ss 18K, 18L, 18M.

[39] The full text of the NPPs are available at http://www.privacy.gov.au. These NPPs are also set out in Schedule 3 of the *Privacy Act* 1988 (Cwlth).

Practice Notes: National Privacy Principles (NPPs) for Handling Personal Information

NPP 1 Collection of personal information. Collection must be necessary for an organisation's activities. It must be collected lawfully and fairly, and as a general principle, with the individual's consent.

NPP 2 Use and disclosure of personal information. Information can only be used or disclosed for its original purpose unless the person has consented to its use or disclosure for another purpose. Exemptions may apply to initial contact for direct marketing if consent was not practicable originally. Other exceptions include law enforcement needs, public safety, need for medical research, need to manage, fund and monitor a health service and where necessary to prevent or lessen a threat to a person's life.

NPP 3 Accuracy of personal information. Organisations must take reasonable care to ensure that they keep personal information accurate, complete, and up-to-date.

NPP 4 Security of personal information. Organisations that collect personal information must be able to document their practices and make the information available upon request.

NPP 5 Openness. Organisations must set out in a document clearly expressed policies on their management of personal information and make that document available to anyone who asks for it.

NPP 6 Access and correction rights. Organisations must give individuals access to their personal information and allow them to correct it or explain something with which they disagree, unless that explanation would invade someone else's privacy. Another exception is where this would compromise a fraud investigation.

NPP 7 Use of government identifiers. Organisations cannot use a government agency's identifier as its identifier. This includes driver's licence numbers, Medicare numbers, TFNs or any other future identity number assigned by a government agency.

NPP 8 Anonymity. Organisations must give people the option of entering into transactions anonymously where it is lawful and practicable. An example of where it would be unlawful, however, would be the opening of a bank account.

Continued on page 223

Continued from page 222

NPP 9 Restrictions on trans-border data flows. Organisations can only transfer the personal information about an individual to a foreign country if they believe that a law or a contract that upholds privacy principles similar to the NPPs will protect the information.

NPP 10 Special provisions for sensitive information. A higher level of privacy protection applies to sensitive personal information. This includes health information, political beliefs, religious affiliation, sexual preferences, membership in political parties etc.[40]

4.3.3 *Data-matching Program (Assistance and Taxation) Act* 1991 (Cwlth)

This Act regulates data matching between specific agencies, namely the Department of Community Services and Health, the Department of Employment, Education, and Training, the Department of Social Security and the Department of Veteran's Affairs. The purpose of this legislation is to prohibit unrestrained matching or creation of comprehensive databanks on citizens. Thus, s 6 of the Act provides for a maximum of nine data-matching cycles in any one year, with only one cycle to be in operation at any one time. Other restrictions include the storage of data, time limits on when the data must be examined and investigated, and what notice needs to be given to persons whose benefits are to be impacted.[41]

4.3.4 *Privacy Amendment (Private Sector) Act* 2000 (Cwlth)

In response to consumer concerns about privacy, as well as international developments such as the European Union Directive noted previously,[42]

---

[40] The *Privacy Amendment (Private Sector) Act* 2000 (Cwlth) defines *personal information* as information, whether fact, opinion or evaluative, about an identifiable individual that is recorded in any form, but excludes information that is generally available to the public such as name, address and telephone number. *Sensitive information* includes information or an opinion about a person's racial or ethnic origins, political opinions, membership of a political association, religious beliefs or affiliations, philosophical beliefs, membership of a professional or trade association, membership of a trade union, sexual preference or practices, criminal record or individual health information.

[41] See Hughes, G 1994 'Government data matching to continue', 68 *Law Institute Journal* 488.

[42] For example, in the E.U. Data Protection Directive of 1995 Art. 25(1) provides that local law must restrict the flow of data to another country unless there is adequate protection within that country. The U.S. is taking a more self-regulatory approach with overall supervision to be provided by the Federal Trade Commission (FTC).

the *Privacy Amendment (Private Sector) Act* 2000 (Cwlth) came into effect in December 2001 and affects the private sector. This includes all businesses having an annual turnover of AUD$3 million or more. The Privacy Private Sector Act applies to bodies corporate and unincorporated and to individuals and sole traders or consultants operating in their business capacity. Government business enterprises previously not under the Privacy Act are also now affected because of the commercial nature of their operations. The new legislation will also apply to the Australian Broadcasting Corporation (ABC), Australia Post, the Commonwealth Scientific Industrial Research Organisation (CSIRO) and Telstra Corporation (a government-owned telecommunications corporation) when operating as business entities.

The Privacy Private Sector Act gives greater protection to sensitive information with stricter limits on how this data may be collected and handled by private sector organisations. Specifically banned is the use of sensitive information such as health information for direct marketing purposes. However, personal information may be used for direct marketing as long as the individual is given an opportunity to opt out of receiving any further direct marketing material.

The Commonwealth Privacy Commissioner has given advice on the use of consultants and outsourcing of IT contracts.[43] This advice suggests privacy clauses to be used in the contracts when outsourcing work by public sector agencies. The advice is also applicable to non-IT contracts where contractors handle personal information. The purpose of these clauses is to impose on a contractor many of the obligations that an agency is subject to, for instance, the IPPs under the *Privacy Act* 1988 (Cwlth) as well as the NPPs, such as reasonable security and the prevention of unauthorised use and disclosure of information. In particular, IPP 4 requires agencies to protect personal information against misuse by taking reasonable security safeguards when outsourcing work to third parties. A key means of compliance with IPP 4 is the inclusion of clauses protecting privacy in outsourcing contracts. Under s 8 of the *Privacy Act* 1988 (Cwlth), while the conduct of agents and employees of an agency is covered, conduct arising from the provision of a service for the agency by a contractor is not covered. Neither the agency nor the contractor is liable under the Act for a failure to comply with IPPs that arises solely from an action of the

---

[43] See Office of the Federal Privacy Commissioner Australia 1994 *Advice for Commonwealth Agencies considering contracting out (outsourcing) information technology and other functions.* See http://www.privacy.gov.au/news/p6_4_19.doc. Office of the Federal Privacy Commissioner Australia 2001 'Information Sheet 14. Privacy obligations for Commonwealth contracts' at http://www.privacy.gov.au/publications/IS_14_01.html [30 June 2004].

contractor. As a result individuals will be unable to assert any rights under the Privacy Act against a contractor.

There are minor exceptions in the legislation; chief among these are small business operators, registered political parties and agencies, authorities and instrumentalities of the Commonwealth, State or Territory. There is also a media exception with *media* defined as acts or practices done in the course of journalism. *Journalism* is defined as including the collection, preparation and dissemination of news, current affairs, documentaries and other information for the purpose of making the material available to the public.[44]

The Privacy Private Sector Act permits related corporate groups to share personal information, but not sensitive information amongst the group. However, all members of the group must comply with either the NPPs or an industry code approved by the Privacy Commissioner. Compliance with the Act thus means that it will allow central data warehousing activities to be undertaken, for instance. Databases existing prior to the enactment of the Privacy Private Sector Act are subject to the new requirements. The Act requires private sector organisations that hold existing databases of personal information to take reasonable steps to ensure that the data are up-to-date, accurate, complete and adequately secured. People should be able to access and correct information collected about them on existing databases when the organisation uses or discloses the information. When pre-existing databases are updated after the 2001 passage of the Act, the owners of the database will be required to comply with all aspects of the new law in updating their information.

Practice Notes: Compliance with National Privacy Principles

What organisations should do to ensure compliance with the NPPs

1. Determine the extent to which the *Privacy Amendment (Private Sector) Act* 2000 (Cwlth) Applies.
   Organisations may need to scope their activities against the application of the Act to determine the extent to which their information handling activities will be governed by its provisions.

Continued on page 226

---

[44] *Privacy Amendment (Private Sector) Act* 2000 (Cwlth) s 6c.

Continued from page 225

2. Conduct a privacy audit.

Organisations whose activities fall within the scope of the Act will need to conduct an audit of all their information handling practices to determine whether those practices are consistent with the requirements set out in the legislation. The audit will include a review of:

- what personal information has been and will be collected by the organisation;
- how that personal information is collected;
- how that personal information is used and disclosed;
- how that personal information is stored and ultimately disposed of.

3. Review results of audit and modify any non-compliant practices.

Once the audit has been completed, any information handling practices that do not comply with the NPPs will need to be reviewed and either modified or abandoned.

4. Implement a privacy compliance program.

The components of an organisation's privacy compliance program will depend on the size of the organisation and the extent to which it handles personal information. This may involve the development of a privacy policy and procedure guidelines, privacy training for relevant staff who handle personal information, consider the appointment of Privacy Managers to monitor privacy compliance and to undertake privacy audits in an ongoing fashion.

5. Decide whether to develop a privacy code or to join others in the same industry where such an industry code has been developed.

This industry code will be in the general language of the NPPs to suit specific requirements and practices of the industry and permits flexibility and sensitivity to industry and market needs.

6. Consider how to address compliant handling.

One benefit in establishing an industry compliant handling process is that it can involve members with industry experience and who may have dispute resolution experience.

The Privacy Private Sector Act does not apply to State and Territory public sector agencies. Some states and territories have introduced their own privacy legislation, for example *Privacy and Personal Information Protection Act* 1998 (NSW), *Invasion of Privacy Act* 1971 (Qld) and the *Australian Capital Territory Government Service (Consequential Provisions)*

Act 1994 (ACT). State and Territory laws dealing with privacy will continue to operate as long as they are not inconsistent with Commonwealth law. Other laws protecting confidential information will continue to operate and these include common law trust principles, breach of confidence, equitable duties of confidence, and confidentiality inherent in particular relationships, as discussed below.

The N.S.W. legislation has permitted the development of private privacy codes subject to the approval of the State Privacy Commissioner. Victoria has introduced the *Information Privacy Act* 2000 (Vic) and the *Health Records Act* 2001 (Vic) and has also passed the *Surveillance Devices Act* 1998 (Vic). The Victorian Surveillance Act regulates the installation and use of data surveillance devices in criminal and civil investigations.

4.3.5 Freedom of Information

One very important aspect of privacy is to be able to find out what information is held by government agencies and private bodies. In the case of governmental records, the *Freedom of Information Act* (FOI) 1982 (Cwlth) aims to give to members of the public rights of access to official documents of the Commonwealth government and its agencies. Each of the states and the Australian Capital Territory (ACT), but with the exception of the Northern Territory (NT), has also passed FOI legislation.[45] In general, a person who wishes to obtain access to a document usually makes a request in writing to the relevant government agency giving all details necessary to identify the subject document. Commercially valuable documents and confidential business information are generally exempt from disclosure.

A majority of the requests made under FOI legislation relate to the personal affairs of the applicant and these for the most part have been granted. A key test, however, is the extent to which access has been granted to non-personal information related to bureaucratic and politically sensitive information. Many observers and commentators are of the view that the FOI legislation has not worked well because of the high costs of getting such information, the long time delay in getting access to the documents, and the use of various devices by governments to keep

---

[45] *Freedom of Information Act* 1982 (Cwlth); *Freedom of Information Act* 1982 (Vic); *Freedom of Information Act* 1989 (NSW); *Freedom of Information Act* 1989 (ACT); *Freedom of Information Act* 1991 (Tas); *Freedom of Information Act* 1991 (SA); *Freedom of Information Act* 1992 (WA); *Freedom of Information Act* 1992 (Qld).

information from the public. One example is that of government contracts which have strict and widely cast confidentiality clauses for the purpose of restricting its release to the public. Another is the claim that documents are 'privileged' because their release would not be in the public interest or the claim by Ministers that the papers have commercial-in-confidence status and thus cannot be released. Even more restrictive is the claim of cabinet confidentiality. Together, these claims for exemption have kept government documents from public scrutiny.

4.3.6 The Common Law and the Disclosure of Personal Information

In contrast to the regulatory means of protecting privacy, the common law controlling the disclosure of personal information is both incidental and limited in its reach. Nevertheless, common law can be both preventative as well as a source of remedy for civil wrongs. The following examines common law protection of privacy through the law of torts, either by way of defamation or negligence or passing off, the law of contract, and the equitable doctrines enshrined in the duty of confidence.

Law of Torts

The law of torts, defamation, negligence and passing off provide some protection of privacy. But these common law theories are narrowly construed and are limited in scope. Defamation, for instance, is concerned with material that is capable of disparaging an individual in the eyes of third parties. Privacy interests may be affected by material about an individual which may be perfectly neutral and true, but which the individual does not want others to know. If these disclosures do not affect either character or reputation it seems that a court will not countenance it as defamation and thus will not be successful in court. Thus, the release of an employee's address, marital status, income and age to some marketing firm does not discredit the employee. Yet, the employee's privacy interests may have been infringed by such a disclosure. The example given thus shows up two competing interests—that of an interest in an individual's privacy on the one hand and the society's interest in freedom of expression on the other.

Negligence, Negligent Advice and Negligent Reports

Negligence, and in particular negligent advice and negligent reports, are further common law actions that may be available if privacy has been infringed. However, the way this particular common law action works is indirect and may require stretching analogous cases to find application. For example, an indirect application of the principles is demonstrated by the leading case of *Shaddock and Associates Pty Ltd v Parramatta City Council*[46] where liability was not limited to persons giving advice within their area of special skill or competence. It was held in this case that the supplier of information has to ensure that the information is correct if the supplier knows that the recipient will rely on its accuracy. But, note that the case dealt with the disclosure of inaccurate information and not with the disclosure of accurate personal information in breach of confidence. Nevertheless the principles that have been established by the case may be applied to the release of seriously incomplete, albeit accurate, personal details.

The principles of tort of negligence, as developed by the courts, to cope with the special problems associated with claims for compensation for economic losses flowing from careless advice, provide only limited protection for the person whose privacy interests have been infringed by an improper disclosure of personal information. This is because only the recipient of the negligent advice or information may sue. The person about whom the personal information pertains has no remedy against the supplier of the negligent misstatements made to third parties.

Negligent Reports

Similarly, in the body of law covering negligent reports, the careless reporting of information does not have the inherent inhibition upon its capacity for incidental privacy protection that is so characteristic of the law relating to negligent advice. With a negligent report, it must be shown that the plaintiff was the recipient of the misinformation and relied upon it to the plaintiff's detriment. One of the leading cases with negligent reports in relation to the protection of information privacy is that of the English Court of Appeal decision in *Ministry of Housing and Local Government v Sharp*.[47] In this case a clerk omitted the plaintiff's land charge from a conclusive certificate issued to a purchaser. The Court held that the clerk was under a duty at common law to use due care. That duty is one that he owed to any person whom he knew, or ought to have known, might be injured if he made a mistake. The principle established in *Sharp* was endorsed in a later Court of Appeal decision in *Dutton v Bognor Regis*

---

[46] (1981) 55 ALJR 713.
[47] [1970] 2 QB 223.

Urban District Council.[48] In *Dutton* LORD DENNING held that a professional man giving advice on financial or property matters such as a banker, lawyer or an accountant, has a duty only to those who rely on that financial or property advice. However, in the case of a professional who gives advice on the safety of buildings, or machines, or material, the duty is to all those who may suffer injury in case the advice is bad.[49]

Nervous Shock

Another category of tort law stemming from negligence is nervous shock as a result of improper disclosure. In the New Zealand case of *Furniss v Fitchett*[50] a doctor was held liable to his patient for disclosing to the patient's husband a confidential account of the patient's illness. The information was later revealed to the patient in the course of divorce proceedings of the patient and her husband, whereupon the patient suffered nervous shock. The doctor's negligence lay in the manner and circumstances in which the report was released. The case reflects the broad principle that where nervous shock and other tangible forms of loss, injury or damage are the foreseeable consequence of carelessness in the handling of personal information, they will be recoverable in negligence.[51] The key to recovery is proof of nervous shock causing actual, that is, organic damage; some tangible and measurable symptom such as a heart attack, a stroke or a diagnosable nervous condition such as depression or schizophrenia. Mere embarrassment arising from the improper release of personal information has never been recognised as a basis in itself for an action for compensation at common law.

Passing Off

Under English and Australasian common law passing off is a tort applying to the sale of goods accompanied by a misrepresentation as to their origin and ownership and calculated to mislead the purchaser. The tort is primarily concerned with the protection of a person's proprietary interests, specifically goodwill. In relation to personal privacy, it offers incidental protection to an individual's privacy from the protection it affords against unauthorised use of a person's name, likeness or life history. Previously,

---

[48] [1972] 1 QB 373.

[49] *ibid.* p. 395.

[50] [1958] NZLR 396.

[51] The decision in *Furniss v Fitchett* follows a long line of similar cases beginning with *Wilkinson v Downton* [1897] 2 QB 57.

privacy protection was limited by the rule that the action was only available to a person who was engaged in some 'common commercial field of activity' with the defendant.[52]

In Australia plaintiffs in passing off actions may also rely on s 52 of the *Trade Practices Act* 1974 (Cwlth) which provides that corporations shall not, in trade or commerce, engage in conduct that is 'misleading or deceptive'. In *World Series Cricket Pty Ltd v Parish*[53] the Federal Court of Australia granted an interim injunction at the suit of the chairman of the Australian Cricket Board. The suit was to restrain the company from referring to its cricket matches as 'Tests' or 'Super Tests' without clearly distinguishing them from matches controlled by the Board. In *McDonald's System of Australia Pty Ltd v McWilliam's Wines Pty Ltd (No 2)*[54] FRANKI J restrained the defendant from promoting its wines as 'Big Mac'. This conduct could be regarded as 'deceptive' within s 52, taking into account the extent to which the phrase 'Big Mac' was identified in the public's mind with the plaintiff's products.

However, in *Henderson v Radio Corporation Pty Ltd*[55] the Full Court of the Supreme Court of New South Wales refused to accept any limitation upon recovery based on the requirement of a common commercial field of activity between the plaintiff and defendant. The Hendersons were well-known professional ballroom dancers. They complained of a gramophone record cover put out by the defendant showing a ballroom scene with them in a typical dancing pose. The appellant claimed that because the respondents were not engaged or likely to be engaged in selling gramophone records it was entitled to appropriate their names and reputations for its own commercial advantage. The majority of the Court held otherwise by stating that, if it was proven that a party had falsely represented his goods as the goods of another or his business to be the same or as connected with the business of another, and in this case names and reputation of the Hendersons, the wrong of passing off had been established and the other is entitled to relief.

Privacy Protection through the Law of Contract

The most obvious means by which a party to a contract may protect privacy is by expressly stipulating terms governing disclosure or use of

---

[52] *McCulloch v Lewis A. May (Produce Distributors) Ltd* [1947] 2 All ER 845.

[53] (1977) 16 ALR 181

[54] 28 ALR 236

[55] (1960) SR (NSW) 576.

personal information that is supplied. If the agreement is breached, damages may be awarded, and an injunction may go to prevent repetition. A further example is that of a contract of employment which stipulates that an employee is not to reveal to outsiders secret information gleaned from employment.[56] There may also be implied terms protecting privacy. Implied terms have been relied upon in the business community to authorise the disclosure of personal information, rather than to reinforce a common understanding of secrecy. Examples from the banking industry may be given where an implied consent is relied upon. This implied consent so authorises the practice of bankers giving opinions as to the financial standing and capacity of their customers.

In relying on a contract as a means of remedying an invasion of privacy, the threshold question is whether there is a valid contract between the record keeper and the record subject. The absence of any contractual relationship may severely limit the effectiveness of relying on contract law to protect privacy and confidentiality of the records. The potential risks to privacy lie in areas where there are no established information privacy guidelines in the handling of personal data and information, hence the need for IPPs and NPPs discussed previously. Thus, for example, in the development of GIS where large databases are involved there is a potential for a breach of privacy to take place whether innocently or otherwise. Also if no privacy protection is written into a contract for the use of such data, the integrity and security of a customer's data may be at risk, especially in the event of a breakdown of the contract.[57]

Duty of Confidence

Proper information handling practices are central to privacy protection. Under the FOI Act there is an exemption to the right of access where it would result in breach of another person's legal right to maintain the confidentiality of the information to which a person is seeking access. Under the common law the issue is not so clear-cut since it has been suggested that an action may be based on contract, property, tort or equity. However, Australian courts have accepted the concept of an equitable duty of confidence and it thus appears well settled that confidential information could be protected in equity in an appropriate case.[58]

---

[56] See Chapter 5, where contractual issues are discussed in detail.

[57] See Lickson, CP 1968 'Protection of the privacy of data communications by contract' (1968) 23 Bus. L. 971.

[58] *Commonwealth v John Fairfax & Sons Ltd* (1980) 32 ALR 485.

It is to be noted that the action for breach of a legal duty of confidence is not directly concerned with protection of privacy interests. The case law in the area has been concerned primarily with the protection of commercial confidences such as trade secrets. However, disclosure of information acquired in confidential relationships such as doctor and patient, lawyer and client, teacher and student, would clearly endanger privacy interests of a personal nature. Courts will hold entirely personal and non-commercial information to be capable of protection. However, because of a dearth of litigation in the area, the impact of the legal duty of confidence in protecting privacy has not been fully elaborated by the courts. But there are clear indications that breaches of the legal duty of confidence may occur when persons disclose or use information acquired by them in confidence *without the consent* of the original communicator or for a purpose inconsistent with the reasons for which the information was originally acquired.

This present discussion is also highly significant in assessing the lawfulness of information brokerage systems and in particular those systems used by the credit reference agencies and direct mail industries. These industries rely upon disclosure by the original recipients, for example, banks, credit agencies, professional organisations, hospitals and hotels of confidential information. However, when the data are collated and collected by third parties that have no direct dealings with the data subjects, there are severe privacy protection implications both in law and in practice. These issues and its ramifications are explored later in this chapter.

The defences to an action for breach of confidence have been stated in and English case in *Tournier v National Provincial and Union Bank of England*.[59] Here it was established that there might be four categories for which disclosure may be permissible, where:

- disclosure is made under compulsion of law;
- the interests of the confidant require disclosure;
- there is a duty to the public to disclose;
- the disclosure is made with the express and implied consent of the confider.

These defences may prove controversial and provide interesting grist for the mill when discussing the use of geospatial technologies embedded in mobile devices such as personal digital assistants (PDA),

---

[59] [1924] 1 KB 461.

mobile phones equipped with global positioning systems (GPS) and other sorts of location identifiers. Whether it is a defence that it is in the interests of the confider or in the public interest will require further exploration. Where the disclosure will avert an apprehended public injury, disclosure of location may be justified. However, the question remains: what is the balance between the public interest in maintaining confidentiality of a person's location as against the rescue of that person in emergency situations. It is self-evident that in the extreme case of tracking of prisoners on a weekend release the use of locating devices such as wrist bands with radio transmitters may be permissible. The justification for the use of these devices may be made on several grounds, including public policy ones. Moreover, the tacit agreement of the prisoner to permit tracking could be a condition of the weekend release.

Thus, as may be noted there exist a number of limitations of the action for breach of confidence as a protector of information privacy interests. Breach of confidence law is concerned with the protection of facts that are confidential, that is, those that are not in the public domain. The circumstances which import a duty of confidence are where the information are given in confidence or acquired in circumstances that require confidentiality. The claim of *information privacy*, however, is concerned with the risks from mishandling of any personal information, whether or not it was acquired in confidence. Thus, information relating to ordinary retail transactions would not ordinarily be secret. It does not appear that there is anything in the seller–customer relationship that would prevent the seller to disclose such retail purchasing details to an information broker. If anything, a plaintiff would have to rely on general principles of equity which would have to be adapted to the particular circumstances of a novel claim.

4.3.7 Industry Codes of Conduct and Self-regulation

While the legislative regimes mandate a minimum level of privacy protection, various industry codes also recommend, and in some cases impose, privacy standards as a precondition to membership of their organisation. One of the most influential and best known of these codes is the Australian Direct Marketing Association (ADMA) code of conduct.[60] The ADMA Code contains specific standards relating to data protection in direct marketing contexts in both the paper and electronic environments. ADMA

---

[60] See ADMA Code of Conduct at http://www.adma.com.au.

has also adopted the latest OECD Guidelines for Consumer Protection in the Context of E-Commerce.[61]

The OECD Guidelines for Consumer Protection articulate a number of general principles that are highly relevant to privacy protection including:

- transparency and effective consumer protection;
- fair business practices;
- on-line disclosures that provide sufficient information to permit consumers to make informed choices;
- clear processes for confirming transactions;
- secure payment mechanisms;
- timely and affordable dispute and redress processes;
- privacy protection following recognised privacy principles; and
- emphasis on consumer and business education.

Depending on the technology used in direct marketing, other codes of conduct also deal with aspects of privacy protection. These codes include:

- Smart Card Code of Conduct;[62]
- Electronic Funds Transfer Code of Conduct;[63]
- Telecommunications Codes;[64]
- Internet Industry Association Code of Conduct;[65] and
- Commonwealth Government's *Building Consumer Sovereignty in Electronic Commerce: A Best Practice Model for Business* based on the OECD Consumer Protection in Electronic Commerce Guidelines[66] cited as Best Practice Model Code.

It seems that governments are moving away from prescriptive 'black letter' law to regulate the market towards industry codes of

---

[61] OECD Council Concerning Guidelines for Consumer Protection in the Context of Electronic Commerce, adopted 9 December 1999 by 29 OECD member countries. See http://www.oecd.org/dsti/sti/it/index.htm.

[62] Asia Pacific Smart Card Forum 1997 Code of Conduct at http://www.smartcardforum.asn.au/.

[63] Electronic Funds Transfer (EFT) Code of Conduct 2002 reproduced in *CCH Australian Consumer Credit Law Reporter*, pp. 95–250 and at http://www.asic.gov.au.

[64] Australian Communications Industry Forum (ACIF) 1999 Telecommunications Code at http://www.acif.org.au.

[65] A draft Internet Industry Association (IIA) code of conduct is to be found at http://www.intiaa.asn.au/codeV2.htm and at http://www.iia.net.au/codes.html.

[66] See OECD Consumer Protection in Electronic Commerce Guidelines at http://www.treasury.gov.au/ecommerce.

conduct as one of a number of ways of ensuring both consumer protection as well as the protection of privacy on an industry-specific basis. Recognition is now accorded to voluntary codes of best practice and of acceptable ethical standards of conduct in the handling of personal information and its protection. The advantages of self-regulation are that the industry standards would be more flexible and less restrictive than those imposed by legislation or regulation. Self-regulation would demonstrate the positive steps that industry is taking and its social responsibilities in addressing privacy concerns. However, to date there has not been any general agreement on what features a code of conduct should contain if it is to produce effective privacy protection outcomes. There are also other concerns. For example, it has been suggested that the protection offered in the various codes may be insufficient to protect data subjects, and recalcitrant members of an industry group may try to gain a competitive advantage over those who comply. Thus, those who 'cheat' the system may undermine self-regulation unless it is perceived by them that there is some benefit to be gained from complying with the code.[67]

As discussed in the next section, the issue of privacy also exemplifies the contrasting approaches of Australia, North America and Europe. In Europe, for example, private industry codes of conduct are tantamount to no regulation or certainly insufficient regulation. In Australia and North America, in contrast, self-regulation is seen as an excellent way to achieve the balance between consumer privacy concerns and business needs. In the employment context, an e-mail or Internet usage policy seeks to balance the interests of employees versus that of the employer.

However, in reality, there is evidence that both systems work just as well. Also, the differences between the two approaches are not as great as they may seem. This is because in the European regime, the Directives are written in broad language that allow for and requires individual industry groups to fill in the detailed rules.[68] In the case of private industry codes, such regulations take place 'within the shadow of the law'. For example, in Australia, private codes must be consistent with the broad privacy protection principles as well as work within the *Trade Practices Act* 1974 (Cwlth).

---

[67] Perritt, JJ 1996 *Law and the Information Superhighway*, New York: John Wiley & Sons Inc.

[68] An analysis of the impact of the various European Directives is found in the final section of this chapter.

4.3.8 The Regime in the United States

The purpose of this brief section on the situation in the U.S. is to provide a comparison to the regulatory framework established in Australia discussed previously. This comparison will serve to demonstrate how the rules have developed in tandem and in some cases how the approaches justifying one course of action have differed because of backgrounds, attitudes, and public policy. At the same time the discussion will be timely because it would lead quite appropriately to contrasting the situation in Europe and in particular the Directives dealing with privacy that have developed there.[69] As with the structure adopted in earlier discussions, there is initially a discussion of Federal and State statutes dealing with the protection of privacy and followed by the common law protection of privacy.

A 'right to privacy' is absent in the U.S. Constitution or the Bill of Rights. However, the U.S. Supreme Court has interpreted a right to privacy to exist for individuals under the First, Fourth, Fifth, Ninth and Fourteenth Amendments.[70] But, many privacy decisions in the U.S. federal courts are based on the Fourth Amendment which generally provides for the right of people to be secure in their persons, houses, papers and effects against unreasonable searches and seizures.[71] In 1965, the U.S. Supreme Court suggested that there may be 'zones of privacy' implicit in the Bill of Rights.[72] This suggestion was made in a case in which the Court invalidated a statute prohibiting the use of contraceptive devices and the giving of medical advice on their use.[73]

---

[69] See further Charlesworth, A 2000a 'Clash of the data titans: US and EU Data Privacy Regulation', *European Public Law* 6, pp. 253–274; and also compare Charlesworth, A 2000b 'Data privacy in cyberspace: Not national vs international but commercial vs individual', in Edwards, L and Waelde, C *Law and the Internet: A Framework for Electronic Commerce*, Oxford: Hart Publishing, pp. 79–124.

[70] The First Amendment guarantees freedom of communications and the expression of ideas; the Fourth Amendment guarantees freedom from unreasonable search and seizure, including (in some cases) electronic, aural, visual, and other types of surveillance; the Fifth Amendment guarantees freedom from self-incrimination, and guarantees due process of the law with regard to the Federal government; the Ninth Amendment recognises that rights not specified in the Constitution are vested with the people; and the Fourteenth amendment guarantees due process and equal protection of the law with regard to the states.

[71] The Fourth Amendment to the U.S. Constitution provides that: The right of the people to be secure in their persons, houses, papers, and effects, against unreasonable searches and seizures, shall not be violated, and no Warrant shall issue, but upon probable cause, supported by Oath or affirmation, and particularly describing the place to be searched, and the persons or things to be seized.

[72] See the penumbra argument put by Scoglio, S 1998 *Transforming Privacy: A Transpersonal Philosophy of Rights*, Westport: Praeger, at 226ff.

[73] *Griswold v Connecticut*, 381 U.S. 479, 14 L.Ed.2d 510, 85 S.Ct. 1678 (1965).

The tensions between privacy interests such as to be left alone and interests for the community good has prompted a court to state these 'zones of privacy' explicitly. The California Supreme Court has stated that '[c]ollectively, the federal cases sometimes characterized as protecting "privacy" have in fact involved at least two different kinds of interests. One is the individual interest in avoiding disclosure of personal matters, and another is the interest in independence in making certain kinds of important decisions'.[74]

In California, the elements of a claim for invasion of privacy based on the state constitution includes a legally protected privacy interest, a reasonable expectation of privacy in the circumstances and conduct by a defendant constituting a serious invasion of privacy.[75] According to the court, there are two aspects of legally recognised privacy interests. The first precludes the dissemination or misuse of sensitive and confidential information and this is generally regarded as 'informational privacy'. The second class is an interest surrounding the making of intimate personal decisions or conducting personal activities free from observation, intrusion, or interference. This is termed as 'autonomy privacy'.

In an indirect way, Federal constitutional rights to privacy, as embodied in the Fourth Amendment, apply to prevent government or state action.[76] However, California added an express right to privacy to its state constitution in 1972.[77] A number of other states have amended their constitutions similarly to provide privacy protection rights that limit both private as well as government action.[78]

In terms of direct legislation, Congress has passed a number of laws that affect the protection of privacy in general and those regulations applying to the on-line world in particular. It seems likely that the Federal courts will permit the laws to define the area rather than expand the contours of implied 'zones of privacy'. Legislation has been passed in the general area of the protection of privacy and in the more specific areas in the protection of the privacy of financial information, education, health, electronic communications, video hiring habits, driver's licence details

---

[74] *Hill v National Collegiate Athletic Association*, 7 Cal.4th 1 at 30; 865 P.2d 633 (1994).

[75] *ibid.* at 35.

[76] *United States v Maxwell*, 42 M.J. 568 (A.F.C.C.A. 1995).

[77] California Constitution, Art. I, § 1.

[78] For example, Alaska Const., Art. I, § 22; Arizona Const., Art. 2, § 8; Florida Const., Art I, § 23; Hawaii Const., Art. I, § 6; Illinois Const. Art. I, §§ 6, 12; Louisana Const. Art. I, § 5; Montana Const., Art. II, § 10; South Carolina Const., Art. 1, § 10; Washington Const., Art. I, § 7.

and protection of children. The following are a selection of specific Privacy Acts that address particular situations for the protection of privacy.[79]

Privacy Act of 1974

The *Privacy Act of 1974*, 5 U.S.C. § 552 provides limited privacy protection for government-maintained databases. In general the Act prohibits any government agency from concealing the existence of a personal data record-keeping system. Each agency maintaining such a system must publish a notice of the existence and character of the system. The *Privacy Act* applies to all collections of spatial data collected by federal agencies. The Federal Geographic Data Committee (FGDC) established under the Office of Management and Budget (OMB) has endorsed a policy on access to public information and the protection of personal privacy in federal geospatial databases.[80] In particular, the policy applies to all federal geospatial databases from which personal information may be retrieved. However, the constraints imposed on the commercial sector are less restrictive, given that most follow self-regulatory guidelines and/or use contracts with individuals as the means of guaranteeing the privacy of the data.

The *Privacy Act* provides for the following types of checks and balances that:

- allow individuals to determine what records pertaining to them are being collected, maintained, or used by federal agencies;
- allow individuals to prevent records obtained for a particular purpose from being used or made available for another purpose without their consent;
- allow individuals to gain access to such records, make copies of them and make corrections;
- require agencies to ensure that any record which identifies individuals is for a necessary and lawful purpose; and
- require agencies to provide adequate safeguards to prevent misuse of personal information.

The *Privacy Act* also contains a number of broad exceptions. The exceptions under this Act include civil or criminal law enforcement activity, circumstances affecting the health or safety of an individual,

---

[79] Other 'privacy-related' acts include: *Privacy Protection for Rape Victims Act of 1978*; and, *Telephone Consumer Protection Act of 1991*.

[80] See Federal Geographic Data Committee (FGDC) Policy on access at http://www.fgdc.gov/ Communications/policies/policies.html.

court orders, and consumer reporting agencies acting within the applicable law. The agency is also required to keep records of disclosures it may have made and make these available to the individual named in the record, who can then review the records and request correction.

Privacy Protection Act of 1980

The *Privacy Protection Act of 1980*, 42 U.S.C. § 2000 establishes safe-guards relating to the privacy of communications and publication materials. A person may possess materials for publication or broadcast in a newspaper, book, or other medium of public communication. Such materials are exempt from search or seizure by a government employee unless there is probable cause to believe that the person possessing the materials has committed or is committing the criminal offence to which the materials relate. Immediate seizure is permitted where necessary to prevent death or serious bodily injury.[81]

Data-Matching Legislation

One of the earliest concerns raised by computers was that of data matching. The *Computer Matching and Privacy Protection Act of 1988*, 5 U.S.C. § 552 was meant, at least in the area of government, to prevent the creation of large dossiers on individuals by making the creation of databases more difficult. There are also other laws that protect the privacy of information transmitted via telecommunication systems. The *Electronic Communications Privacy Act of 1986* (ECPA), 18 U.S.C. § 2510–22, 2701–11 is the primary federal legal protection against unauthorised interception and disclosure of electronic communications, including e-mail, while in transit or in storage. The ECPA contains a number of significant exceptions, however. For example, while operators of electronic communication services for the public are generally barred from disclosing the content of a mes-sage in storage, operators of purely internal e-mail systems are not covered by this prohibition. Companies operating internal e-mail systems, say on an Intranet, do so for purely for their own purposes and these systems do not normally connect to the outside world. However, so long as this is known to all staff and by not establishing a custom, practice, policy or procedure that may give rise to a reasonable and legally enforceable

---

[81] *Steve Jackson Games v United States Secret Service*, 816 F.Supp. 432 (W.D. Tex. 1993), *aff'd*, 36 F.3d 457 (5th Cir. 1994).

expectation of employee's privacy in their inter-office e-mail and other files, then the provisions of the ECPA would not apply.[82]

Privacy of Financial Information

Several federal laws protect the privacy of financial information. The *Fair Credit Reporting Act of 1970*, 15 U.S.C. § 1681 regulates information maintained by credit bureaus. Such bureaus are required to implement and maintain procedures to avoid reporting obsolete or inaccurate information. This is because credit reports may be furnished only for credit, insurance, employment, to obtain government benefits, or other legitimate business needs involving a business transaction. In such instances it is imperative that the reports are as accurate as possible. The *Fair Debt Collection Practices Act of 1977*, 15 U.S.C. § 1692–92 prevents debt collectors from disclosing information about a consumer's debt to third parties.

The *Right to Financial Privacy Act of 1978*, 12 U.S.C. § 3401–22 regulates the federal government's access to financial information held by a financial institution. Such institutions are prohibited from providing the government with 'access to or copies of, or the information contained in, the financial records of any customer except in accordance with the provisions of the Act'. On the other hand, financial institutions are permitted to notify a government authority that it has information that may be relevant to a possible violation of any statute or legislation. In addition, the institutions are permitted to release records in order to ensure that a security interest is properly validated, to prove a claim in bankruptcy has merit, to collect a debt or to process an application for a government loan or give a loan guarantee. Customers, of course, may authorise disclosure to the government as well as to revoke this authority. The *Gramm–Leach–Bliley Act of 1999*, 15 U.S.C. § 6801–10 requires financial institutions to provide consumers with notice and an opportunity to opt out before certain types of personal financial information to non-affiliated third parties may be disclosed.

Transactional Information Privacy

The *Telecommunications Act of 1996*, 47 U.S.C. § 222 includes a provision that protects transactional information concerning telephone

---

[82] See generally 'ECPA and Online Computer Privacy', 4 *Federal Communications Law Journal* 17, 39 (1989).

calls, including amount of usage and the destination of calls. The *Cable Communications Policy Act of 1984*, 47 U.S.C. § 551 requires cable operators to disclose to their customers what types of personally identifiable information (PII) they collect, and how they use and disclose such information. The Act also limits the permissible purposes for collection of PII from subscribers; limits disclosures of subscribers' PII; gives subscribers the right to access and correct PII concerning them that a cable operator holds; and requires cable operators to destroy PII that is no longer needed for its original purpose.

Children, Video, Health and Education Privacy Legislation

The only federal statute that specifically addresses the collection of information on-line is the *Children's Online Privacy Protection Act of 1998* (COPPA), 15 U.S.C. § 651–05 which regulates the collection, use and dissemination of personal identifying information obtained on-line from children under 13.

The *Video Privacy Protection Act of 1988*, 18 U.S.C. § 2719 limits disclosure of records of videotape rentals. Apparently, this statute was enacted in response to the public outcry over the disclosure of the video rental records of a judge during the Senate's consideration of his nomination to be a Justice of the Supreme Court.

The *Health Insurance Portability and Accountability Act of 1996* (HIPPA), 42 U.S.C. § 1320 limits the disclosure of individually identifiable health information by health care providers, health plans, and health care clearinghouses, without the consent of the individual, except in defined circumstances. The Department of Health and Human Services (DHHS) is required to establish *Standards for Privacy of Individually Identifiable Health Information.*

The *Family Education Rights and Privacy Act of 1974*, 20 U.S.C. 1232 regulates the handling of student records by educational institutions that receive public funds. This statute limits the disclosure of student records if parental consent is not given, and provides parents with a right to access their children's records and correct inaccuracies.

The *General Education Provisions Act of 2002*, 20 U.S.C. § 1232h(c)(2)(C)(i) includes a provision requiring school boards to provide parents with a notification on an annual basis. That notification is about the collection, disclosure, or use of student personal information that may be used for marketing purposes or the selling of that information. Parents should be given an opportunity to opt out if they did not wish to participate in the scheme. This requirement, which applies to both on-line and off-line collections of information, was inspired partly by the unease to which technology companies would provide schools with free computers or Internet access in exchange for the right to monitor students' activities on-line.

The *Drivers Privacy Protection Act of 1974*, 18 U.S.C. § 2721 regulates the disclosure and resale of personal information contained in records maintained by state Departments of Motor Vehicles. The Act was enacted partly in response to the murder of actress Rebecca Schaeffer, whose killer was reported to have obtained her address from records made publicly available for a two-dollar fee by the California Department of Motor Vehicles.

Location Privacy Protection

The *Location Privacy Protection Act of 2001* is designed to protect the privacy of individuals who use Internet-ready devices that can pinpoint a person's location.[83] The Act will require companies to notify users about location data collection, their consent as to the collection of the data, and prohibit the sale and use of the information collected without consent. There are also safeguards that permit individuals to correct errors in the data. In the findings leading up to the passage of this Bill it is interesting to note that Congress has recognised the right to privacy of location information. This right is based on the fact that customer proprietary network information is subject to s 222 of the *Communications Act of 1934*, thereby preventing use or disclosure of that information without a customer's express prior authorisation.[84] The *Wireless Privacy Protection Act of 2003* has also amended the *Communications Act of 1934* 'to require customer consent to the provision of wireless call location information'.[85]

Freedom of Information Act

The *Freedom of Information Act of 1966* (FOIA) U.S.C. 5 § 552 and the Open Records Laws of the individual states create a balance between the right of citizens to be informed about government activities and the need to maintain confidentiality of some government records. These laws generally support a policy of broad disclosure by government. Hence, if a data set held by a federal agency is deemed to be an agency record, that

---

[83] See http://www.allnetdevices.com/wireless/news/2001/07/13/location-privacy.html and http://www.techlawjournal.com/cong107/privacy/location/s1164is.asp.

[84] 47 § U.S.C. 222.

[85] See http://www.theorator.com/bills108/hr71.html; and see Schilit, B, Hong, J and Gruteser, M 2003 'Wireless location privacy protection' at http://www.computer.org/computer/homepage/1203/invisible/rz135.pdf.

record must be disclosed to any person requesting it, unless the record falls within one of the nine narrowly drawn exceptions contained in the FOIA. Provision of records is on the basis of cost recovery at the cost of duplication, even for those citizens requesting an entire geographic data set produced by a U.S. government agency.

Many States in the U.S. have a general privacy act that mirrors the federal government's Privacy Act. In general the state acts are designed to control the information that a state agency or local government may gather on individuals and how these may be used. Also, most states have separate acts that address the protection of privacy in specific situations. A selection of privacy-related legislation in the U.S. is summarised in Table 4.1.

Table 4.1 Summary of selected 'privacy' protection legislation in the U.S.

Legislation	Subject matter
Privacy Protection Act of 1980, 42 U.S.C. § 2000	Protects the privacy of communications and publication materials[86]
Computer Matching and Privacy Protection Act of 1988, 5 U.S.C. § 552	Prevents the creation of large dossiers on individuals by making of government databases difficult
Electronic Communications Privacy Act of 1986 (ECPA), 18 U.S.C. § 2510–22, 2701–11	Federal legal protection against unauthorised interception and disclosure of electronic communications, including e-mail
Fair Credit Reporting Act of 1970, 15 U.S.C. § 1681	Regulates information maintained by credit bureaus
Fair Debt Collection Practices Act of 1977, 15 U.S.C. § 1692–92	Prevents debt collectors from disclosing information about a consumer's debt to third parties
Right to Financial Privacy Act of 1978, 12 U.S.C. § 3401–22	Regulates the federal government's access to financial information held by a financial institution
Gramm–Leach–Bliley Act of 1999, 15 U.S.C. § 6801–10	Requires financial institutions to provide consumers with notice and an opportunity to opt out before certain types of personal financial information to non-affiliated third parties
Telecommunications Act of 1996, 47 U.S.C. § 222	Protects transactional information concerning telephone calls, including amount of usage and the destination of calls

---

[86] *Steve Jackson Games v United States Secret Service*, 816 F.Supp. 432 (W.D. Tex. 1993), *aff'd*, 36 F.3d 457 (5th Cir. 1994).

Cable Communications Policy Act of 1984, 47 U.S.C. § 551	Requires cable operators to disclose to their customers what types of personally identifiable information (PII) they collect, and how they use and disclose such information
Children's Online Privacy Protection Act of 1998 (COPPA), 15 U.S.C. § 651–05	Regulates the collection, use and dissemination of personal identifying information obtained on-line from children under 13
Drivers Privacy Protection Act of 1974, 18 U.S.C. § 2721	Regulates the disclosure and resale of personal information contained in records maintained by state Departments of Motor Vehicles
Video Privacy Protection Act of 1988, 18 U.S.C. § 2719	Limits disclosure of records of videotape rentals
Health Insurance Portability and Accountability Act of 1996 (HIPPA), 42 U.S.C. § 1320	Limits the disclosure of individually identifiable health information by health care providers, health plans, and health care clearinghouses, without the consent of the individual
Family Education Rights and Privacy Act of 1974, 20 U.S.C. 1232	Regulates the handling of student records by educational institutions that receive public funds
General Education Provisions Act of 2002, 20 U.S.C. § 1232	Provision requires school boards to provide parents with an annual notification on the collection, disclosure, or use of student personal information that may be used for marketing purposes or the selling of that information
Location Privacy Protection Act of 2001	To protect the privacy of individuals that use Internet-ready devices that can pinpoint a person's location[87]

4.3.9 Common Law Privacy in the United States

In the U.S. the common law of tort is more developed and formalised. Four activities are said to give rise to liability for the invasion of privacy. These are: (1) intrusion upon seclusion; (2) appropriation of name or likeness; (3) publicity given to private life; and (4) publicity placing a person

---

[87] Interestingly, it appears that Congress is now recognising a right to privacy of location information. See Schilit, B, Hong, J and Gruteser, M 2003 'Wireless location privacy protection' at http://www.computer.org/computer/homepage/1203/invisible/rz135.pdf. See also http://www.allnetdevices.com/wireless/news/2001/07/13/location-privacy.html and http://www.techlawjournal.com/cong107/privacy/location/s1164is.asp.

in a false light.[88] Some states, however, do not recognise such claims; for example, New York does not have a false light claim provision.[89] Other states protect a larger class of persons and private persons such as California and New York laws on misappropriation of name or likeness.

However, sometimes the public interest may override private interests. In *Stern v Delphi Internet Services Corporation*[90] the controversial radio talk show host Howard Stern who announced his candidacy for governor of the state of New York, sued Delphi on-line systems for using his photograph without permission in an advertisement for a subscriber-participation debate on his political candidacy. Stern claimed that this was in violation of New York privacy laws. The court, however, held against Stern because it argued that the 'Delphi bulletin board, like a letter-to-the-editor column of a newspaper, is a protected First Amendment activity' and '[t]he newsworthy use of a person's name or photograph does not give rise to a cause of action under [New York's privacy law] as long as the use is reasonably related to a matter of public interest'.[91]

An invasion of privacy is thus a '... wrongful intrusion into one's private activities, in such a manner as to cause mental suffering, shame or humiliation to a person of ordinary sensibilities'.[92] Individuals or corporations who wish to withhold themselves or their property from public scrutiny is supported in equity if there is no remedy at law.[93]

4.3.10 Evolving Fair Information Privacy Principles

In the U.S. it appears that the law is slowly coming to recognise information privacy as an important interest. Privacy *per se*, serves several valuable functions and more generally it is the ability to control what other people can know about you. The right to keep identity information secret serves to help protect the individual from stalkers, abusive ex-spouses, and others whose company that individual may wish to avoid. In the on-line world, privacy makes identity theft—the wrongful use of a person's identifying information to obtain goods and services fraudulently—less likely, and anonymity enables an individual to blow the whistle on wrongdoing without fear of retribution. Thus, anyone handling personal

---

[88] *Restatement (Second) of Torts* § 652A.

[89] See *Howell v New York Post Co.*, 596 N.Y.S.2d 350; 612 N.E.2d 699 (Ct.App. 1993).

[90] 626 N.Y.S.2d 694 (Sup.Ct. 1995).

[91] *ibid.* at 698.

[92] *Shorter v Retail Credit Co.* D.C.S.C., 251 F.Supp. 329 at 330.

[93] *Federal Trade Commission v American Tobacco Co.* 264 U.S. 298, 44 S.Ct. 336, 68 L.Ed. 696.

information would need to respect the emerging 'right' to privacy and its protection, and other competing interests that may require the free flow of information.

Alpert and Haynes (1994)[94] believe that because GIS as a tool can be used in so many different contexts it might be difficult to evolve meaningful rules or remedies. For example, fair information practices that underpin many of the privacy laws may be inadequate to combat the data manipulation of the different databases that takes place within GIS. Where the parameters may be strictly construed there may be no problems, however, when there are no strict controls privacy problems are more than likely to arise. Alpert and Haynes (1994) use as their example the intersection of GI and intelligent transportation systems (ITS). While no personally identifiable data may be generated from the ITS applications personal privacy may be implicated. For example, when the spatially arrayed data on, say, travel patterns of vehicle origins and destinations are matched with geographic coordinates, such as a zip code of the travel points, together with geodemographic data, a reasonably accurate profile of a distinct demographic segment of the travelling population may be obtained.

Over the past quarter of a century, governments in Australia, U.S., Canada and Europe have examined and analysed their 'information practices' and the safeguards for the collection and use of personal information, and the adequacy of privacy protection. Fair information practice principles were comprehensively articulated in the U.S. Department of Health, Education and Welfare's seminal report entitled *Records, Computers and the Rights of Citizens* in 1973. Since then fair information practice principles have evolved in various government and quasi-government agency studies and reports. These include the Privacy Protection Study Commission's 1977 report *Personal Privacy in an Information Society*; the Organisation for Economic Cooperation and Development *OECD Guidelines on the Protection of Privacy and Transborder Flows of Personal Data* (1980), the Information Infrastructure Task Force, Information Policy Committee, Privacy Working Group *Privacy and the National Information Infrastructure: Principles for Providing and Using Personal Information* (1995); the U.S. Department of Commerce *Privacy and the NII: Safeguarding Telecommunications-related Personal*

---

[94] Alpert, S and Haynes, KE 1994 'Privacy and the intersection of geographical information and intelligent transportation systems' at http://www.spatial.maine.edu/tempe/alpert.html. See also Alpert, S 1994 'Privacy on Intelligent Highway: Finding the right of way', *Santa Clara Computer and High Technology Law Journal*, v. 11, p. 1.

Information (1995); the *European Union Directive on the Protection of Personal Data* (1995); and the Canadian Standards Association *Model Code for the Protection of Personal Information: A National Standard of Canada* (1996).

Core Principles of Fair Information Privacy Protection

Five core principles of fair information privacy protection may be gleaned from the various codes noted above; and these are Notice–Awareness; Choice–Consent; Access–Participation; Integrity–Security; and, Enforcement–Redress.[95]

1. *Notice–Awareness* is a fundamental principle for fair information privacy protection. Individuals should be given notice of an entity's information practices before any personal information is collected from them. Without notice, an individual cannot make an informed decision as to whether and to what extent to disclose personal information.

2. *Choice–Consent* is the second widely accepted core principle. Choice means giving an individual an option as to how any personal information collected from them may be used. More particularly, this choice relates to the secondary uses of the information and is beyond what may be necessary for the purposes in the first instance. In general, there are opt-in and opt-out regimes. In the former an affirmative step is required to permit the collection and/or use of information; opt-out regimes require affirmative steps to prevent the collection and/or use of such information. The distinction between the two regimes is the default rule when no affirmative steps are required of the individual.

3. *Access–Participation.* The third core principle refers to an individual's ability both to access data and to correct the accuracy and completeness of the data. Both access to and participation in the collection of the data are essential in ensuring that the data are accurate and complete. The access processes must be timely, inexpensive, simple with a mechanism for verification of the

---

[95] See Information Infrastructure Task Force Privacy Working Group at http://www.iitf.nist.gov/ipc/ ipc/ipc-pubpubs/niiprivprin_final.html; see also Federal Trade Commission (FTC), *Privacy Online: A Report to Congress* (1998) at http://www.ftc.gov/reports/privacy3/index.htm; and FTC 2000 *Privacy Online: Fair Information Practices in the Electronic Marketplace*, May at http://www.ftc.gov/os/ 2000/05/index.htm#22.

data, and a means by which corrections and objections may be recorded.

4. *Integrity–Security.* The data need to be accurate and secure. To ensure data integrity, collectors must take reasonable steps by using reputable sources and cross-checking data against multiple sources, providing individuals access to the data, and destroying stale data. Security of the data would include both the management of and technical measures to protect against loss, unauthorised access, use, and disclosure of the data.

5. *Enforcement–Redress.* It is generally accepted that these principles of privacy protection can be effective only if there are mechanisms in place to enforce them. Without such mechanisms the code become merely ideals rather than prescriptions for a course of action and hence will not ensure compliance with the core principles. Several methods to achieve enforcement include industry self-regulation, legislation that creates private remedies for individuals, and/or regulatory schemes enforceable through the courts.

Given below is an OECD Privacy Policy Statement Generator that gives guidance on conducting an internal review of personal data practices that are in use in an organisation and on developing a privacy policy statement. A questionnaire is included to learn about personal data practices in an organisation and a Help Section provides explanatory notes and practical guidance. Needless to say, use of the Generator does not imply any seal of approval or endorsement by the OECD of the privacy policy and statement developed by users.[96]

Practice Notes: Design Your Own Privacy Policy Statement

How to develop a privacy policy

Step 1. To ensure that answers to the questions contained in the Generator are accurate you need to know what your personal data practices are. Before completing the questionnaire, it is essential to undertake an extensive internal review of current personal data practices in your organisation. For example:

Continued on page 250

---

[96] See OECD Privacy Policy Statement Generator at http://cs3-hq.oecd.org/scripts/pwv3/pwhome.htm.

Continued from page 249

- do you collect personal data?
- what kinds and how are they collected?
- who is responsible for deciding what personal data are collected and why do you collect them?
- how are the data used, who controls the data after collection and are the data disclosed to third parties?
- how and where are the data stored, what standards apply to the data collection and use of data and do visitors have access to the personal data you have about them?
- how do you deal with visitor queries about their personal data?

Step 2. After reviewing your current personal data practices:

- you should review laws and self-regulatory schemes that may apply to your collection and use of personal data;
- it is recommended that you review current practices against such regulations and amend where necessary to ensure compliance.

Using the Generator to create a Privacy Policy Statement

Step 3. You are now ready to complete the Generator questions. The Help Section provides explanations of terms used, guidance on what is consistent with the OECD Privacy Guidelines and where appropriate additional information on national and international instruments. After completing the questionnaire, a draft privacy policy statement is automatically generated with proposed pre-formatted sentences based on answers you have given.

Assessing the Draft Privacy Policy Statement

Step 4. Make sure that:

- the draft privacy statement accurately reflects your organisation's personal data practices;
- the draft privacy statement complies with applicable national and international laws and self-regulatory schemes;
- errors are corrected and that the privacy statement reads smoothly.

Continued on page 251

Continued from page 250

Making your Privacy Policy Statement Available

Step 5. Once you are satisfied that your privacy policy statement accurately reflects your personal data practices and complies with applicable regulations, you need to consider how to make your statement publicly available.

Remember: once your privacy statement is publicly posted, you may be legally liable if you fail to abide by your privacy policy statement or if that statement does not comply with local laws.

Source: OECD Privacy Statement Generator at http://cs3-hq.oecd.org/scripts/pwv3/pwhome.htm.

Conclusion

This discussion of the protection of privacy developments in Australia and elsewhere point to a number of conclusions. First, as shown by the introduction of a legislative regime to cover the private sector, there is a clear indication that there is a fundamental shift and structural change in relation to the protection of privacy. In Australia, as in the U.S., it seems that judges have been loath to expand privacy tort law to apply to the area of data gathering, information warehousing, and data matching. It appears that the preferred forum is to have legislature determine whether, and to what extent, further rights should be determined for protecting the information privacy of individuals.

Second, the shift towards a legislative regime also suggests a different mindset. No longer will the issues revolve primarily on the collection of information by governments. Privacy is now an issue that is of concern for both public and private sector organisations who need to comply with consumer expectations as well as consumer protection.

Third, the rapid pace of technological developments make it imperative that both the legislators and corporations will need to work together to set rigorous standards for the protection of personal and informational privacy.

Fourth, the standards and codes of practice will be enforced by sanctions and penalties either emanating from the Privacy Commissioner or from industry codes of conduct and industry sanctions. But here again there is a divide between what happens in the U.S. and what happens elsewhere, say in the E.U. or in Australia. In the U.S. there is the tendency to restrict the personal information that government may collect and there are significant safeguards against privacy intrusions by government agencies. However, the commercial sector is given greater leeway in privacy protection of its customers and in what they are allowed to do with the information they have collected. In the U.S. individuals are responsible for protecting their own privacy interests relative to the commercial sector; there is a belief that imposing greater privacy restrictions on corporations will stifle economic efficiency. Moreover, there is a greater distrust of government power *vis-à-vis* commercial power. In addition there may be an inability to overcome industry resistance to privacy legislation proposed at state and federal levels.[97]

In the E.U. on the other hand there is a strong privacy protection mandate given by the Data Directive. This Directive may lead to greater harmonisation of data protection legislation across all member states when compared with the situation across individual states in the U.S. Furthermore, in Australia, the U.K. and Canada the respective Privacy Commissioners oversee the protection of privacy. This has been achieved by permitting and sanctioning industry codes, especially where data subjects are given access rights to correct faulty data about themselves and to legal redress in cases of loss or injury.

Finally, it needs to be acknowledged that the aim of achieving consistency in privacy protection across all jurisdictions may be an impossible task. What might be achieved are some form of compromise between jurisdictions—such as the Safe Harbour mechanism that exists between the U.S. and the E.U.—that may generate greater protection of privacy for all citizens at large.

4.4 Geospatial Technologies and Privacy Implications

As previously intimated this section examines the class of geospatial technologies that are in common use that rely on both spatial attributes and other information that 'populate' this data, namely personal and other feature information. Geospatial technologies may be used for tracking people, their shopping and travel habits, the places that they go to for recreation, for what duration and in some instances making an inference on the purposes of that event. In particular we will consider location-based

---

[97] See Onsrud, HJ 1998 'Access to geographic information in the United States' in *Free Accessibility of Geo-information in the Netherlands, the United States, and the European Community, Proceedings,* Delft, The Netherlands, 2 October, pp. 33–41.

services (LBS) that rely on the key ingredients of time and space. LBS may be considered to be no different from geodemographics, an information technology that enables marketers to predict behavioural responses of consumers based on statistical models of identity and residential location.[98]

LBS have become commonplace because of the use of geocodes and GPS and other mobile communication and tracking technologies. LBS inferentially involve the tracking of people through the use of credit card data that may result in profiling exercises, statistical modelling, and pattern analysis. More generally, GI science using such technologies may be misleading as to who we are, where we are, and what we have been doing either by way of speech, purchases or simply being at a location. There is one view that without legislation to curb the (mis)use of such data there would be chaos both in space and cyberspace. However, an equally compelling but opposing view is that there should be no legislation, but rather just self-regulation by industry itself.[99]

The basic ingredients of geospatial technologies include the spatial aspects of data and databases, maps and visual presentations, and, statistical modelling and inferential statistics. Together these produce 'information' that has been synthesised and manipulated to result in new insights and inferences that previously were either not thought of or which may have led to chance discoveries because of the unusual combination of facts. Purveyors of LBS and geodemographics would put these discoveries down to the 'science' and the predictive insights of the data in the systems that have been used.[100] But, precisely because of the integrative capabilities of geospatial technologies and the perceived strategic insights into consumer behaviour, there is the danger that the privacy of individuals may be compromised, either vicariously or by association. Such invasions of privacy may arise in a number of ways as highlighted in the discussions that follow.

4.4.1 Data Aggregation and Databases

Private sector commercial applications of both GIS and LBS are perhaps the fastest growing areas of business and this has fed the need for more

---

[98] Goss, J 1995 'We know who you are and we know where you live: The instrumental rationality of Geodemographic Systems', *Economic Geography*, v. 71(2) April, pp. 171–198 at p. 171.

[99] Westin, AF 1967 *Privacy and Freedom*, New York, NY: Atheneum; and Westin, AF 1971 *Information Technology in a Democracy*, Cambridge, MA: Harvard University Press.

[100] See Goss JD 1994 'Marketing the new marketing. The strategic discourse of Geodemographic Information Systems' in Pickles, J (ed.) *Ground Truth: The Social Implications of Geographic Information Systems*, New York: Guildford Press, pp. 130–170.

data. In order to maintain a competitive edge marketers need good databases to make their decisions, but also need to be able to handle geographical data efficiently.

Patterns, relationships and trends show up much better when depicted visually in graphs, charts and maps than just columns of numbers or text. EQUIS, developed by the National Decision Systems 'maintains a database of financial information for over 100 million Americans on more than 340 characteristics including age, marital status, residential relocation history, credit card activity, buying habits, credit relationships (by number and type), bankruptcies, liens. This information is updated continuously at a rate of over 15 million changes per day'.[101]

In the early 1990s Equifax National Decision Systems announced the introduction of Infomark-GIS – a fully integrated GIS specifically designed for marketing applications and decision-making. At that time it was noted that:

> Infomark-GIS ... provides sophisticated marketing analysis and mapping capabilities that enable marketers to easily integrate their own internal customer, sales, and operations data with more than 60 national marketing databases. With an intuitive point-and-click interface, Infomark-GIS users can quickly and easily overlay proprietary information on current market data, analyze options, evaluate 'what if' scenarios, and visualize results that are key to making better and faster decisions. In addition, Infomark-GIS offers optional automated applications that are customized to solve specific marketing problems common with the retail, restaurant, consumer products, utilities, and financial services industries.[102]

There are other U.S. companies that are engaged in the collection, processing and storage of data pertaining to individuals.[103] These firms obtain consumer information from credit bureaus, public records, telephone records, professional directories, surveys, customer lists and other data aggregators. The data are cleansed, using information on changes of names or address, whether opt-in or opt-out processes were part of the collection procedures and the data are used to develop products, including household lists and specialty lists. For example, arguably the three largest

---

[101] Curry, DJ 1992 *The New Marketing Research Systems: How to Use Strategic Database Information for Better Marketing Decisions*, New York, NY: John Wiley & Sons. Inc. p. 264.

[102] ESRI 1993 'GIS System for Marketing Applications Introduced', *ARC News*, v. 15(3) Summer 1993, p. 6; see also Equifax and National Decision Systems (1993) *InfoMark-GIS: Tomorrow's Technology for Today's Business Success*, Atlanta, GA: Equifax, Inc.

[103] A search of those firms classified under the Standard Industrial Classification (SIC) code 7374 for companies engaged in marketing and business research services yielded approximately 50 companies.

credit-reporting agencies in the US – EQUIFAX, EXPERIAN and Trans-Union – maintain information on approximately 190 million individuals.[104] The data are obtained from credit-granting businesses such as banks, credit card companies and other lenders; and to this is added information that is publicly available from public records. The data maintained include name, address, social security number, phone numbers, date of birth and employment information plus a detailed credit history. All this information in a credit report can be bought and sold under the *Fair Credit Reporting Act* 1970 because there is a 'legitimate business need'.[105] U.S. law prohibits the release of credit information to entities that do not meet the criteria and there appear to be additional limits on the dissemination of personal credit data under the *Gramm–Leach–Bliley* Act 1999.[106]

Information is also gathered in other ways. For example, Naviant Technologies captures purchasing habits and demographic information by providing services to companies whose customers register products on the Internet.[107] With a database of more than 17 million Internet-using households, Naviant helps clients develop traditional and Web-based direct marketing campaigns. There are also specialised providers of information. A company called ChoicePoint, for example, allegedly maintains the largest database of physicians, chiropractors, dentists, and orthodontists in the world.[108] The database is available on-line, where for a small fee it is possible to obtain information about a practitioner, including malpractice information and patient ratings. Gale's *Directory of Databases* is also a reference source for different kinds of databases that are commercially available for those companies that are either seeking to purchase a database or intending to build their own.[109]

In the U.S. and Canada reports of break-ins into commercial databases have raised the real risk of legal problems. Due diligence and security practices, if found wanting, can lead to legal liability and prosecutions. The recent theft of more than 13 000 confidential records from the Costa Mesa California-based EXPERIAN company, a major credit-reporting agency, highlights the liability issues for database

---

[104] See Associated Credit Bureau website for details at http://www.acb-credit.com.

[105] See definition under s 604 *Fair Credit Reporting Act* at http://www.acb-credit.com.

[106] Discussed previously *supra*.

[107] See Naviant's website at http://www.naviant.com/Products/elist/hthh.asp.

[108] See Choicepoint's website at http://www.choicepointinc.com.

[109] Gale Group 2001 *Gale Directory of Databases*, 1993–2001, Farmington Hills, MI: Gale Group.

companies.[110] In another report more than 1400 Canadians, mainly living in the provinces of British Columbia and Alberta have been notified of a major security breach at EQUIFAX Canada Inc. a national consumer-credit-reporting agency. Unauthorised access had been gained to personal credit files containing social insurance numbers, bank account numbers, credit histories, home addresses and job descriptions.[111]

Clarke (1999c) has reported on the development of InfoBase in Australia by Publishing and Broadcasting Ltd (PBL) in conjunction with an American firm Acxiom, a data warehouse that has personal data on 'almost every Australian'.[112] The announcement of the venture caused major concerns among journalists and politicians. The worry revolved around the consent requirements before using and/or disclosing any of its data. The proposal also highlighted the growing importance of the need for a privacy legislation covering the private sector that will ensure data quality, data security, and openness in the use of these kinds of databases containing personal information.

4.4.2 Regulation and Use of Databases

In view of the commercial market for data, databases and data aggregation services, Curry (1994) has suggested that the use of geospatial technologies will produce multi-faceted problems that would similarly require multi-dimensional solutions.[113] The concerns raised include the fact that the technologies consist of and promote the widespread availability of unregulated data. This leads to the difficulty of regulating data matching that must take place if the geospatial tools are to produce meaningful results. Further, geospatial technology is inherently visual, but this strength also exposes a major weakness in that it may produce map inferences that may be both statistically and ecologically fallacious. Finally, there is the altered expectation of privacy rights because the case law may promote an erosion of those aspects of life where a person can feel safe, secure from

---

[110] Vijayan, J 2002 'Recent breaches raise spectre of liability risks' at http://www.computerworld.com/printthis/2002/0,4814,71609,00.html.

[111] Suppa, C 2004 'Credit agency reports security breach' at http://www.computerworld.com/printthis/2004/0,4814,91319,00.html.

[112] Clarke, R 1999c 'The Packer/PBL/Acxiom InfoBase' at http://www.anu.edu.au/people/Roger.Clarke/DV/InfoBase99.html; and Grayson, I 1999 'Packer sets up Big Brother data store' at http://technology.news.com.au/news/4277059.htm.

[113] Curry, MR 1994 'In plain and open view: Geographic information systems and the problem of privacy' at http://www.spatial.maine.edu/tempe/curry.html.

search and surveillance, and most importantly keep private. For example, when individual pieces of information are combined into a spatially coded dossier or where Courts use criteria for judging violations that may change along with technological developments.

Safeguards are already in place to ensure that the information that governments and their agencies keep on individuals are handled properly following fair information privacy practices. Similarly, in the private sector such safeguards are governed by specific industry codes of conduct and more recently the *Privacy Act* was amended to apply to the private sector in Australia. It is an accepted fact that inaccuracies in personal information in databases are bound to occur whether kept in paper or electronic form. In most cases access to correct such inaccuracies are afforded to individuals *when* they find out that the records kept on them are in error. However, as information is distributed, re-distributed, re-packaged, traded and stored in ever greater quantities, it would be very difficult for individuals to find out first who or which agency may be holding and distributing inaccurate information about themselves and second how to dispute and correct the errors. An even more significant problem is the proliferation of databases so that the data on individuals become hidden in a greater number of sub-directories and sub-files that the chances of discovery that the data about them exists become even more remote.

Geospatial Technologies

Geospatial technologies are said to do best at the intersection of location, time and content. But each of these elements of location, time and content tends to produce tensions of their own. For example, in regard to letting people know where they may be and to keep this fact hidden because they do not wish to be found. Here we have a technology that is employing the power of the 'place', since this power has been proven in other applications. GIS-based geostatistical models using locations and space have studied the home range of animals, for instance. Similarly, geographical profiling may be applied to the 'home range' of predatory humans that they may have inadvertently patterned when stalking their potential victims. Leipnik *et al.* 2001 have reported on the important role of geospatial technologies in investigating and gathering evidence of a locational nature in order to convict a serial killer.[114] When the serial killer Robert Lee Yates was

---

[114] Leipnik, M, Bottelli, J, von Essen, I, Schmidt, A, Anderson, L and Cooper, T 2001 'Coordinates of a killer' at http://www.geoinfosystems.com/1101/1101spokane.html.

257

arrested in April 2000 in Spokane County, Washington State, he was reported to have told his wife to 'destroy the GPS receiver'. This was because there were incriminating data on it showing the 72 waypoints associated with several journeys that he had made in disposing the bodies of the victims. Indeed, this example shows the extent to which GIS and GPS technologies are permeating society and their use—both by law enforcement agencies and by criminals. However, a court appeal could challenge the validity and use of the GPS data. The argument could be that using the GPS to track suspects without their knowledge might involve an invasion of privacy rights and might not meet the legal test of finding the 'least obtrusive means' for police to gather information about a suspect. In counter-argument, sometimes consent needs to be conspicuously absent in cases where suspicions are not to be aroused prematurely.

The question may be asked: is there any special data protection or privacy issue associated with locational data? 'Sensitive personal data' is regarded as data that identify, among others, a person's ethnic background, religion, political affiliations or sexual habits. However, the location of a person is not considered sensitive personal data. Yet, when processing data on persons, especially when locations are involved, say in terms of visiting synagogues on a regular basis or particular areas of ethnic concentrations, it is arguable that sensitive data are being processed and unintended inferences made about a person's religion or ethnicity.[115] Where then do the data protection laws play a part in ensuring that a person's privacy is preserved? There is no notice given to the data subject that the data are being collected on location and are going to be used and analysed in a particular way. What the present discussion does illustrate, however, is that the issues raised are complex and multifaceted and are special to this particular technology's preoccupation with space.

4.4.3 Some Definitions: Location, Tracking and Dataveillance

Location

Most geographers will understand location to mean the relative positions of events and entities in space and in time. Location gives a description of a person's or entity's whereabouts, in relation to other known objects or

---

[115] Rowe, H and McGilligan, R 2001 'Data protection. Location technology and data protection', *Computer Law and Security Report*, v. 17(5), pp. 333–335.

reference points. Common examples are the x- and y-coordinates of that location in relation to the Earth's reference points according to a particular projection. The location of events and entities in time will often have important meanings when analysing the spatial distribution of entities or the tracking of events through space over time. A consideration of time has become even more important in the information technology (IT) age because the location of entities can now be determined to be both absolute, fixed in one spot, and relative, variable over space as a result of movement and changes in location. The data may be referenced to the 24-hour clock in the first instance. But, measures of location may also be available with varying degrees of timeliness, that is, the lag between an event taking place and the availability of that information to the entity undertaking the monitoring of that event.

Tracking

By tracking is meant the plotting of the trail or sequence of locations within a space that is taken by an entity over a period of time. The 'space' within which the entity's location is tracked is generally physical or geographical space, although it may be virtual, in that a person may have had successive interactions with different organisations. Tracking of entities may take place in real-time when a device is attached to the entity that emits a continuous pulse of transmissions, such as coordinates from a GPS unit. However, the data on tracking locations can also be captured and stored in some system for retrospective analysis of the paths taken sometime in the past. Data about an entity's past and present location may provide the necessary information to make inferences and projections of likely behaviour patterns sometime in the future. A good example is the case of traffic planning where the estimated volumes of flows over a period of time is based on past and present performance over a stretch of roadway.

Dataveillance

Data surveillance, abbreviated to dataveillance, is the systematic use of personal data systems in the investigation or monitoring of actions or communications of one or more persons. Clarke (1999a) identifies two separate classes of surveillance. *Personal surveillance* is the surveillance of identified persons for various purposes, including investigation or monitoring, or as a means of deterrence against particular actions by the person or particular behaviours of that person. *Mass surveillance* is the surveillance of large groups of people, again for the purpose of investigation or monitoring and aid in the identification of individuals who belong to a class of interest that the surveillance organisation has cause for concern. Mass surveillance may also be used for its deterrent effects. In general, therefore dataveillance involves the monitoring of data about individuals and their actions.[116]

In the following the focus is on those technologies that have been designed for location and tracking and its by-products such as those that have been co-opted or subverted for that purpose, hence dataveillance. The discussion will focus on home locations and addresses, movement data, transactional information and electronic communications.

4.4.4 Geospatial Technology Applications: Home Location

While geospatial applications based on remote sensing of the earth on regional scales, and the use of GIS in city planning are relatively well known, the applications of such technologies to home location are less prominent. In general, home location data may be used by utility companies to track usage of power, gas, and water, whereas benefit providers and licence administrators make use of a home address tag for administrative purposes. In Europe, as in some large Asian cities, such as Jakarta and Beijing, inhabitant registration systems and locality identity cards are employed to prevent large scale in-migration into the cities and as a means of monitoring the residential population within the city. On a micro-scale, building access is regulated by identity tags that permit access to only certain areas of a building depending on the 'security clearance' of that person.

Telecommunication services to the home via the double-twisted copper wire attached to the wall socket provide home location information to the Public Switched Telephone Network (PSTN) operated by the phone utility. All phone communications to and from the home address are recorded, stored, analysed and made available to others. Telephone traffic data may be analysed as call records and are used for billing and invoicing purposes. The data also give paired locational information of the origin and destination of calls that may be analysed for particular purposes. Given the physical connection between parties to a phone conversation, there arise also the possibilities of telephonic interception, eavesdropping

---

[116] See Clarke, R 1999a 'Introduction to dataveillance and information privacy, and definition of terms' in http://www.anu.edu.au/people/Roger.Clarke/DV/Intro.html.

and real-time call tracing. While these activities are normally court-sanctioned through a warrant to the police to help in their investigations, sometimes these activities could connote illegal activity by criminals.

Phone numbers are usually found in the White Pages Directory that gives details of names and addresses of the registered person. However, some numbers could be ex-directory or a 'silent' number that is unlisted with the names and addresses of the registered persons known only to a select few. More recently home telephone services have facilities such as caller id, calling line id (CLI), and calling number display (CND) that give information of callers, their numbers and the time of their call. Caller, line, and number identifications are now generally available and are used by telecommunication companies, law enforcement agencies and consumers as a means of screening calls to domestic users. These facilities provide pertinent information that is displayed on a handset to enable individuals to decide whether to pick up the telephone and answer the call or to screen such calls for later response. Also available are directory services known as 'reverse' White Pages that can find phone numbers and addresses simply on the basis of just providing the name of a person and the general area/suburb where they live during an inquiry. As will be noted the privacy implications of these developments can have wide impact where home address and location information are concerned.

Australia on Disk (AOD) is a directory of every residential and business phone number in the country. The disk has all 55 National Phone Directories with a total of 1.3 million business listings and 6.9 million residential listing and phone numbers, updated every six months. While there are obvious environmental benefits in not having to produce paper copies of telephone directories, there are equally potent privacy implications, especially the disk containing residential information.[117] However, as the product does not permit reverse searching, that is, if you have a phone number and you want to search to find out who owns it, the product is unable to answer the query. On the other hand, the product permits searches and printing of whole lists by whatever criteria have been specified, such as by post codes or target markets. Individually addressed letters can then be sent out or a phone marketing campaign launched. An additional product, the *AOD Mapper* presents data and maps with areas of interest colour-coded to show their significance. The maps could identify hotspots where the

---

[117] Each year Sensis/Telstra produces about 55 000 tonnes of directories or 18 million sets of Yellow Pages and White Pages of which 80% is recycled. But that still means about 11 000 tonnes go into land fills each year. See Lowe, S 1994 'Indecent disclosures', *The Sydney Morning Herald*, 21 March, p. 47.

residents are most likely to match particular profiles, for example, suit-buying yuppies, high-income earners with young families, or low-risk, single, professionals.[118]

4.4.5 Tracking Movements of Individuals in Space

The tracking of individuals over borders has normally been controlled and monitored at immigration desks by checking and stamping national passports and the use of identity cards. More recently such travel documents have electronic chips embedded in them that permit electronic scanning and give permission to enter or pass the checkpoints. The data may yield patterns of entry and departure of citizens and visitors alike. At a micro-scale movements within buildings may be monitored using video surveillance equipment. In combination with video evidence movement over time and over space within buildings, analysis using pattern recognition and/or pattern matching algorithms may yield greater insights to both the most trafficked areas as well as identifying strategic locations for the placement of information or notice boards. While the system may provide greater effectiveness to the security company securing this particular building, it has significant implications for privacy because of the technology's intrusiveness.

The U.S. Department of Transport (DOT) Intelligent Transportation Systems (ITS)[119] program has made extensive use of GIS technologies, along with surveillance and other computer technology, to provide in-vehicle mapping and both lagged and real-time transportation management services. 'ITS represents perhaps one of the most ambitious applications of GIS and other technologies interacting simultaneously to make for an improved transportation system.'[120] As reported in Alpert and Haynes (1994) the DOT's vision for ITS is 'a future of safer and better informed travelers, improved traffic control systems, and systems aimed at increasing the efficiency of commercial vehicle and transit operations. The safety of

---

[118] Due to an ongoing litigation between Sensis (Telstra) and Australia On Disc, both parties reached an out of court settlement which has forced the publisher of AOD to immediately stop distributing the product as of 29 December 2003. See http://www.bsgi.biz/aodspecial.htm; and Hull, C 1994 'Privacy question in phone-book CDs', *The Canberra Times*, 1 August, p. 13.

[119] Also known as Intelligent Vehicle Highway Systems (IVHS), and used interchangeably here.

[120] Alpert, S and Haynes, KE 1994 *op. cit.* See also Alpert, S 1994 *op. cit.*; and Agre, PE and Harbs, CA 1994 'Social choice about privacy: Intelligent Vehicle Highway Systems in the United States', *Information Technology & People*, v. 7(4), pp. 63–90 and at http://www.mcb.co.uk/services/articles/liblink/ipt/agre.htm.

highway travel will be significantly increased through products which ensure the driver's state of fitness, enhance driver perception, warn of impending danger, and intervene with emergency control to prevent accidents from occurring'.[121]

The ITS concept is one where there is an interactive link of a vehicle's electronic system with roadside sensors, satellites, and a centralised traffic management system to constantly monitor each vehicle's location and the traffic conditions. More advanced systems would include the receipt of alternative road information in real-time via two-way communications, on-board video screens and mapping systems. While few will argue with the efficacy of this system in regard to public safety and convenience, there are other unintended outcomes that must be considered. Policy makers and developers of such a system must contend with its impact on individual privacy. While the data culled from ITS may provide individual travel patterns it has the potential to know where individuals travel, what routes they take, and, their travel duration. However, it may also be possible to predict 'when someone is or is not at home; where they work, spend leisure time, go to church, and shop; what schools their children attend; where friends and associates live; whether they have been to see a doctor; and whether they attend political rallies'.[122] While ITS can perform surveillance of travellers on highways, the collection and the use of ITS information may be seen to threaten the privacy interests of road users.

Roger Clarke reported the 1998 case where several residents of Sydney were charged with the murder of a diplomat in Canberra.[123] A key factor was that the person's car had passed along the Hume Highway between the two cities on the night in question and had been detected on the Safe-T-Cam system operated by the N.S.W. Roads Traffic Authority (RTA). A further example is the completion of a segment of Melbourne's inner-city ring road network in 1999 and implementation of a toll-road (Clarke 2000a). A corporation, Transurban that collects fees using an electronic tag affixed to a car's windscreen, operates the Melbourne CityLink. The passage through the toll-gates is recorded and a fee charged to the tag-owner's account. Thus, a previously anonymous journey on the toll-road has now been converted into an identified one because the only

---

[121] DOT Plan 1992 'IVHS Strategic Plan—Report to Congress', 18 December 1992 as cited in Alpert and Haynes, 1994 *op. cit.*

[122] *The Privacy Bulletin* 1990 Special Issue, August, v. 6(2), Sydney, Australia, p. 2.

[123] Clarke, R 2000a 'How to ensure that privacy concerns don't undermine e-transport investments' http://www.anu.edu.au/people/Roger.Clarke/EC/eTP.html.

way of obtaining the e-tag is to open an account with an identifiable individual or company. This now means that it is possible for highway surveillance and pattern recognition to be valuable supervisory tools for regulators as well as investigative and forensic tools for law enforcement agencies.

4.4.6 Tracking Transactions

A variety of methods may be used to capture transaction data. These include cheques that carry data in MICR (magnetic ink character recognition) and turnaround documents where the form has already been filled-in automatically for the client to authenticate and return, for example, OCR (optical character recognition). Other means of data capture are magnetic strips, embossed data codes, bar coding, RFID (radio-frequency identity tags) or a device with a location identity such as a phone socket. The main applications are in financial transactions from deposits, to loan repayments, salaries, cash withdrawals at automatic teller machines (ATMs), use of credit/debit cards and electronic funds transfer-point of sale (EFTPOS) and others. The observation to be made is that unrecorded and/or anonymous activities are now being converted into recorded/identified transactions. Even more important is that the data are being aggregated in a far more intense manner than before and these data have a location tag associated with it as well as a data trail. There are also real-time locater mechanisms built into these electronic transaction tools so that passive monitoring and surveillance can be turned into law enforcement.

4.4.7 Tracing Communications

While locational information and address tags may be readily available when using the PSTN, mobile telephony services, including the use of pagers, analogue or digital phones, and satellite phones make the tracing of persons and locations difficult. With mobile telephony the tracing of a device and its usage, either in real-time or logged in message banks is more dynamic. Here surveillance is more difficult, but not impossible and there are developments to replace location-identifying numbers with personalised numbers, as a service to phone customers. When this happens the auto-reporting by mobile devices of their location can provide much more exact and timely surveillance data. Inevitably, the developments in telephony will facilitate person location and person tracing to an even more intrusive extent than before.

A global plan to give home phones its own e-mail address has raised privacy concerns. An alliance between the telecommunications industry and the Internet Engineering Task Force (IETF) is seeking to create a standard called *ENUM*.[124] ENUM is a protocol that is used to relate telephones, facsimile machines and other devices from a phone number to the Internet. The plan is to use the domain name system (DNS) for the storage of E.164 numbers. E.164 is the International Telecommunications Union (ITU) standard that defines the format of telephone numbers, specifically for international subscriber dialler numbers (ISDN).[125] Thus, for example +61 2 6201 5650 is interpreted as an international dial-code 61 for Australia, followed by the subscriber trunk dialling code of 2 for Canberra, followed by the local telephone number for that domain. Thus, with ENUM you have a single electronic access point or address. However, Clarke (2002) is not sanguine with this prospect because 'the effect of ENUM would therefore be to establish a single unique contact number for each individual. If it were successful, it would represent a unique personal identifier, with all of the threats to privacy and freedoms that this entails. Worse still, ENUM is to support mobile devices, and hence represents a location and tracking tool. Rather than just each person's data being subject to consolidation and exploitation, each person themselves would thereby be directly subjected to surveillance and interception'.[126]

A combination of national security, law enforcement and corporate marketing interests is pushing for the implementation to make handsets locatable to within a few metres; just as in the case of the location of callers to the emergency number 911 in U.S. or 112 in Europe.[127] In the U.S. when a user punches 911 from a traditional phone the screen at the answering centre displays the number of the caller, street address and a map of where that phone is located. However, cell or mobile phones

---

[124] See Internet Engineering Task Force website at http://www.ietf.org.

[125] IETF 2000 'E.164 Number and DNS' RFC 2916 September 2000 at http://www.ietf.org/rfc/rfc2916.txt.

[126] Clarke, R 2002 'ENUM' at http://www.anu.edu.au/people/Roger.Clarke/DV/enum.html; and at International Telecommunications Union (ITU) 2001 'ENUM' at http://www.itu.int/osg/spu/enum/index.html.

[127] The 112 emergency number is incorporated in international global system for mobile communications (GSM) and can be dialled from anywhere in the world where there is GSM coverage with the call transferred to that country's primary emergency call service number. The 106 emergency calls service number is for the exclusive use of text-based telecommunication users especially the hearing or speech impaired. See Australian Communications Authority (ACA) 2004 *Location, Location, Location*, January, Melbourne: ACA.

may be on the move and change locations and the phone numbers do not correspond to a fixed address. There was a reported case of a Florida woman who died in February 2001 after her car sank in a canal off a Florida turnpike. It was reported that she had dialled 911, but was unable to tell the operator where she was.[128] The U.S. Federal Communications Commission (FCC) has now imposed requirements on wireless carriers to provide emergency call services with far more precise locational identifiers. Handset-based solutions must be identified to within 50 m of a call while for network solutions the identification must be within 100 m.[129]

In contrast, the Australian Communications Authority (ACA) has adopted an interim Mobile Location Indicator (MoLI) that requires mobile carriers to provide emergency service location information about the caller's location to be in compliance with the draft ACIF Code.[130] The code specifies the identification of the call location to be within the 'standard mobile service area' (SMSA)—this may range from 2000 to 500000 km^2! However, a new 2004 draft code for the future use of location information to enhance the handling of emergency mobile phone calls proposes that the SMSA locational accuracy range between 50 and 500 m.[131] The tension here is to let rescuers know where they are on the one hand in emergency situations and concealing this fact from a stalker on the other.

4.4.8 Convergence of Locational and Tracking Technologies

By far the technology of greatest relevance is GPS, depending on a constellation of satellites to give positional information in four dimensions—latitude, longitude, altitude and time. With the Presidential edict of turning off selective availability—the purposeful degrading of positional information—users of GPS are now able to poll satellites for positional information and be given references to within a metre of their location.[132]

---

[128] See Ho, D 2001 'Cellular phone carriers buy more time to develop emergency location technology' at http://www.nando.net/technology/v-text/story/124854p-1312017c.html.

[129] See FCC requirements at http://www.fcc.gov/911/enhanced/.

[130] Australian Communications Industry Forum (ACIF) 1999 'Mobile location indicator for emergency services' DR ACIF G530.

[131] See ACA 2004 *op. cit.*

[132] The White House, Office of the Press Secretary 2000 'Statement by the President regarding the United States' decision to stop degrading global positioning system accuracy', 1 May at http://www.ostp.gov/html/0053_2.html.

Differential GPS (DGPS) uses the same technology, except that locations are determined as a differential to the data received relative to a surveyed point on the ground. Hence, the accuracy obtained by DGPS methods is quite precise. Assisted GPS (AGPS) technology, on the other hand, has been developed in conjunction with information communication technology which uses a server at a known geographical location in the network. This information reduces the time, complexity and power required to determine location.

Recent technologies include the use of smart cards, transponders, biometrics, GPS/DGPS and video surveillance with close circuit TV (CCTV). Smart cards now have computer chips that provide electronic storage of information and in some cases minor processing capabilities. Transponders are devices that, on receiving a particular signal, respond automatically with details of any kind of a message. The transponder may simply re-transmit a signal, a predetermined message, or a dynamic message response, depending on context upon instructions given by a program. Identifiers and locational indicators are typical transponder transmissions. Biometrics depend on measures of some aspect of a person, or some behaviour of the person, in order to either identify or authenticate their identity.

With the advent of wireless communications location information has come under greater scrutiny. In the U.S. the *Telecommunications Act of 1996* included location information as Customer Proprietary Network Information (CPNI) that gave time, date, and duration of a call, and the number dialled. The regulation does not specify what kind of customer consent—express prior consent to use (opt-in) or not to use (opt-out)—is required for CPNI. However, the *Wireless Communication and Public Safety Act of 1999* (WCPSA or E911 Act) rectified the omission by requiring opt-in for location information used for any non-emergency purpose. In the E.U. the *Directive on Privacy and Electronic Communications* (2002/58/EC) established a technology-neutral legal standard for privacy protection in the processing of personal data for all electronic communications.[133] Article 9 requires informed opt-in consent for the provision of telecommunication services based on the use of location information. Subscribers must be able to withdraw consent for the collection or processing of their location information without charge at any

---

[133] Directive 2002/58/EC of the European Parliament and of the Council of 12 July 2002 concerning the processing of personal data and the protection of privacy in the electronic communications sector at http://europa.eu.int/comm/internal_market/privacy/law_en.htm.

time. By the end of 2003 only four countries: Denmark, Sweden, Finland and Spain, have implemented the directive.[134]

RFIDs is an abbreviation for radio frequency identification, a technology similar to bar code identification. With RFID the electromagnetic or electrostatic coupling in the radio frequency portion of the electromagnetic spectrum is used to transmit signals. RFID systems can be used just about anywhere, from clothing tags to missiles to pet tags to food—anywhere a unique identification system is needed. The tag can carry information as simple as a pet owner's name and address or the cleaning instruction on a sweater and as complex as the instructions on how to assemble a car.[135]

4.4.9 Privacy Risks with Location and Tracking Technologies

There used to be a saying that 'a man's home is his castle'. However, the home as a bastion of privacy is slowly being eroded not only in regards to a 'physical' loss, but also in terms of loss of privacy of the airwaves, thermal emissions from the property as well as visual monitoring. In Real Property Law, the Latin saying *cujus est solum ejus est usque ad coelum* [whose is the soil, his is also that which is above it] suggests that the absolute owner of real property has domain over the land, to the centre of the earth and up to the sky above it. But today that is no longer the case since in common law countries of the English tradition the Crown has claim over any water, minerals and precious metals that lay beneath the land. In addition, owners in possession of land no longer have any claim over the air space above their land parcels, and are subject to any restrictions contained in a lease, covenant or other agreement.[136]

With geospatial technologies there has already been litigation on the basis of such 'trespass' to land properties. The basis for this litigation include the idea of an 'objective' expectation to privacy as may be found,

---

[134] In Japan the Guidelines on the protection of personal data in telecommunications business established a clear standard for consent to use of location information. In May 2003 the Diet passed a package of bills known as the Personal Data Protection Law codifies the requirement for informed opt-in consent. See Ackerman, L, Kempf, J and Miki, T 2003 'Wireless location privacy: Law and policy in the U.S., E.U., and Japan' at http://www.isoc.org/briefings/015/index.shtml.

[135] See Webopedia definition of RFID at http://www.webopedia.com.

[136] See Bradbrook, AJ 1988 'The relevance of the *cujus est solum* doctrine to the surface landowner's claims to natural resources located above and beneath the land' (1988) 1 Adel LR 462. See also reference to this issue discussed in Chapter 1 Section 1.4.4 in relation to 'corner jumping' and the use of GPS in the U.S.

for instance, in the Fourth Amendment to the U.S. Constitution or a 'subjective' expectation as provided by case law.

In *Dow Chemical v United States*[137] Dow Chemical claimed that the capture and collection of aerial photo imagery over their site was an invasion of its privacy and a violation of their Fourth Amendment rights. Although the District Court held that the aerial photography was a 'violation of Dow's reasonable expectation of privacy and an unreasonable search in violation of the Fourth Amendment', the U.S. Supreme Court held that the 'open field' doctrine applied to the case, and therefore there was no invasion of privacy. Similarly in *California v Ciraolo*[138] the Supreme Court found that it was acceptable for the police to fly over fenced-in backyard at an altitude of 1000 feet to undertake monitoring. Also in *Florida v Riley*[139] a court approved the use of a helicopter, hovering at 400 feet, to observe marijuana plants through a hole in the roof of Riley's greenhouse. Here Justice O'CONNOR interpreted the 'reasonableness' in any expectation of privacy to mean that 'if the public can generally be expected to travel over residential backyards at an altitude of 400 feet, [then] Riley cannot reasonably expect his curtilage to be free from such aerial observation'.[140]

The remit to observe from the air has also extended to monitoring emissions. In *United States v Penny-Feeny*[141] a Hawaii District Court endorsed the police use in a helicopter of a forward looking infrared (FLIR) device to discern heat emissions from a garage to gather information on illegal activities. Beepers and identity tags have also been endorsed to track the location of individuals in motor vehicles and then use GIS to map and trace routes.[142]

However, sometimes the permission to monitor is more narrowly defined. In *Smith v Maryland*[143] which clarifies the landmark 1967 case of *Katz v United States*,[144] the case at hand concerned the installation of a pen register by the telephone company that recorded the telephone numbers that had been dialled from the petitioner's home. Justice HARLAN's concurring opinion argued that a search may be carried out under the terms of the

---

[137] 106 S.Ct. 1819, 90 Led 2d 226 (1986).

[138] 106 S.Ct. 1809 (1986).

[139] 488 U.S. 445 (1988).

[140] *ibid.* at 455. Compare this with *Streisand v Adelman* Case No. SC 077 257. Cal. W.D. 31 December 2003 at http://www.californiacoastline.org/streisand/slapp-ruling.pdf discussed in Chapter 1 *supra*. part III on the nexus between GI and law and the subsection on aerial photographs and images.

[141] 773 F.Supp. 220 (D. Haw. 1991).

[142] *United States v Knotts* 460 U.S. 276 (1983); *United States v Karo* 468 U.S. 705 (1984).

[143] 442 U.S. 735 (1979).

[144] 389 U.S. 347 (1967).

Fourth Amendment when two requirements are met. First, that a person has exhibited an actual (subjective) expectation of privacy; and second, that the expectation be one that society is prepared to recognise as 'reasonable'. The Court concluded, 'we doubt that people in general entertain any actual expectation of privacy in the numbers they dial'.[145] Also, in a similarly narrowly defined case of *United States v Place*[146] the court concluded that because the drug-sniffing dogs in question were capable of discovering only one thing, here cocaine, the privacy interests of the people involved were not threatened. This was because the information, as in the pen register case, was so limited.

Finally in *United States v Smith*[147] a U.S. Court of Appeals for the Fifth Circuit has ruled that technological advances may be capable of expanding the legally protected range of privacy that individuals enjoy. In 1990 David L. Smith lived with his girlfriend in Port Arthur, Texas. His next-door neighbour had experienced problems with break-ins to his garage and cars, and suspected Smith. Using a Bearcat scanner designed to scan and intercept 400 'radio-type' channels the neighbour began to intercept Smith's home phone conversations made over a cordless phone. While not finding anything to suggest the break-ins the neighbour discovered Smith was involved in drug dealing. Police then participated in the interception and recording of Smith's calls. Smith was convicted on the facts, but on appeal argued that the interception of his telephone calls violated the Fourth Amendment, which bars unreasonable searches and seizures. The court stated that in order to establish such a violation he had to show 'that a government activity intruded upon a reasonable expectation of privacy in such a significant way that the activity can be called a "search"'. The Appeal Court held that while the trial court had erred in assuming that there could never be a reasonable expectation of privacy for a cordless telephone conversation, nevertheless the court affirmed Smith's conviction. This is because he had failed to discharge his burden in showing that his Fourth Amendment rights were violated.

4.4.10 Privacy-invasive Technologies (PIT): Privacy-enhancing Technologies (PET) and Privacy-sympathetic Technologies (PST)

The literature suggests that there may be three different types of technologies that either invade or enhance the privacy of individuals. Some are

---

[145] *ibid*. at 743.
[146] 462 U.S. 696 (1983).
[147] No. 91–5077 5th Cir. Nov. 12, 1992.

expressly privacy-invasive technologies (PIT) including 'data-trail generation through denial of anonymity, data-trail intensification as in identified phones, stored valued cards (SVC), and intelligent transport systems (ITS), data warehousing and data mining, stored biometrics, and imposed biometrics'.[148] Then, in the last decade or so technologies have emerged that are expressly designed as privacy-enhancing technologies (PET), seemingly as a reaction and bid to reverse trends in technological developments that have hitherto been privacy-invasive.

Tools that have been developed to assist the protection of privacy interests include those that make individuals genuinely untraceable and anonymous. Examples include the Electronic Privacy Information Centre's (EPIC) list of privacy-enhancing tools, Cranor (2001) and Cranor *et al.* (2003) who report on a World Wide Web Consortium's (W3C) Platform for Privacy Preferences Project (P3P).[149] P3P enables websites to express their privacy practices in a standard format that can be retrieved automatically and interpreted easily by user agents ('robots' or 'bots'). Users will then be informed of the privacy practices at that site and then to automate decision-making based on these practices when appropriate. Thus, users need not read the privacy policies at every site they visit.

The third type of technology, labelled privacy-sympathetic technologies (PST) is capable of delivering genuine anonymity. Such PSTs are further subdivided into anonymity services, pseudonymity services and personal data protection.[150] These services use various devices that provide outright anonymity to those that encrypt identity tags which may be shared with selected parties only. In between there are pseudo-anonymous devices that give a semblance of anonymity, but from which identities may be traced. The warning for using such technologies is that

---

[148] See Clarke, R 1999b 'The legal context of privacy-enhancing and privacy-sympathetic technologies' at http://www.anu.edu.au/people/Roger.Clarke/DV/Florham.html.

[149] See EPIC Online Guide to Practical Privacy Tools at http://www.epic.org/privacy/tools.html; Cranor, LF 2001 'The role of privacy enhancing technologies' at http://www.cdt.org/privacy/ccp/roleoftechnology1.shtml; Cranor, LF, Byers, S and Kormann, D 2003 'An analysis of P3P deployment on commercial, government, and children's websites as of May 2003', Technical Report prepared for the Federal Trade Commission (FTC) at http://www.research.att.com/projects/p3p; and World Wide Web Consortium (W3C) http://www.w3.org/P3P/.

[150] Anonymity services make use of identifiers for the purpose of records or transactions whereas pseudonymity services are those which cannot, in the normal course of events, be associated with a particular individual. When a person uses a pseudo-identifier then a digital persona, e-pers or a nym is born; this is the model of an individual's published personality, based on data, maintained by transactions and intended for use as a proxy for that individual. See Clarke, R 1999b *op. cit.*

one should always be conscious of maintaining a balance between privacy and other interests, such as accountability.

To assuage consumer privacy concerns both legislators and those in the geospatial technologies industries are battling to find solutions that provide PET and PST while at the same time preserving the functionality of the technology, even if it may have the effect of PIT. The effects of a perceived invasion of privacy by a technology may have a 'reputation' effect on a company and potentially a loss of business. New technologies are becoming available to consumers to assist in enhancing privacy such as the P3P protocol. The protocol will provide standardised information in machine-readable form about a website's privacy policy and individuals may be able to configure their own browsers when transacting business on a website.[151]

Two concluding observations may be appropriate here. First, location information should be used only to provide services to users. It may be inappropriate for location information to be used for secondary purposes, such as marketing based on an individual's location or by government for law enforcement and national security purposes as it may create a society of 'dataveillance' without the necessary investigatory mandate. Second, while voluntary standards and codes of conduct may work well in Australia, U.K., Canada, the U.S. and elsewhere, the preferred route in the E.U. to stem the erosion of privacy is through the use of legislation and mandatory standards. However, within any jurisdiction the strong message is that laws and regulations specifying how consumers authorise access to their privacy need to be clear, consistent and technology-neutral. A summary table showing privacy risks in the various jurisdictions and parallel legislation is given as Table 4.2.

4.5 Emergent Policy and Practice

The use of legislation as a means of protecting privacy and the soft-touch self-regulatory approach to such protection may reflect differences in cultures, histories and philosophies. It may be that for economies such as Australia, U.K., Canada and the U.S. there is reluctance for governments to interfere with market forces and so opt for minimal legislation. Most of these countries use sectoral approaches that rely on a mix of legislation,

---

[151] Two commercial on-line Privacy Policy Editor services to help sites meet the current P3P specifications are available at http://www.p3pwriter.com and http://policyeditor.com.

Table 4.2 Summary of privacy risk status and geospatial technologies

Geospatial technologies related to	Applications	Privacy risk status	Legislation/jurisdiction	Significant references: fair information Practice
Location:				
Home address: Fixed	Utilities, benefit providers, licence administration	N/I	*Australia* *Privacy Act 1988* *Freedom of Information Act 1982*	U.S. Department of Health, Education and Welfare 1973 *Records, Computers and the Rights of Citizens* U.S. Privacy Protection Study Commission 1977 *Personal Privacy in an Information Society*
	Inhabitant registration systems	I	*Data-Matching Program (Assistance and Taxation) Act 1991*	OECD 1980 *OECD Guidelines on the Protection of Privacy and Transborder Flows of Personal Data.*
	White Pages, ex-Directory, silent numbers	E	*Privacy Amendment Act 1990*	U.S. Information Infrastructure Task Force Privacy Working Group 1995 *Privacy and the National Information Infrastructure: Principles for Providing and Using Personal Information*
	Reverse White Pages	E	*Privacy Amendment (Private Sector) Act 2000*	E.U. 1995 *Directive on the Protection of Personal Data*
Mobile	GPS/DGPS/AGPS	I	*U.S.* *Privacy*	Canada Standards Association 1996 *Model Code for the Protection of Personal Information: A National Standard of Canada*
	Auto-reporting mobile devices	I	*Privacy Act 1974* *Privacy Protection Act 1980* *Electronic Communications* *Privacy Act 1986* *Computer Matching and Privacy Protection Act 1988*	

Table 4.2 (Continued)

Geospatial technologies related to	Applications	Privacy risk status	Legislation/jurisdiction	Significant references: fair information Practice
			Financial/credit information	
			Fair Credit Reporting Act 1970	
			Fair Debt Collection Practices Act 1977	
			Right to Financial Privacy Act 1978	
			Gramm-Leach-Bliley Act 1999	
Movement			Telephony	
Individual	Trans-border (passports)	N/I	Cable Communications	
	Credentialed building access	I	*Policy Act 1984*	
			Telephone Consumer Protection 1991	
	Biometrics	I	*Telecommunications Act 1996*	
Mass	Transponders	N/I	*Wireless Privacy Protection Act 2003*	
	CCTV	I		
	Pattern recognition; pattern matching	N	Children	
	Intelligent Transport Systems and imposed identifiers (travel cards)	N/I	*Children's Online Privacy Protection Act 1998*	

		Video	Video Piracy Protection Act
Transactions			
Financial		Health	Health Insurance Portability and Accountability Act 1996
Cheque data with magnetic ink character recognition	N		
Turnaround documents with optical character recognition	N	Education	Family Education Rights Privacy Act 1974
Retail			General Education Provisions Act 2002
Encoding, magnetic strips	N		
Bar coding	I	Drivers Licence	Drivers Piracy Protection Act 1974
ATMs, EFTPOS, Credit/debit cards	I		
Loyalty schemes	E	Freedom of Information	Freedom of Information Act 1966
Smart cards, stored valued cards	I		
Real-time locater mechanisms		Location	Location Privacy Protection Act 2001

Table 4.2 (Continued)

Geospatial technologies related to	Applications	Privacy risk status	Legislation/jurisdiction	Significant references: fair information Practice
Communications				
Fixed	PSTN traffic data, call records	N/I	*U.K.*	
			Data Protection Act 1998	
	Real-time tracing, interception	I	*Canada*	
Mobile	Mobile telephony, pagers, analogue/digital/satellite phones/PDAs	N/I	*Personal Information Protection and Electronic Documents Act 2001*	
	Personal phone numbers, ENUM	N/I	*New Zealand*	
			Privacy Act 1993	
	Caller id, CLI, CND	I	*Japan*	
			Personal Data Protection Law 2003	
Convergent technologies				
Information technology	Voice data TCIP/VoIP	N/I		
	RFID computer wear	I		
	EPIB emergency beacon	N/I		
Mobile, on person	Identity tags (prisoners, children)	I		

Privacy risk status: E enhancing; I intrusive; N neutral

regulation, and self-regulation while the European approach is one of strict regulation. Moreover, a sceptical citizenry may be suspicious of, and loath to permit governments to have a greater control of private information than is necessary. The soft-touch approach will permit a middle ground with industry regulating itself and giving governments some control over privacy protection. Whether the marketplace or legislation will decide what is best for the protection of personal informational privacy is still an open question; but with globalisation as well as the commodification of personal data, the E.U. Data Protection Directive portends a different future. This Directive, discussed in the next section, marks a major shift in the shaping of global privacy protection and indirectly heralds the imposition of a foreign protective schema on the rest of the world.

4.5.1 European Union Data Protection Directive

The European Union promulgated a comprehensive privacy legislation that became effective on 25 October 1998, entitled the Directive on the Protection of Individuals with regard to the Processing of Personal Data and on the Free Movement of such Data[152] (E.U. Data Protection Directive). Among others, Art. 25 of the regulations requires that transfers of personal data take place only to non-E.U. countries that have an 'adequate' level of privacy protection. This is designed to prevent the circumvention of the Directive and the creation of 'data havens' outside the E.U.

Provisions of the Data Protection Directive

The Directive places limitations on the 'processing' of 'personal data'. Defining *personal data* broadly as any information relating to an individual or legal person, the Directive aims to develop a high level of data privacy protection within the E.U. But at the same time the E.U. seeks to promote the free flow of personal data by harmonising privacy laws member states. The focus of the E.U. Data Protection Directive is primarily on the private sector with a limited number of exceptions, including public security, defence, state security, law enforcement and personal or household activities.

---

[152] European Union Data Protection Directive 1995 *Directive on the Protection of Individuals with regard to the Processing of Personal Data and on the Free Movement of such Data*, Brussels: European Commission; Directive 95/46/EC; *Official Journal of the European Commission* (L 281) 31 and at http://europa.eu.int/eur-lex/en/lif/dat/1995/en_395L0046.html.

Several types of limitations have been placed on the processing of personal data by a 'data controller', defined as any person with a role in determining 'the purposes and means of the processing of personal data' (Art. 2(d)). First, there are limitations on what data may be collected and how it must be maintained. Personal data may be collected only for specified purposes, and may not be processed 'in a way incompatible with those purposes'. The quantity of data collected must not be 'excessive' in relation to those purposes. Data must be accurate and kept up-to-date. And data may be retained in a form that makes them identifiable to a person only for as long as necessary to accomplish the purposes for collection (Art. 6).

Second, there is a general consent requirement: processing of personal data is allowed only if 'the data subject has unambiguously given his consent' (Art. 7(a)). The requisite consent is strictly defined.

Third, there is a general prohibition against the processing of certain types of sensitive personal information, namely that 'revealing racial or ethnic origin, political opinions, religious or philosophical beliefs, [and] trade-union membership' (Art. 8(1)).

Fourth, the data controller must furnish certain information to the data subject. Information that is required to be disclosed includes the identity of the data controller, the purposes for processing the data, and whatever additional information is required 'to guarantee fair processing in respect of the data subject' (Arts 10, 11).

Fifth, data subjects have the right to obtain from the data controller 'at reasonable intervals and without excessive delay or expense' confirmation whether data concerning the subject are being processed, and the purposes of such processing (Art. 12(a)).

Sixth, data controllers must assure the confidentiality and security of the data they process (Arts 16, 17).

Seventh, a data controller must notify the national supervisory authority, which the Directive requires each E.U. member state to set up, before commencing any automatic processing of data (Art. 18).

Eighth, the Directive places limitations on a controller's transfer of personal data to a country outside the E.U. Transfers of personal data to a 'third country' are allowed only if the 'third country in question ensures an adequate level of protection to the data' (Art. 25(1)).

Before entering into an agreement with a foreign country to allow the circulation of personal data outside the E.U., an evaluation of the adequacy of data and privacy protection in that country has to be undertaken. This has already been completed with regard to Australia,[153] Switzerland, Hungary, the U.S.[154] and Canada.[155] In the case of the U.S. it has been a protracted decision since it related to a specific system applied in that country known as the 'safe harbour' principle discussed below.[156]

4.5.2 European Union–United States Safe Harbour Framework

To protect American business interests against possible interruptions in data transfer when dealing with European counterparts and to avoid potential prosecutions under European privacy laws, the U.S. Department of Commerce negotiated a 'safe harbour framework'. This system permits U.S. companies to satisfy the European 'adequacy' standard while maintaining their traditional self-regulatory approach to data protection. In July 2000 the European Commission approved the safe harbour framework as meeting the 'adequacy' standard. This framework provides some measure of predictability and continuity for U.S. companies transmitting personal information from the E.U. To qualify for safe harbour protection a self-certified business entity must notify the Department of Commerce in writing annually and declare publicly its published privacy statements that adhere to the Safe Harbour Principles summarised below.[157] The Department maintains a publicly available list of organisations that have self-certified.[158] Self-certification services have been offered by many companies, one of which is TRUSTe. In particular the company offers a third party the E.U. safe harbour privacy standard which consists of a verification that the company's privacy policy and practices

---

[153] See *Privacy Amendment Act* 2000 (Cwlth) approved 22 December 2000; http://www.privacy.gov.au.

[154] *Official Journal of European Commission* L 215 of 25 August 2000, pp. 1, 4, 7 respectively.

[155] See http://europa.eu.int/comm/external_relations/canada/summit_12_99/e_commerce.htm and *Official Journal* L002, 04/01/2002 pp. 0013–0016 and see also the E.U. 'adequacy' standard agreement at http://europa.eu.int/comm/internal_market/en/dataprot/wpdocs/wp39en.pdf.

[156] Safe Harbour Principles are available at http://europa.eu.int/comm/internal_market/en/dataprot/news/shprintiples.pdf.

[157] See Yu, P 2001 'An introduction to the EU Directive on the Protection of Personal Data', http://www.gigalaw.com/articles/2001/yu-2001-07a-p1.html. See also Harvey, JA and Verska, KA 2001 'What the European Data Privacy obligations mean for U.S. Businesses' at http://www.gigalaw.com/articles/harvey-2001-02-p1.html.

[158] See U.S. Department of Commerce list of agencies and companies that have self-certified at http://www.web.ita.doc.gov/safeharbor/shlist.nsf/web-Pages/safe+harbor+list.

are compliant with safe harbour principles. Companies that participate in TRUSTe's program are entitled to display TRUSTe's E.U. Safe Harbour privacy seal.[159]

Practice Notes: E.U.–U.S. Safe Harbour Principles

1. Notice. The business must clearly tell its customers why it collects their personal information, how it plans to use their personal information, whom they can contact with inquiries and complaints, the types of third parties to which the business intends to disclose their personal information, and the choices and means through which they can restrict the use and disclosure of such information.

2. Choice. If the business wants to disclose to a third party the personal information of its customers or to use such information in a way that has not been previously authorised, the business must give the customers an opportunity to opt out of such disclosure or use. For sensitive information—such as data revealing racial or ethnic origin, political opinions, religious or philosophical beliefs, trade union membership and information concerning health or sex life—the business must also provide an affirmative or explicit opt in procedure.

3. Onward transfer. To disclose personal information to a third party, the business must comply with the notice and choice principles. In addition, the business must limit its disclosure to third parties that subscribe to the Safe Harbour Principles or that are subject to the E.U. Directive or an 'adequacy' finding. Contracts ensuring that the third party will offer the same level of protection as required under the Safe Harbour Principles will satisfy this principle.

4. Security. The business must take reasonable precautions to protect personal information from loss, misuse and unauthorised access, disclosure, alteration and destruction.

5. Data integrity. All personal information must be relevant for the purposes for which it is to be used. The business must not process

Continued on page 281

---

[159] See TRUSTe *The TRUSTe EU Safe Harbor Privacy Program* at http://www.truste.org/webpublishers/pub_eu.html.

Continued from page 280

any personal information in a way that has not been previously authorised and should take reasonable precautions to ensure that data are reliable for their intended use, accurate, complete and current.

6. Access. Each customer should have reasonable access to the stored information about him or her. Customers should also have the opportunity to correct, amend or delete any inaccuracies, except where the burden or expense of providing access would be disproportionate to the risks to the customer's privacy or where the rights of other persons would be violated.

7. Enforcement. The business must provide an independent, readily available and affordable dispute resolution mechanism for investigating and resolving customers' complaints and disputes. It must also institute a procedure for independently verifying its compliance with the Safe Harbour Principles. In addition, the business must be committed to remedy problems arising out of its failure to comply with the Principles. To ensure compliance, the dispute resolution body must be able to impose sanctions that are sufficiently rigorous. Examples of such sanctions include publicity for findings of non-compliance, deletion of data, suspension from membership in the privacy program, injunctive orders and damages. A privacy seal program that incorporates and satisfies the Safe Harbour Principles will satisfy this principle.

Source: Yu, P 2001 'An introduction to the EU Directive on the Protection of Personal Data', http://www.gigalaw.com/articles/2001/yu-2001-07a-p1.html; see also http://www.export.gov/safeharbor.

European Union–United States Fair Information Practice Principles

The E.U. Data Protection Directive goes beyond the FTC's formulation of the fair information practice principles in four significant respects.[160] First, the E.U. Data Protection Directive places substantive limitations on the quantity of personal data that is collected, requiring that it be 'not

---

[160] See Radin, MJ, Rothchild, JA and Silverman, GM 2002 *Internet Commerce: The Emerging Legal Framework, Cases and Materials*, New York: Foundation Press, p. 565.

excessive in relation to the purposes for which [the data] are collected and/or further processed' (Art. 6(1)(c)). Second, personal data must be 'kept in a form which permits identification of data subjects for no longer than is necessary for the purposes for which the data were collected or for which they are further processed' (Art. 6(1)(e)). Third, the processing of certain sensitive types of data is off limits: with certain exceptions, it is impermissible to process 'personal data revealing racial or ethnic origin, political opinions, religious or philosophical beliefs, trade union membership, and...data concerning health or sex life' (Art. 8(1)). Fourth, the onward transfer of personal data is subject to the limitation that in the hands of the transferee the data must retain 'an adequate level of protection' (Art. 25(1)).

A reading of the articles given above shows that the approaches to data protection between the U.S. and the E.U. are at different ends of a spectrum.[161] First, in the E.U. privacy is a fundamental human right and hence legislation is the chosen means to protect the privacy of personal information. To achieve this databases need to be registered with government data protection agencies, including those developed pre-Directive, and there is a need for approvals prior to processing of personal data, if only to monitor compliance with the Directive. In contrast, the U.S. regime relies on the free market in data and personal information and the limits placed on government intrusions in such activities in the private sector. As a result in the U.S. a sectoral approach is used that combines legislation, administrative regulation and industry self-regulation through codes of conduct as an alternative to government monitoring.

The second major difference is that in the U.S. the First Amendment to the Constitution has imposed strict limits on the ability of the government to regulate the flow of information including personal data. A comprehensive set of rules such as embodied in the E.U. Data Protection Directive may undermine significant interests that have so far been protected by the First Amendment. Discussion of the various privacy related laws in a previous section above has shown that each of the laws are very narrowly focussed on specific information—financial records, health information, video rental activities, driver's licence details and credit ratings.

Finally, the U.S. does not have a specific government data protection agency or a Commissioner as such. A large number of agencies oversee the

---

[161] See Yu 2001 *op. cit.*

protection of privacy information and data: the Department of Commerce; the Department of Health and Human Services; the Department of Transport; the Federal Reserve Board; the Federal Trade Commission; the Internal Revenue Service; the National Telecommunications and Information Administration; the Office of the Comptroller of the Currency; the Office of Consumer Affairs; the Office of Management and Budget; and the Social Security Administration.

4.5.3 European Union Data Protection Directive and Implications for Australia, Canada and United Kingdom

Australia

Australia's *Privacy Amendment (Private Sector) Act* 2000 (Cwlth) discussed previously, has put in place regulations for the use and handling of personal information by individuals and private companies and in particular NPPs. The principles oblige companies to secure all personal and sensitive electronic data that is stored, processed or communicated in their software, systems and networks. The provisions here and all other NPPs would appear to satisfy the requirements of the E.U. Data Protection Directive.[162] While a European Commission decision on whether the Australian Privacy Private Sector Act is sufficient to meet E.U. 'adequacy' standards is as yet available, in the interim there is enough flexibility to allow data to continue to flow unhindered. This is provided that adequate safeguards are put in place either in the form of contractual agreements or approved industry codes.[163]

Canada

The Canadian *Personal Information Protection and Electronic Documents Act* 2001 (PIPEDA) (Canada) came into effect on 1 January 2001 and regulates the use and collection of personal information.[164] The Act applies, not only to Canadian companies, but also potentially to any entity that collects personal information in Canada and/or personal information from

---

[162] See Handelsmann, A 2001 'Strategies for complying with Australia's Privacy Principles' at http://www/gigalaw.com/articles/2001/handelsmann-2001-11-p1.html.

[163] Hughes, A 2001 'A question of adequacy?' The European Union's approach to assessing the *Privacy Amendment (Private Sector) Act 2000* (Cwlth) [2001] University of NSW Law Journal 5, also at http://www.austlii.edu.au/au/journals/UNSWLJ/2001/5.html.

[164] See Krause, B 2001 'An Overview of the Canadian Personal Information Protection and Electronic Documents Act' at http://www.gigalaw.com/articles/2001/krause2001-02.html.

Canadian residents. Private sector entities that are covered by this Act are the so-called federally regulated companies within Canada, including financial services, telecommunications, broadcast media, and air transport industries. Thus, all federally regulated Canadian firms and all Canadian firms transferring personal information for a fee, must take steps to ensure that the firms to whom they transfer personal information abide by the Act's ten principles of Fair Information Practices after the data is sent from Canada.[165] The legislation is designed to build trust in the use of technology by implementing clear, predictable and standard rules that ensure the protection of personal information in digital and traditional forms. With a Privacy Commissioner to oversee the Act, the rules are meant to be workable and flexible so as not to unreasonably impair the growth of e-commerce, and more importantly the rules are designed to put Canadian businesses in compliance with European and American privacy regulations. This enactment ensures that Canada complies with the E.U. Data Protection Directive's standard for 'adequacy' for the protection of privacy.

United Kingdom

In the U.K. existing legislation focuses on two aspects of data protection. The first involves giving rights to individuals about whom personal data are stored. The second regulates the organisations that collect and process the data to ensure compliance with certain principles and rules.[166] The *Data Protection Act* 1998 (UK) replaces the *Data Protection Act* 1984 (UK) and took effect from 1999. The new Act governs the collection, processing, and use of data in the U.K. by any organisation that are required to register with the Data Protection Commissioner. The legislation is wider in scope than the previous provisions and gives greater powers of enforcement to the Data Protection Commissioner.[167]

The Act covers only personal data, that is, data which relate to a living individual and from which the identity can be derived directly or indirectly. Data includes that which is processed automatically, recorded

---

[165] The Canadian Fair Information Practices Principles include: accountability, identifying purposes, consent, limiting collection, limiting use, disclosure and retention, maintaining accuracy, establishing safeguards, openness, access and challenging compliance. See Winer, J 2001 'What the Canadian Privacy Act means for U.S. companies' at http://www.gigalaw.com/articles/2001/winer-2001-02.html.

[166] See Westell, S 1999b 'New data protection legislation. How will this affect a geographic information business?' *Mapping Awareness*, April and at http://www.geoplace.com/ma/1999/0499/499gis.asp.

[167] For details of principles, legal obligations, codes of practice, individual rights, international transfers, and exemptions see http://www.informationcommissioner.gov.uk/eventual.asp.

with the intention of automatic processing and data recorded as part of a 'relevant filing system' as in health records and certain public authority records. This final category represents an extension of the scope of the legislation and includes both manual as well as computer records. The individual about whom information is held is the 'data subject' and the person who decides how the data are to be used is the 'data controller'. A new definition of 'processing' of data under the Act will mean that a wider range of activities are likely to come under the jurisdiction of the regulation from merely holding, adding, deleting through to recording.

For organisations, on the other hand, before they may collect or otherwise process personal data, there are a number of pre-conditions that must be satisfied. The first of which is to ensure that the consent of the data subject has been obtained; a fact to be established by the data controller. There are other pre-conditions that have been set out as alternatives, but which may be more difficult to establish. For example, the data controller would need to show that the processing is necessary for the performance of a contract to which the data subject is a party. Alternatively, that the controller was taking steps at the request of the data subject, with a view to entering into a contract. There are other specified reasons that may be used to justify particular actions. The personal data that may be obtained should be for fair and lawful purposes and it must then be used fairly and lawfully. This means that data subjects must be aware of the use of the data. Thus, in obtaining a mailing list from a third party, a data controller must identify him/herself to the data subject and to inform the latter that it is going to either process or use the data in a particular way. The Act also sets out a series of fair information practice principles that must be observed by data controllers, including accuracy, adequacy, relevancy, quantity, timeliness and purposes of data collection. Steps to be taken to ensure compliance include regular updating of the database to ensure the personal data that are no longer relevant or out-of-date are removed and purged from the records system.

A new information category for 'sensitive' personal data has been included. Sensitive data relate to racial or ethnic origin, political opinions, religious or other beliefs, trade union membership, physical or mental health, sexual preferences, and offences committed or alleged to have been committed by the individual. There are restrictions on the processing of such data and who may control this type of data. Data subjects must give explicit consent to the processing of the information since a general consent is insufficient. In the alternate, the processing of such data must be necessary for certain specified purposes.

Transfer of data or permitting its access by anyone outside the E.U., for example, via the Internet, is heavily regulated and data controllers have to ensure compliance with the regulations governing such transfers. These are in place to ensure that the data are not transferred to and subsequently misused in third countries where no data protection laws may exist.

Data subjects are also given extended rights under the Act. These include the right:

- to prevent certain types of processing;
- to object to processing for the purposes of direct marketing;
- to query those rights related to decisions which may significantly affect the data subject where the decision is based solely on automated processing;
- to obtain court orders to correct inaccurate data;
- to seek compensation for damages arising from a contravention of the Act.

Failure to comply with the provisions of the Act is a criminal offence. The Data Protection Commissioner has been given wide powers including a right to issue an 'information notice', requesting information relating to the use of data by any organisation; and, the ability to search premises and ask questions in order to determine whether the data protection principles have been broken.

Postcript

Without doubt, the E.U. Data Protection Directive has shaped global privacy protection by imposing a rigid approach on other trading partner countries and producing a domino effect on other jurisdictions in the area of data protection. In particular the approach may seem to impose on some countries a protective scheme that may be inconsistent with that country's traditions, for example, the U.S. tradition and the protection afforded under the various Amendments to its Constitution. But at the same time the E.U. Parliament is conscious that it may need its partner countries for trade and economic development, and that the Directive should not stifle such activities nor intrude upon autonomy and sovereignty of other nations.

Summary

This chapter has canvassed the issues of personal and informational privacy and the use of geospatial technologies. The underlying theme was that 'spatial data is special' given that locational information is the key ingredient in the privacy cake. The question of how invasive GI technology has been or can potentially be is answered in the negative, that is, the technology is not personal data invasive. Even so, there is a need for vigilance as well as the ethical use of such technologies, even if there are social benefits. The legal and regulatory framework governing the issues of privacy was then discussed using the Australian jurisdiction as the backdrop and then contrasting this with the U.S. regime. While there are no constitutional impediments in preserving a right to privacy in Australia, there are four so-called privacy laws supported by the common law in protecting the confidentiality and disclosure of personal information. Supplementing the law are industry codes of conduct and self-regulation that complete the privacy package.

The U.S. privacy jurisdiction is sectoral, piecemeal and fragmented with no overarching Federal privacy Act, but rather a number of acts targeted at specific sectors. While the Constitution provides limited protection through the various Amendments there are over twenty or more 'privacy' type federal acts. The tradition in self-regulation in the U.S. is mixed, as there are quite different approaches between the states and the federal jurisdictions. One example is the unremarkable success of the FTCs 'Do Not Call Register' which suggests that the American public are increasingly cynical about self-regulatory regimes that do not have some legal backing. Other models include third party certification via privacy seals, empowerment of individuals through technological tools, and the commodification of privacy.[168]

A discussion of the legal and regulatory framework is necessary since it provided the rules under which geospatial technologies were employed. In examining such technologies, which depend on data aggregation and databases as the fuel for the marketing and advertising engine, various privacy implications has been highlighted. The analyses of geospatial applications were undertaken in regards to home location, the tracking of individuals over space, tracing financial transactions and communications. The identified privacy risks in relation to locational and tracking technologies were categorised as invasive, enhancing or sympathetic.

The implications for user organisations is that geospatial technology applications is but one of the array of different kinds of surveillance and in particular that of dataveillance. Organisations need to consciously be

---

[168] See Radin *et al. op. cit.* pp. 594–623.

aware of PIT and to encourage the adoption of PET and PST in their technology use policy. Equally, the technology providers should be aware of these sorts of issues and to genuinely strive for anonymity in the use of personal information when marketing their products. For developers of standards on the other hand, there is a dire need to develop standards for technologies that have substantial privacy implications. These standards should be soundly based on an appreciation of privacy concerns, a reconciliation of conflicting interests between the developers, users and the people affected by the technology. International and national standards bodies should also be involved and some model codes developed for general application.

The final section on emergent policy and practice has examined the impact of the E.U. Data Protection Directive and how other jurisdictions have responded to this by implementing Safe Harbour principles or fulfilling the adequacy requirement. Such a principled solution to the issue of trans-border data transfers has demonstrated the willingness of various jurisdictions to resolve their differences in a simple, yet effective manner.

The implications for policy makers, including privacy and data protection commissioners is one where the tensions between economic rationalism and the social good is stretched and seemingly irreconcilable. But this need not be the case if governments are focused on law and order, as well as striving for stability, consistency and sensitivity that are supportive of privacy protection.

Geospatial technologies such as LBS may 'push' content but at the same time 'pull' in locational information. Use of these should not have a chilling effect on personal behaviour or actions. That effect may only be apparent where there is the danger of the acontextual use of personal information and data. Hence, it is imperative that the idea of a 'zone of privacy' around one's personal and private affairs should be fostered and encouraged so that the onus is on those who intrude into the zone to justify their conduct. This zone will then draw a boundary to a private zone and a 'public' zone where everyone can interact and relate with each other and for technology to be freely used. Privacy need no longer be 'too indefinite a concept to sire a justiciable issue' (Tapper 1989: 325).[169]

While technology will continue to be both a problem and a solution, technological advances such as LBS, informatics, and GI science,

---

[169] Tapper, C 1989 *Computer Law* (4th edn), London: Longman.

will continue to challenge privacy boundaries. Also, new technology will come to the aid of consumers and enhance the level of privacy protection. As the power of information technology strengthens, grows and becomes more flexible and adaptable to different kinds of uses, this growth may be accompanied by an increasing threat to privacy rights. It seems that it is becoming easier than ever before to monitor the private, lawful activities of people, without their knowledge or consent. Equally, the more we rely on technology the more it seems we have to give up and the more it is looking like the Orwellian world of 1984. However, this time it is not someone watching you but something is... 'the thinking software, described anthropomorphically as a neural network but known also as a "bot"—a non-physical robot that can hunt down specific information in a computer system and "learn" what is required as it goes along' (Millar *et al.* 1997).[170]

But technological means alone cannot help manage and enhance privacy protection, legislation, corporate policy, and social norms may, in the final analysis, eventually dictate the use of location information generated from tracking devices and geospatial technologies.

Fair information practices are the cornerstone of many privacy laws today. However, these practices may be found wanting, especially when dealing with data manipulation using disparate databases joined together in geospatial technologies such as a GIS. The policy and technical solutions suggested, with a mix of technical and legal remedies may minimise the risk of a loss of privacy. But where the data are stretched and used in situations beyond the original intent when they were collected, then the remedies suggested might perhaps be insufficient to deal with the privacy problems that will inevitably arise. The solutions may lie in a mix of international standards, self-regulation, legislation and government policy. While the harmonisation of laws and regulations and getting consistency of privacy protection, especially across all jurisdictions is very difficult to achieve, yet, international standards must of necessity emerge. One way forward would be to keep canvassing for a global convergence of privacy regulation. It may not be desirable for each country to impose a separate privacy regime. 'The prospect of protecting personal privacy given the capabilities of GIS will be one of the most difficult privacy protection challenges public policy has had to accommodate.'[171]

---

[170] Millar, P, Grey, S and Rufford, N 1997 'Prying eyes', *The Sunday Times*, 15 June 1997, p. 16 (U.K.); and described as 'the all-seeing, all-knowing eyes of a bunch of Windows NT servers, humming quietly to themselves in a cool, dark room' cited in Rubin and Lenard 2001 *op. cit.* p. 34.

[171] Alpert and Haynes 1994 *op. cit.*

Practice Notes: Aide Memoire for GI Professionals

- Collect *only* the data that you need and use it *only* for the stated purpose.
- Ensure that the formal administrative steps required by law are *fulfilled*, for example to register, collect and use of personal data.
- Individuals must be *informed* of what, why, when and how the data collected about them are to be used.
- The goal is to *protect* the identity of any person who might be identified by the project or research.
- Be *privacy conscious* because it can impact on how you collect the data, your funding source, the storage and repository of the final database and its intended uses.
- Where relevant *comply* with Information Privacy Principles (IPPs).
- Check *compliance* with National Privacy Principles (NPPs) for the handling of personal information.
- Ensure adoption of the five core principles found in the *Fair Information Privacy Protection Principles* to maintain best practice standards.
- Design your own *Privacy Policy Statement*.

Review Safe Harbour Principles to ensure E.U. *'adequacy'* requirements are met.

Chapter 5
Geographic Information and Contract Law

Learning Objectives

After reading this chapter you will:

- Get a grasp of legal constructs of the role of the law of contract in geographic information (GI) science.

- Know why a mastery of contract law will lead to sound business practice as well as a foundation for the building of good business relationships.

- Master the elements of contract law as both necessary and sufficient conditions for relationship building.

- Be able to identify the differences between contracts *for* service versus contracts *of* service.

- Be familiar with the problems associated with using licensing as a means of protection of information products against misuse, loss of proprietary interests, privacy issues, confidentiality, and the minimisation of liability.

- Recognise the need for the privity of contract and third party interests.

- Gain knowledge of the different techniques for dealing with the execution of contracts, including those that have been discharged,

Geographic Information Science: Mastering the Legal Issues   George Cho
© 2005 John Wiley & Sons, Ltd   ISBNs: 0-470-85009-4 (HB); 0-470-85010-8 (PB)

or failed and the kinds of remedies available with the breakdown of relationships.

• Master Web-based contracts and identify the differences from conventional contract formation.

5.1 Introduction

This chapter is about the law of contract and its role in geographic information (GI) science. Geographic information scientists need to have a working knowledge of contract law because such knowledge will come into use immediately on the first contact between information providers, software consultants, and end-users. The law of contract binds the major players as well as provides the platform for establishing relationships between the major players. Knowledge of the law of contract is vital for building sound business practices and providing a foundation for good business relationships.

Contract law is about relationship building rather than simply attempting to either drive a hard bargain or to get out of a dispute. It will indeed be too late if the parties were to face each other in a court of law as this may signal a breakdown in the relationship. If information product and service businesses wish to sustain their presence in the marketplace then it is vital to put in place preventative measures to avoid any breakdown. To do so there is a need to recognise the real nature of those contractual elements that bind and to strive to use the best features of these elements in every agreement to guarantee successful outcomes.

However, the nature of spatial data is so special that there may be times when a contract may not be the preferred course of action. This is because in the first place there may be problems in defining the data as digital objects as opposed to the appearance of this data on a derived map— that is, data taken as one whole entity as against data taken as single bytes of information. The derived map referred to here may be quite different from the original map because of the manipulation of data elements, the addition of other information and new data so that the resultant product is a reconstituted 'new' map. Secondly, defining each of the alterations, additions and editorial deletions is no mean task, let alone writing these in formal language in a contract. Finally, the user should be given interpretative rules for that particular map and these rules have to be clearly communicated either on the map itself or as additional information accompanying the map.

In sum, therefore, because spatial data are special it may mean that a letter of agreement may be a more appropriate way to proceed than to using a formal contract. The letter of agreement could also spell out the lineage of the data, its characteristics as well as notes on rules of interpretation and use. Such an agreement may be endorsed on each page by the recipient and gives the feel of a legal document, even though both parties will be generally aware that it may not sustain close legal scrutiny by way of its enforcement by the courts.

This chapter is thus about those elements of a contract that are both necessary and sufficient conditions for relationship building. A necessary condition for a contract to be formed is one whereby its non-occurrence, prevents the contract being formed. Just as for a lost bushwalker, stranded in the bush, having access to adequate water is a necessary condition to stay alive. However, having water does not assure that the bushwalker will survive, because other elements such as exhaustion, starvation, or snakebite may cause death. Similarly, even an offer to perform certain GI services will not create a contract without the requisite acceptance and consideration moving from one party to the other. A sufficient condition refers to those elements in a contract that determine unequivocally that a contract will be formed. Thus, a sufficient condition of any contract is the set of elements, including offer, acceptance, consideration as well as criteria for a valid contract. These are discussed below.

The purpose of this chapter is thus twofold. First, it is intended that the chapter illuminate the context of agreements, and in particular the relationship of such contracts in GI science based primarily on common law principles. It is of little consequence that there are small, but subtle differences between common and civil law countries, since, in general the laws achieve the same ends.[1] The second purpose is to provide some precedents as templates for both users and producers of GIS and information products and services employing spatial data. In practice it has been found that many research-oriented users of spatial data do not understand the true nature of the terms of contracts and the warranties for the data or software that they incorporate in their projects. Equally, very few laypersons ever bother to read the terms and conditions in their purchase of data and services. This state of affairs is exacerbated when doing business over the Internet where there is even greater potential for carelessness and inattention

---

[1] For a review of the major differences between civil and common law jurisdictions in regard to the enforcement of contracts see Chance, C 1996 (ed.) *European Computer Law: An Introductory Guide*, Current Issues Publication Series. London: The Computer Law Association, Chapter 2.

to the finer legal details. Experience suggests that 'far more disputes are caused by vague contracts than are averted by them' (Knight and FitzSimons 1990: 137).[2]

This is an important guiding chapter for producers and users of spatial information who either disregard legal niceties to their peril or who may need to take into consideration the contractual terms under which they are collecting and/or providing spatial data or services incorporating such data. Here, practical guidelines based on 'better practice' case examples are given for the benefit of persons using this text, especially those in academia and the research community. The commercial vendor community, national mapping agencies, and similar quasi-government bodies, on the other hand, may find such examples illuminating as well.

Equally important is the recognition that the 'boilerplate' contract will contain issues of indemnification and disclaimers of warranties as part of a checklist. The provider of data may use these either to limit or prevent liability in the event of errors in the data. The creation of spatial data, whether in the form of maps or databases, brings to the fore the issues of different types of employment contracts. The use of consultants and independent contractors or employees in the organisation to produce spatial data may raise collateral concerns, including liability and the assignment of intellectual property rights. While these latter two issues are the subject of separate chapters in this volume it suggests the strong interaction between the various legal theories and that the legal issues cannot be considered separately and in isolation.

In writing this chapter, the picture that is kept in mind is that of a GI specialist, whether user or data provider, or the legal adviser with little or no experience with both GI and information technology (IT). Any one of these actors may be expectedly or otherwise asked to 'look over the contract' in case there may be 'problems'. It is suggested that the pitfalls, disputes and problems arise chiefly from the use of the technology and industry practices and rarely from the law of contract. This may be because the foundations of the law of contract have been built on strict formalities.

Perritt (1996: 386) has observed that contract law has reduced the universe of likely disputes over offer and acceptance and interpretation by defining certain formalities.[3] As hallmarks of contract formation, these formalities are included in any or all of the following: the requirements

---

[2] Knight, P and FitzSimons, J 1990 *The Legal Environment of Computing*, Sydney: Addison-Wesley Pub. Co.
[3] Perritt, JJ 1996 *Law and the Information Superhighway*, New York: John Wiley & Sons. Inc.

for writing and signatures, the *Statute of Frauds*, parol evidence rule, requirements that certain contracts can be under seal with its accompanying unique qualities, attestation and requirements for notarisation.

It has further been postulated that formalities such as signatures serve defined legal purposes (Fuller 1941) such as an evidentiary function, a cautionary function, a channelling function, and a protective function.[4] The evidentiary function is to ensure that contract documents can be used as probative evidence should disputes arise on what was agreed to. The channelling function provides the demarcation between the intent to act within the law and the intent to act otherwise, while the cautionary function ensures that there is deliberation and reflection before action. Finally, the protective function is there to give protection to both parties should some aspects of the contract to fail, even perhaps, due to the fault of neither party.

Organisation of this Chapter

In the first section traditional law of contract is discussed in terms of a 'meeting of the minds' where an offer is accepted and sealed with some sort of consideration. It is here that well-constructed contracts will help avoid problems in the future, as it will be quite expensive for the parties to meet in court. There is also a discussion on the express and implied terms to a contract.

A second section addresses the important issues of understanding and distinguishing between a contract *for* service and a contract *of* service. There is a distinction with a difference as between a consultant and an employee and the legal implications that flow from such differences need elaboration.

A third section discusses whether the tools of GI science deliver a product or a service. Vending a product requires different contractual terms to offering a service to the public at large.

The fourth section discusses licences and the use of this contractual device for both software and hardware systems driving the information technology.

A fifth section discusses liability implications and the importance of the privity of contract. This privity is vital if there is to be sheeted home liability, responsibility and the requirements to make good the damage and loss suffered.

---

[4] Fuller, L 1941 'Consideration and form', 41 *Columbia L. Rev.* 799 (1941).

A sixth section deals with the execution a contract, its discharge and the end of a contract. In addition, this section also examines failed contracts and what remedies may be available to the respective parties.

The final section addresses the increasingly pervasive Internet-based contract where the problems raised in the preceding discussions on largely paper-based contracts are revisited in an electronic environment. While seemingly similar there are subtle differences both in treatment and in solutions when compared to the paper-based 'traditional' contract law formulations that have been built up on precedents and case law.

5.2 A Contract is a Meeting of the Minds

Why Contract?

In Australia, a contract is a legally enforceable agreement.[5] In our daily lives we make contracts almost at will and sometimes without realising it. But such contracts are merely agreements to do something or to refrain from doing something. In law, however, contracts signify that an agreement has been reached and is strong indicia that two or more parties intend it to be binding. This means all parties to the contract agree to be bound to the agreement. Again, the agreement may be that one party will do something in return for the other party to do something else, sometimes thought of as a bargain, and this may often be in the form of a payment in return for service or goods rendered. Also, in law the contract defines the respective roles of the parties and the terms under which each of the participants agrees to be bound and the manner of their binding.

While this part of the law of contract can be formulistic and formal, nevertheless it may have been arrived at after lengthy discussion, negotiation, bargaining, and jockeying of their relative positions *vis-à-vis* each other. One presumption here is that each of the parties is in more or less similar bargaining positions with the other, in experience and in financial or economic strength.

Hence, a simple legal definition of a contract may read as follows: 'A contract is a promise or a set of promises which the law will enforce'.[6] Such a definition is applicable in English and Australian law.

---

[5] Hocker, PJ and Heffey, PG 1994 *Contract: Commentary and Materials* (7th edn), Sydney: The Law Book Company Ltd, p. 2.

[6] Guest, AG (ed.) 1994 *Chitty on Contracts*, London: Sweet & Maxwell, 27th edn, p. 1.

Since contracts can take different forms, depending on the context, it is imperative that the various terms used are carefully understood. Such terms as preliminary discussions need to be contrasted with negotiations; agreements and understandings differ; and binding contracts are dissimilar to 'subject to contract'. These terms are not considered 'legalese', but rather are somewhat important and serve to enforce a precision in the use of language, without which, there may be more litigation that may be necessary. Rather than discuss each of these in isolation, these terms will be illuminated by way of examples and in the course of answering a series of questions that mark the pathways through the law of contract.

In GI science there may be many instances where an agreement may be necessary, indeed, fundamental to the whole basis of the business relationship between a person and agency seeking GI services or products and a vendor proffering such services and products. However, at various stages before any agreement and before contracts are signed a person or agency may need to explore any of the following avenues in the development of a product or service:

- Request for information (ROI)
- Request for proposals (RFP)
- Licensing and leasing of software and/or provision of programming services
- Maintenance of databases and mapping activities
- Developing a spatially-based information system
- Outsourcing of mapping functions
- Delivery of mapping services on the Internet

In many ways these points may be considered as invitations to treat. An invitation to treat is one that merely initiates negotiations or provides information. These may ultimately lead to an agreement. Hence, these may be invitations to the other party to make an offer. The invitation allows for some possibility of negotiation.[7] Whether what amounts to an offer or to an invitation to treat depends on the circumstances and courts have developed certain rules and presumptions for common situations. Thus, advertising brochures sent in a GIS journal or advertisements and the display of products in an exhibition at a GI conference will normally amount to invitations to treat rather than offers. The offer is made only

---

[7] Seddon, NC and Ellinghaus, MP 1997 *Cheshire and Fifoot's Law of Contract* (7th Australian edn), Sydney: Butterworths, pp. 90–91.

when a potential customer responds to the invitation by offering to purchase the goods or services advertised or displayed.

Once the customer approaches the vendor these preliminary discussions and understandings made at the initial stages may then be written into a contract so that these become both the terms and conditions of the contract when agreement is reached. Even when the discussions lead to some agreement there may remain the view that all these discussions and agreements are 'subject to contract'. This means that the parties are nearly in agreement and need only to go through the formal motions of setting out the terms of the contract that is to be signed. Hence, when these have been set out and both parties are in agreement, then the contract comes into being when both parties put their respective signatures to the agreement. A binding contract is now said to exist when both parties have signed the contract and then proceed to perform their respective roles.

In the following discussion, the rules refer to Australian law. Where relevant laws from other jurisdictions are given as a contrast to show differences or similarities.

What are the Elements of a Valid Contract?

This question may be rewritten to similarly ask what it necessary before the law will enforce a promise or set of promises. Does the agreement need to be in writing? Must it be witnessed by anyone? In truth, none of these is required to form a valid contract. There are however, six requirements for a legally enforceable contract and these requirements are equally relevant whether the contract is made between persons sitting across a table or whether it is made over the Internet without written agreements other than the click of the mouse. These requirements are not alternative tests and each of the six must be satisfied (necessary conditions) for a contract to be valid.

- Is there an intention to create legal relations? Do the parties intend to be legally bound by the contract?
- Is there an agreement? Offer and acceptance make up an agreement when connected to each other.
- What is the consideration? The agreement must be reinforced by way of an exchange of sorts—'something for something', or to give something of value.
- Do the parties to the contract have legal capacity? If one party lacks capacity, for example, an under-aged child, then there is no contract.

- Is there genuine consent? In the absence of coercion or unconscionable conduct on the part of any party, what was actually agreed to?
- Are the objectives legal? The question is whether a contract formed for an illegal purpose may be upheld in a court of law? An example in a GI context is the contract to overfly security installations and taking aerial photographs where there is a prohibition against overflying, let alone taking such photos.

Must a Contract be in Writing?

Under Australian law, some contracts need to be in writing whereas other contracts may be those that should be evidenced in writing under the *Statute of Frauds* and its equivalents, or else the contract would be unenforceable in court. There may be other contracts that need no writing. Thus, an oral contract or 'parol' contract entered into by word of mouth is perfectly valid and creates enforceable rights and obligations between the parties.

In practice it might be prudent and sensible business practice to keep a written record of what takes place in relation to a contract. It is always preferable to reduce contracts or arrangements and agreements to a written document where possible and signed by each party. This act alone has the benefit that the courts will carefully look at the document to guide the judgement as to which party may have better rights than the other in cases of a dispute.[8] In GI, documentation is made all the more simple and important since it is possible to precisely define what is being agreed to and what is to be delivered by one party to the other, whether it be raw data, databases, layers of spatial information, and maps.

Some contracts must be in writing and will be unenforceable if they are not. Most of these requirements are statutory in nature and are unaffected by any repeal of the *Statute of Frauds*. Contracts that *must* be in writing are usually financial in nature, such as bills of exchange, promissory notes, cheques, insurance policies, consumer credit, and submission to arbitration. A contract for the sale of land must be in writing. In

---

[8] Where the contract is reduced to writing then parol or oral evidence is inadmissible to vary, contradict, add to or subtract from the words of a written document. This is a rule of evidence—the so-named parol evidence rule. See *Mercantile Bank of Sydney v Taylor* (1891) 12 LR (NSW) 252 per INNES J at p. 262 (affirmed [1893] AC 317).

GI, however, an agreement that must be in writing is the assignment of copyright.[9]

Practice Notes: Five Reasons Why Written Contracts Are Better Than Oral Contracts

1. The process of writing down a contract's terms and conditions and signing the contract forces both parties to think carefully and precisely the obligations they are undertaking.
2. When the terms of a contract are written down, the parties are likely to create a more complete and thorough agreement than by way of an oral agreement. Oral contracts tend to have gaps that may have to be resolved later when the relationship may have deteriorated.
3. With oral contracts the parties may tend to have different recollections on what has been agreed to. Written contracts will eliminate disputes over what was promised by each party.
4. Some types of contracts must be in writing in order to be enforceable.
5. If one is taken to court to enforce a contract or to obtain damages, a written contract will mean fewer disputes about the contract's terms and conditions.

Why must there be an Intention to Create Legal Relations?

Irrespective of what the parties to an agreement might promise each other, a legally enforceable contract will not be formed unless the parties intended to create legal relations.[10] This is to distinguish social and domestic situations where a court will normally presume that the parties did not intend to enter into legal relations. However, these presumptions can be rebutted by clear evidence to the contrary.

Can an Offeror Prescribe the Method of Acceptance?

An offeror may prescribe the method of acceptance which means that acceptance must comply with the requirements before an agreement is

---

[9] *Copyright Act* 1968 (Cwlth) s 196(3). This issue is discussed in greater detail in Chapter 3.
[10] Seddon and Ellinghaus 1997 *op. cit.* pp. 180–181.

completed. The difficulty arises in whether acceptance sufficiently complies with the terms of the offer. Where the communication between parties is instantaneous (such as person-to-person, telephone or telex), acceptance must be received by the offeror to conclude the agreement; and the contract is made at the place where the acceptance is received.[11] This instantaneous communication rule applies equally to communication by facsimile—there is no contract if a faxed acceptance is not received.

Acceptance by E-mail

A message is deemed to have been mailed by e-mail when sent, but the offeree's acceptance by e-mail is not received until the offeror has logged-in to the mailing system at the offeror's end. Hence, e-mail is not a form of instantaneous communication. The act of logging-in is the equivalent of opening mail.[12]

Termination of Offer

An offer once made can be terminated by various means to prevent acceptance by the offeree, including revocation, lapse of time, non-occurrence of a necessary condition and death. Termination of the offer is different from termination of the contract, the former not leading to an agreement, whereas the latter refers to a contract being terminated, revoked or at an end.

Revocation of Offer

The party who made the offer at any time before acceptance may revoke the offer. However, the revocation will not be effective unless it has been communicated to the offeree, or someone who is authorised by the offeree to receive such communications. Any words or conduct that clearly indicate that the offer has been revoked will suffice.[13] Where an offer is made to the public at large or to an indeterminate number of persons, and the offeror does not know the identity of the persons who may be contemplating acceptance, the offer may be revoked. The revocation is by giving a similar notice in the same way as when the offer was initially made.

---

[11] per DENNING LJ in *Entores Ltd v Miles Far East Corporation* [1955] 2 QB 327 at pp. 332–334; [1955] 2 All ER 493 at pp. 495–496.

[12] See infra under Part VII on the discussion of web-based contracts and the *Electronic Transactions Act* 1999 (Cwlth) in Australia.

[13] *Financings Ltd v Stimson* [1962] 3 All ER 386.

Lapse of Offer

Where offers are open for acceptance until a certain time, a late acceptance will clearly be ineffective, although in some instances it may be considered as a counter-offer. If there is no time limit specified in which the offer is to be accepted, the offer must be accepted within a reasonable time. What constitutes a 'reasonable time' will depend on the circumstances of the case. This issue of timeliness is of particular relevance to offers made via the Internet. Websites are often unchanged for periods of time and so it may be possible for a customer to seek to accept an offer that was made months earlier but which the offeror may no longer wish to be bound by. To avoid such a situation it may be prudent for Web masters who are responsible for the sites to specify that offers will lapse on a particular date. If a date has not been specified, then a Web master may have to advertise a termination of the offer on the website. Simply removing the offer from the website would not meet the legal requirement that the termination of the offer be communicated to the offeree.

Terms in a Contract

In attempting to enforce a provision or entitlement in a contract that same party must establish that the provision which they are seeking to rely on constitutes a term of the contract. There are basically two ways that terms may be imported into a contract—expressly or implicitly.

Express Terms

The acceptance of and incorporation of written (and sometimes oral) terms by signing a contract is a usual way that parties create contractual obligations. Parties are generally bound by an agreement that they have signed, and sometimes, regardless of whether or not they have read the document.[14] Hence, obtaining a party's signature is normally an important step for securing a contractual obligation from that party.

In Internet dealings where a customer may accept the terms and conditions by merely clicking on an '*I accept*' icon, one party may require the other to download terms and conditions from a website, print a copy, and then accept these terms by signing the paper copy and returning it by post. This very act creates a level certainty in the acceptance of an electronic transaction.

---

[14] *L'Estrange v F Graucob Ltd* [1934] 2 KB 294.

Parties seeking contractual certainty are advised to expressly agree to written terms and conditions. Many business-to-business (B2B) procurement website portals use this as a matter of course where participation is contingent on the parties accepting detailed rules of participation.

Implied Terms

While most terms in a contract will be expressly written there are instances in which terms may be implied into a contract. A term is implied into a contract to give business efficacy to the particular transaction and this includes contracts formed over the Internet or via e-mail. There are five criteria to be satisfied if a term is to be implied into a contract.

1. The term to be implied in fact must not be unreasonable, as it must operate reasonably and equitably as between the parties.[15]
2. The term must also be necessary in order to make the contract work.[16]
3. The particular term implied must be so obvious that 'it goes without saying'.[17]
4. The implied term must be capable of clear expression, and be reasonably certain in its operation.[18]
5. The term sought to be implied must not be inconsistent with any express term of the contract nor deal with a matter that the contract already sufficiently deals with.[19]

A term may also be implied into a transaction by law. Of significance here in GI services and products are the terms implied into contracts relating to the execution of work and the supply of materials. As opposed to the 'sale of goods', a contract will relate to the execution of work and supply of materials. Such is the case where the substance of the contract is the skill and labour involved in the production of the article—the subject of the contract. In such instances the transfer of materials is merely ancillary to the skill involved.[20] In these contracts, there will be an implied term requiring the contractor to execute the work using reasonable care, and to

---

[15] *Peters American Delicacy Co. Ltd v Champion* (1928) 41 CLR 316.

[16] *Codelfa Construction Pty Ltd v State Rail Authority of NSW* (1982) 149 CLR 337.

[17] *ibid.* at 374 per AICKIN J.

[18] *B P Refinery (Westernport) Pty Ltd v Hastings Shire Council* (1977) 180 CLR 266 per VISCOUNT DILHORNE, LORD SIMON of GLAISDALE and LORD KEITH of KINKEL.

[19] *Ansett Transport Industries (Operations) Pty Ltd v Commonwealth* (1977) 139 CLR 54.

[20] See *Robinson v Graves* [1935] 1 KB 579.

supply materials that are of good quality and fit for the purpose for which they are supplied.

Case Note: Caslec Industries Pty Ltd v Windhover Data Systems Pty Ltd[21]

In this case a computer installation worth AUD$25 000 under a contract was in dispute. Caslec was in the electrical contracting business and in 1986 decided to automate its manual costing and invoicing system. Caslec signed a software licence and maintenance agreement with Windhover in 1988. Caslec bought its own hardware directly from a different supplier and Windhover installed the software before it was delivered to Caslec. After installation of the new system problems were discovered, as interim invoices were not generated as planned, double invoices sent, incorrect accounts received and paid and incorrectly generated 'fatal error' messages were found which also corrupted the data. Windhover responded that the hardware was at fault and disagreed with the terms of the software agreement.

The Court held that Windhover had contractual obligations to rectify faults in the software. The obligation is an express term of the contract as well as an implied obligation under s 74(2) of the *Trade Practices Act* 1974 (Cwlth). The legal obligation here is to produce software that was fit for the purpose for which it was sold. The court chose to characterise the transaction as a supply of services and not as goods. Compared with the agreed price of about $25 000 Caslec was awarded almost $100 000 in damages. It is to be noted that, even in the case of commercial suppliers, who deal only with businesses, consumer protection legislation was applied in this case.

In the case of the supply of IT products and services, the *Trade Practices Act* 1974 (Cwlth) will require certain implied terms in a contract including:

---

[21] Federal Court of Australia No. N G627 of 1990 FED No 580 Trade Practices, 13 August 1992; 43 KB.

- software errors will at the very least be corrected;
- incorrect or incomplete data in a database may need to be remedied, depending on whether the data are considered a good or service;
- failure of the technology to meet the consumer's reasonable functional requirements may amount to a lack of fitness for purpose.

Margaret Calvert (1995)[22] has suggested that suppliers may minimise the risks in IT contracts associated with such problems by drawing customer's attention to a notice that may read something like the following.

Practice Notes: Notice to Customers

Customers Please Note the following carefully:

Customers please note that the technology that is used is not error-free and steps may need to be taken to remedy errors when identified.

Customers please note that the information and data supplied is either accurate or complete only to a certain standard or to no standard at all, but inaccuracies and incompleteness will be rectified where possible following certain specified procedures.

Customers are obliged to assess whether the technology intended for use is suitable for particular purposes and uses.

Customers please note that documentation has been provided that specifies what the technology is capable of doing and may be silent on what it may be unable to do.

Terms Implied by Law

Legislative provisions will imply certain terms and conditions into particular types of contracts. Parties engaged in electronic commerce (e-commerce) for example, in the sale of goods, may find provisions relating to terms that may be implied into contracts. E-commerce is a term used to refer to business transactions on, or using facilities provided by, electronic networks, including the Internet.[23]

---

[22] Calvert, M 1995 *Technology Contracts*, Sydney: Butterworths, p. 193.
[23] Latimer, P 1999 *Australian Business Law* (18th edn), Sydney: CCH Australia Ltd, p. 245.

Sale of Goods Acts have been enacted in various Australian jurisdictions to protect consumers by implying particular conditions, warranties and terms into contracts for the sale of goods.[24] In relation to the supply of goods by corporations, state legislative provisions are supplemented by Commonwealth legislation. The provisions deal with implied terms relating to ownership and legitimate title, correspondence of goods with description, fitness for purpose and merchantable quality.[25]

Where services provided by a corporation to a consumer during the course of business, there will be an implied warranty that the services will be rendered with due care and skill. There is also an implied warranty that any materials supplied in connection with those services will be reasonably fit for the purpose for which they are supplied.[26] Accordingly, corporations who are providing services over the Internet or electronically must ensure that the service is delivered with due care and skill.

Where a consumer has made known any particular purpose for which the services are required or the result that a certain desired outcome is to be achieved, a provider is duty bound to give such services that have been agreed to. In such circumstances, there is an implied warranty that the services under the contract and any accompanying materials in connection with those services will be reasonably fit for the purpose or are of such a nature and quality that they might reasonably be expected to achieve that result.

A failure to comply with the implied warranties as to due care and skill and fitness for purpose will result in the consumer being able to claim damages for breach of warranty.

Internationally, contract law has been given some degree of uniformity. In 1980 the United Nations sponsored Uniform Laws on the Formation of Contracts for the International Sale of Goods (Vienna Convention). This Convention sets out minimal rules for the formation of, and mutual obligations in, contracts for parties whose countries are signatories. The chief features of the Convention are that it:

- relates only to the sale of goods and excludes the supply of services;
- restates the rules on 'offer' and 'acceptance';
- requires that a price term be specified for there to be a contract;

---

[24] See *Sale of Goods Act* 1923 (NSW); *Sale of Goods Act* 1896 (Qld); *Sale of Goods Act* 1895 (SA); *Sale of Goods Act* 1896 (Tas); *Sale of Goods Act* 1895 (WA); *Sale of Goods Act* 1972 (NT); and *Sale of Goods Act* 1954 (ACT).

[25] See Pt V, Div 2 of the *Trade Practices Act* 1974 (Cwlth).

[26] *Trade Practices Act* 1974 (Cwlth) s 74. See also State legislation: *Goods Act* 1958 (Vic), s 89(2); *Fair Trading Act* 1987 (WA), s 33(2); and *Consumer Affairs and Fair Trading Act* 1990 (NT), s 61(2).

- provides for a number of minimum obligations on the part of the seller and buyer;
- sets up rules on the passing of risk;
- provides for dealing with breach of contract and damages, including *force majeure* clauses.

These provisions are entirely voluntary with no obligation for a country to accede to or adopt these requirements into its domestic law. Also parties to international contracts are free either to exclude or modify any of these terms as they wish.

In international trade, there is also the General Agreement on Trade in Services (GATS) treaty developed by the World Trade Organization (WTO) that came into being in 1995.[27] The GATS provides a framework of rules for international trade in services and a timetable for progressive liberalisation on a multilateral basis. The Agreement contains some general obligations, which apply to all WTO Members. One of the most important obligations is a requirement not to discriminate between service suppliers of different countries in regulatory frameworks for services and transparency obligations. Members also assume specific obligations relating to market access for foreign suppliers and non-discrimination between foreign and domestic service suppliers. The GATS covers almost all service industries except air transport services for which there is a separate agreement.[28]

5.3 Contract *for* Service and Contract *of* Service

Under Australian common law, an important issue is to distinguish between a contract *for* service and a contract *of* service. This distinction is really one concerned with personnel contracts—whether it is that of employee or an independent consultant. An employee contract is one under which the employer has the right to directly control the activities and services of the employee. In an employment contract, the employer has the vested right to give directions as to the manner of performance of the work.

An independent consultant or contractor may contract for a result. Such a result may culminate in a product or the performance of an activity. The consultant or contractor may not agree to accept the directions of the

---

[27] For the full text of GATS see http://www.wto.org/english/tratop_e/serv_e/gatsintr_e.htm.

[28] See http://www.dfat.gov.au/trade/negotiations/services/index.html.

employer as to the execution of the work. Hence the distinction is made where the consultant performs 'for service' as opposed to an employee contract where it is one 'of service'.

5.3.1 Personnel Contracts

Many personnel contracts are about the performance of services over a set period of time that may or may not be renewed. In the case of the maintenance of computer systems, it appears that there may be few contentious issues that can arise. However, where personnel are hired for program development work then an employer should be more conscious of several legal implications that can arise.

The consultant/contractor may well wish to contract on a time-limited basis in order to avoid being committed to the performance of a particular task. There may also be the prospect of a fixed fee and a limit as to the risks that the contractor may wish to be responsible for. In such cases the customer needs to scrutinise the contract carefully to discover what is being promised. Even though a customer may terminate the contract at any time, there are other risks, such as incomplete software development, the lack of documentation, preventative maintenance manuals and source codes and whether someone else may be able to complete the task should the first consultant fail to do so.

Where the developer has assigned the copyright to the customer in writing, the latter should require the delivery of the source code. If no assignment is made, an express licence in comprehensive terms should be made mandatory and desirable with the source code put in escrow.[29]

From the customer's perspective, the only way the customer can legally require the escrow agent to release the material is to sign a contract with the escrow agent. This can be either a two-party contract as between the customer and the escrow agent, or a tripartite contract as between the program developer, the customer and escrow agent. If the only contract that exists is between the program developer and the escrow agent, then the customer is unlikely to be able to enforce disclosure.

---

[29] '*Escrow*' is an ancient legal term with origins stemming from the law relating to deeds. A deed is binding upon delivery; so an escrow agent is therefore an independent third party who holds a sealed deed and guarantees its delivery to the intended party under certain conditions. The use of escrow is a useful device in cases where, say, a software developer were go to out of business and the customer needs access to the source code in order to either modify certain steps or to update the software itself.

It may also be prudent to tie payments to such consultants to milestones of significant events in program development where this is feasible. When this staged payment plan is agreed to there will be incentives for the consultant to complete tasks on time, and the customer has the advantage of ensuring that program development is on time and on budget.

Software development contracts do not assign copyright, nor are source codes given to the contractee. On a literal interpretation of the contract the user who is paying for the software development has neither copyright nor general usage rights to the program, other than as a licensee. However, there is always an implied term that the user may have a right to the program, but this right is accompanied by a duty to keep the source code confidential. Furthermore, another implied term may be that the user will not be permitted to sub-licence that right to other parties to use the program.

In regard to 'bespoke' or customised software written to the requirements of a client the situation may be entirely different from that described in the previous paragraph. By definition, bespoke software is written pursuant to a direct contract between a programmer and the client (Smith 2003: 74).[30] The software may not exist at the time of the contract. Hence, two legal implications may be read into this statement. First, a bespoke contract may vest intellectual property rights of the software to the contractee including copyright and confidentiality, along with patent rights. This is contrary to the law on copyright that vests intellectual property rights (IPR)in an author, even though commissioned by another.

In England, it has been held that if there were no express provision as to ownership, it would be open to a court to imply that notwithstanding the general rule, in equity the copyright belongs to the user. However, in order to reach such a conclusion, there would have to be some evidence that this was the intention of the parties. In *Saphena Computing Ltd v Allied Collection Agencies Ltd*[31] a company that commissioned software to be written claimed that it owned the software by reason merely of the fact that it commissioned it. The claim was denied by the learned Recorder who stated that 'the commissioning of a computer program by a person is not of itself sufficient to vest copyright in that program in that person'. However, in *John Richardson Computers Ltd v Flanders*[32] an

---

[30] Smith, G 2003 'Chapter 1: Software contracts' in Reed, C and Angel, J (ed.) *Computer Law* (5th edn), London: Oxford University Press.

[31] [1995] FSR 616.

[32] [1993] FSR 497 at 516–519.

English court held that the computer program written by independent contractors belonged in equity to the company that commissioned it. This holding has since been reviewed in *Cyprotex Discovery Ltd v The University of Sheffield*[33] where an English Court of Appeal has upheld the finding at first instance that the claim of copyright by a university had merit, even though the researcher in question was employed by an external company.

A second legal issue arising from bespoke software revolves around the nature of the contract. Sometimes the agreement could either be a *sale* or a *hiring* of the software. A bespoke software contract could also be one either for services alone if no materials are transferred or for work and materials, the major component of which the provision of services forms a large part of the entire contract. The terms to be implied from either of these will depend on the circumstances of each case. The provision of services implies that it will comply with the particular requirements or achieve a given result. The provision of materials in the form of software and documentation implies that the software will be reasonably fit for its purpose. In most cases, however, the user may seek to impose express warranties in order to ensure the desired outcomes are achieved.

The issue of ownership therefore is not equivocal for it depends very much on what was agreed to and how the software developed as a result of the contract is both to be used and owned.

There are other features of personnel contracts that should also be considered. For example, in the U.S., such features as place of work, number and quality of personnel involved, rates of remuneration, overhead costs and formula for charging and insurance and liability issues should be in the forefront of personnel contracts. Also it should be made absolutely clear the extent to which a customer may be liable for the service provider's employees should a contract be terminated prematurely.[34]

The appraisal by consultants of the GIS needs of an organisation, for example, is another form of a personnel contract. The problem here might be how to access the competence of a consultant and how to avoid situations where the consultant may only be interested in 'promoting' a particular product or software system. As the information industry has

---

[33] [2004] EWCA Civ. 380 also at http://www.courtservice.gov.uk/judgementsfiles/j2444/cyprotex-v-sheffield.htm.
[34] See *Telecomputing Services Inc.* 1 CLSR 953 (1968).

grown so rapidly there has been no measure of competence and skill except from the benefits of experience.[35]

The issue of 'poaching' and the raiding of employees by aggressive firms from their competitors are a further information industry issue. If the 'raider' can show that it is only interested in an employee and not the training, experience or the contacts that that employee can bring, then there can be little or no difficulty. However, such is rarely the case in the IT industry. The former employer may have two possible avenues of relief. First, in contract a former employer may claim a breach by the employee of the terms that governed employment in a similar industry after termination of employment. Such a restriction may be possible in 'closed' industries where there may be trade secrets involved, but such restraints must be reasonable in all circumstances, otherwise a court may not uphold such a restriction. Second, an employer may also rely in particular circumstances on torts such as conspiracy, inducement for breach of contract, interference with contract and the tort of 'passing off'.[36]

Also in the U.S., a remedy in tort may also be available if it can be shown that the employee has misappropriated trade secrets and confidential materials.[37] However, where contracts of employment contain burdensome clauses, a court could strike these down as unreasonable restraint of trade.[38] On the other hand, an agreement not to actively compete with the former employer should not be stricken down merely because there would be an overly burdensome effect on the employee.[39]

Under the English common law system, contracts restricting a person's right to pursue a trade or occupation is against public policy.

---

[35] It is likely that the URISA-led Institute for GIS Professionals and their examinations system and GIS Code of Ethics may provide criteria for the assessment of GIS Professionals and their competence in future. See details at http://www.urisa.org/ethics/code_of_ethics.htm and the URISA-led GIS Certification Institute at http://www.gisci.org/why_certification.htm. In Australia the Spatial Science Institute (SSI) initiative may similarly provide such evaluations on competence and skills of GIS practitioners. The SSI Code of Ethics and mission statement may be found at: http://www.spatialscience. org/Prospectus.htm.

[36] See *Craig Carnahan v Alexander Produdfoot Co. World Headquarters & Alexander Proudfoot Co. of Australia* (Fla DC App 1991) 3 CCH Comp. Cases.

[37] Compare *Sperry Rand Corporation v Kinder* 360 F.Supp. 1044 (ND Tex. 1973), and *Republic Systems and Programming Inc. v Computer Assistance Inc.* 322 F.Supp. 619 (D. Conn. 1970). Also *Electronic Data Systems Corp. v Kinder* 360 F.Supp 1044 (1973), and *Kelsey-Hayes Company v Ali Malehi* 765 F.Supp. 402 (1991).

[38] *Booth v Electronic Data Systems Corp.* (DC Kan. 1992), 4 CCH Computer Cases 46,801.

[39] *Pinch-a-Penny of Pinellas County v Chango* [1990–1 Trade Cases 68–961] 557 SO. 2d 940 (Fla 2d. DCA 1990). For 'Employment Contracts' see generally CCH (1992) *Guide to Computer Law*, CCH ¶21,100.

Preparing to leave an employer, either to join a competitor or to set up in competition oneself, does not constitute such a cause if there is no solicitation of the employee by the prospective employer or inducing existing customers to transfer their custom to the competitor.[40]

In the IT industry a common practice in employment contracts is to have a term that restrains employees from exercising their profession in a certain area or for a designated period of time. However, given that such contracts impact upon a person's livelihood, the courts have tended to interpret such clauses narrowly. Such clauses are in addition to breach of confidence provisions that prevent an employee from disclosing a secret process or using a client list or other confidential information once they have terminated their employment with that particular employer. It is unfortunate that many contracts put together confidentiality provisions with restraint of trade provisions, thus inviting confusion and increasing the likelihood that such provisions may be struck down by the courts.

Another scenario is a restraint against a current employee to stop them from competing against their employer by doing work for a competitor on the weekend or evenings. A court is more likely to prohibit such activity, even if there is no restraint of trade clause, because an employment contract involves a duty to be faithful to one's employer.

In Australia, the law against contracts that unreasonably restrain trade in a profession is classically stated in *Esso Petroleum Co. Ltd v Harpers Garage (Southport) Ltd*[41] A more recent celebrated case in the U.K. is *Panayiotiou v Sony Music Entertainment (UK) Ltd*[42] that involved a dispute between the singer, George Michael, and his record company. Under a restraint of trade doctrine, a court will refuse to enforce provisions in a contract that unreasonably prevents carrying on a trade. In general, a restraint of trade is where it affects a person's future ability to carry on his or her trade, business or profession. Such restraints are against the public interest unless shown to be reasonable in the interests of both the parties and the public.

---

[40] *Computer Services Corporation v Ferguson* 74 Cal. Rptr. 86 (Cal. 1968). See also *Craig Carnahan v Alexander Proudfoot Co. World Headquarters & Alexander Proudfoot Co. of Australia* (Fla DC App 1991) 3 CCH Comp. Cases.
[41] [1968] AC 269.
[42] [1994] ECC 395.

Case Note: Shroeder Music Publishing Co. Ltd v Macaulay[43]

This English case involved a publishing contract in which a young songwriter was locked into a contract, which was one-sided in favour of the publishing company. The company received the songwriter's exclusive services for a period of five years with an automatic extension for another five if the royalties exceeded £5000 sterling. Copyright to all compositions was to reside in the company with no obligation on its part to publish or promote the songwriter. The company was entitled to terminate the employment with one month's notice. The company could assign its rights under the contract with no reciprocal rights to the songwriter. The Court held that as a matter of public policy the contract was unreasonable. The Court placed particular emphasis on the lack of bargaining power between the parties.

5.3.2 Academics and Researchers as Employees?

In a University or research centre environment it has always been asked who is the owner of the copyright such as maps in a GIS. In an Australia-wide survey on this issue of copyright ownership conducted by the Australian Vice-Chancellors Committee (AVCC) there appears to be a diversity of practices.[44] A majority of universities responded that the university staff member is the owner of copyright in books and journal articles. On the other hand, the ownership of teaching materials, including electronic material is regarded as the property of the university. Students are regarded as the owners of the research data, inventions and publications resulting from their enrolment in the university, even in circumstances where they may have been engaged in collaborative research projects.

However, there are specific legal principles that are applicable to the assignment of rights in different categories of intellectual property. For example, in the copyright for literary, dramatic, musical and artistic works, the rule is that the author is the first owner. But, the *Copyright Act* 1968 (Cwlth) does not define the term 'author'. Judicial interpretation of

---

[43] [1974] 1 WLR 1308.
[44] Australian Vice-Chancellors Committee (AVCC) 2002 'Ownership of Intellectual Property in Universities: Policy and Good Practice Guide', Canberra: AVCC, p. 6.

the term indicates that it refers to the person who reduces the idea or information into the relevant material form including electronic versions or formats. Again, this basic rule about ownership is subject to special provisions. For *employees*, where the work is produced by the employee in the course of his or her contract of employment, the *employer* is the owner.

Section 35(6) of the *Copyright Act* 1968 (Cwlth) provides that where an author produces literary works 'in pursuance of the terms of his or her employment... under a contract of service' the employer is the owner of any copyright in that work. This is a crucial point to note since academics and researchers may alter the relevant terms and conditions of their contract of employment when they begin work at the institution. Thus, the legal position in initial ownership of intellectual property rights can be modified by agreement, either expressly or by implication from conduct. Express modification to the terms of employment may be written into the contract. Implication by conduct may become apparent through a course of practice where the institution has consciously permitted staff to deal with certain intellectual property rights as though they were the owners. In time this question of ownership would come to be an implied term in staff contracts of employment and they will own those rights.[45]

5.4 Geographic Information Systems: Product or Service?

GIS analyses may be seen as the provision of products rather than one of services, even though the latter may also be contemplated. The products of GIS analyses include maps and written reports and the accompanying recommendations and expert opinion on the problem at hand, whether these concern environmental, logistical or merely organisational issues. A common feature of commissioned works is that one party can expect a document or series of documents providing both analyses as well as strategies to assist in the decision-making process. In doing so, it appears that there is an implied warranty of some form. That warranty would include the fact that the task was undertaken with all due diligence, skill and care and that the final product was delivered in a workman-like, professional manner. Furthermore, it is implied that the work was executed using all

---

[45] Comparative information for the U.K. may be found in the University of Strathclyde study on 'Policy Approaches to Copyright in Higher Education Institutions' at http://www.strath.ac.uk/ces/projects/ jiscipr/report.html.

reasonable care, and that the materials supplied are of good quality and fit for the purpose for which they are given.

If, on the other hand, the GIS were to be considered to be the provision of services, then the legal issues that might arise would include indemnities and exclusions from liability. Such exclusions may be necessary to avoid liability arising from negligence, and subsequent damages and loss arising as a result.

Where the GIS contract was for the provision of software—the mapping system itself—there may be difficulties in distinguishing between the supply of goods and/or the provision of services. For example, in the U.K. the *Sale of Goods Act* 1979 (UK) relates to goods alone whereas the *Supply of Goods and Services Act* 1982 (UK) relates to both goods and services. But in both these Acts there is an implied condition that the goods will be reasonably fit for the purpose for which they were intended and that they were of 'merchantable quality'. For the provision of services, it is implied that these were provided with reasonable skill, care, and performed in a timely manner. In both Acts, whether it was a supply of goods or services depends on the particular facts of the case.

In the U.S., contracts for the sale of goods are subject to Article 2 of the Uniform Commercial Code (UCC) which generally applies to all goods offered for sale in interstate commerce. Most states have adopted this code, either in its entirety or large parts of it. If there is no explicit contract between the supplier of goods and a consumer then the law applies the provisions of Article 2 to the sales relationship. GIS software, hardware, and datasets are classed as 'goods' and hence the transaction would be subject to UCC provisions.

In the N.S.W. Supreme Court, the supply of both software and hardware as constituting the sale of goods was considered. In *Toby Constructions Products Pty Ltd v Computa Bar (Sales) Pty Ltd*[46] the court held that the sale and installation of software and hardware and the training of the purchaser's staff constituted a sale of goods. The judge stated that the 'system, software included, whilst representing the fruits of much research and work, was in current jargon off-the-shelf, in a sense mass-produced'. In a similar fashion, an English court in *Micron Computer Systems Ltd v Wang (UK) Ltd*[47] appears to have treated the supply of a computer system as comprising both the hardware and software as a sale of goods.

---

[46] [1983] 2 NSWLR 48 per ROGERS J at p. 51.
[47] Unreported 9 May 1990 (QBD).

Furthermore, a later English court in *St Albans City and District Council v International Computers Ltd*[48] considered the development of a computer package for a consortium of local authorities to be used for calculating and administering the community charge (poll tax). The software contained an error and St Albans suffered a loss of revenue because of this error. The judge in the case held that International Computers Ltd was liable for the loss because it was in breach of its obligations under the contract to supply software which would calculate the figures correctly, and it was held that the company was in breach of its obligations to provide a service with due diligence.

The underlying facts of each case might suggest whether a good or a service or even both goods and services are being provided for in a contract.[49]

5.5 Licensing

Licence agreements have been designed to provide a non-exclusive licence to use public (government) information for specific purposes that are subject to a set of conditions. For example, most licensing agreements are designed to protect:

- the information from misuse when it is transferred to others;
- the government's proprietary interests in its information asset;
- privacy and confidentiality in the information;
- the government from being liable for the degree of accuracy of the supplied information, or any subsequent amendments or misuse of the supplied information.

From information supplied by one respondent in a questionnaire survey undertaken by the author, one agency has developed three types of licensing agreements when sharing information.[50] The Western Australian Land Information System (WALIS) uses the following types of agreements:

---

[48] [1995] FSR 686 (QBD), unreported 26 June 1996 (Ct. Appeal).

[49] See also a comparative discussion of the different jurisdictions throughout the common law world in Commerce Clearing House Inc. (CCH) 1992 *Guide to Computer Law*, 'Computer hardware and software as goods' (CCH ¶7120), 'Hybrid transactions: Contracts including system support services' (CCH ¶7140); and 'Time-sharing and service bureau contracts' (CCH ¶7160). E.U. 1986 *Council Directive on Self-Employed Commercial Agents* 86/653/EEC.

[50] Personal communication. Western Australian Land Information System (WALIS) 1999 in response to an Australia Research Council-funded questionnaire survey on GIS and the Law.

- *Level 1 agreement*: a letter of understanding for the transfer and use of digital information between WALIS agencies;
- *Level 2 agreement*: a licensing agreement for the use of digital information acquired from the State of Western Australia for non-commercial purposes;
- *Level 3 agreement*: a licensing agreement for the commercial use of digital information acquired from the State of Western Australia.

The three tiers of agreements are designed for use for different types of users. Level 1 is mainly for use in sharing information within WALIS agencies, while levels 2 and 3 are distinguished as to whether the usage is for non-commercial or commercial uses. An example is the sharing of data and information between WALIS and other governments, including local government, research institutions and the private sector. As formal contractual agreements, action may be taken where a party breaks any terms or conditions. Commercial use is taken to mean purposes relating to trade and the buying, selling or exchange of products for profit.

As GIS users are well aware, many of the tools of trade are in the form of software embedded in systems. Originally software licenses were used as a means of protecting the owners of the software where it was thought that copyright laws did not cover computer software. The owner 'lent' or gave 'on licence' or 'bailed' the software for the use of the licensee in return for a fee. The grant of the licence typically is a non-exclusive, non-transferable right to use the data or software. Users need to treat the data as proprietary since the vendor remains the exclusive owner of the data.

In terms of intellectual property rights, for instance, a licence can protect both the content and arrangement of the data. Also, the licence may also be a means of keeping the data out of the public domain. More recently, software developers and owners have introduced licences as a means of reminding users of 'fair use' governing copyright as well as rules under consumer protection laws. Thus, 'box-top' licences and 'break-seal' licences purport to bind consumers to a contract immediately upon opening a box containing software. In opening the box the consumer signifies acceptance of the terms and conditions of the licence.

In Australia and similar common law jurisdictions elsewhere require that all terms and conditions of the contract be placed outside the box in order to overcome the rule against introducing terms and conditions *after* a contract has been formed.

As a general rule no contract terms may be of any effect if introduced after the contract has been formed. However, in the purchase and sale of IT products such as computer software, it may appear that contract terms can be introduced after the point of sale. 'Shrink-wrap' licences, for example, are contained inside packages, out of sight of the purchaser. The terms and conditions of the licence are only discovered on unpacking the product. On the strength of prevailing legal opinion such a licence may be invalid and of little worth.

In the Australian common law context, the issue revolves around whether reasonable notice of the terms was provided to the customer prior to or contemporaneously with the formation of the contract. English law considers this as the classic so-called 'parking ticket' case where the customer has been presented with sufficient notice of the pre-printed terms and conditions prior to entering into the contract in a car parking station. The same pre-printed terms and conditions will bind the customer in accepting the contract.[51] Thus, in Australia and the U.S. where a party is seeking to rely on shrink-wrap terms and conditions, it is important that these terms are brought to the attention of the customer prior to or during contract formation. This is because such developments in electronic contracting have given rise to two concerns associated with mass-marketing agreements. The first is that the public cannot or will not ordinarily be able to negotiate their own terms and conditions. Second, the standard terms may override exceptions granted in an Act or may impose other undesirable restrictions.[52]

Below is an example of software licence agreement used by an English company for licensing its GIS software.

Practice Notes: Example of a Licence Agreement

Important: Read Before Opening

This software is supplied on licence upon the terms set out below by AAA [the publisher] and all copyright and other intellectual property rights in the software and accompanying documentation are owned by BBB [the software developer] absolutely. If you do not wish to use the software on the terms of this notice, you may return this sealed package unopened with your sales receipt to your supplier for a full refund.

Continued on page 319

---

[51] *Thornton v Shoe Lane Parking Ltd* [1971] 2 QB 163; 1 All ER 686; *Olley v Marlborough Court Ltd* [1949] 1 KB 532.

[52] Founds, G 1999 'Shrink-wrap and click-wrap agreements: 2B or not 2B?' 52(1) *Federal Communications Law Journal* 100.

Continued from page 318

AAA provides the software within the rules of the copyright and licences its use to the purchaser.

If you have ordered a site licence, please refer to your site licence agreement for licence terms and conditions. Please complete and return the enclosed licence agreement to register your copy of the software.

If you have ordered a single user licence:

(a) you may use the software on a single computer, or move the software and use it on another computer, but you may not use the software on more than one computer at a time;

(b) you may make one copy of the software for back-up purposes;

(c) you may not make any copy of the documentation.

Please complete and return the enclosed reply paid registration card to register your copy of the software.

This product is subject to continual development and improvement. Information may be made available to registered users from time to time regarding updated and/or enhanced versions of the software.

Technical support can only be provided to registered users.

Disclaimer

No warranty is made with respect of the software, its quality or performance, its merchantability or fitness for any particular purpose. The software is sold 'as is'. You, the user, assume responsibility for the selection of the software to achieve your intended results and for the installation and use of the results obtained from the software.

This example of a licence agreement provides several noteworthy features. There is an initial section, spelling out property rights and the reservation of ownership rights. The grant of the licence is given in regards to its permissible and non-permissible uses, the terms of the licence and its termination, limitation as to warranties and a disclaimer. Here the license also states what remedies are available and a limitation of liability and a statement on quality, performance, merchantability and fitness for particular purposes. Note also that the terms and conditions of use in licence agreement displayed above are set out by the publisher, AAA while the intellectual property rights are owned by BBB absolutely.

On the strengths of the wording and the prior notice given by the licence agreement it would appear that this is a 'take it or leave it' agreement and that the customer has been so advised that breaking the seal signifies acceptance of the terms and conditions of the licence.

In the U.S., laws have been introduced to give effect to shrink-wrap licences by requiring the words such as 'read this before breaking the seal' be displayed prominently on the outside of boxes containing software.[53] Otherwise, as shown in several cases, courts in the U.S. have held that a contract had formed when the customer negotiated the order for software over the telephone, and hence the shrink-wrap contract contained in the packaging were additional terms to which the licensee had not consented.[54] In both instances the licensor had failed to inform the customer of the shrink-wrap terms and conditions during the original telephone conversation when the contract was being formed.

In recent times the U.S. courts have been more willing to find that contract formation occurs when the pre-packaged terms and conditions have been delivered to the customer (see Case Note on *Hill v Gateway 2000 Inc.*).

Case Note: Hill v Gateway 2000 Inc.[55]

In 1997 the Hills family bought a Gateway 2000 computer by phoning a company. The Gateway sales representative did not read the text of the agreement on any exclusions or conditions over the telephone and the Hills remained ignorant of them until the computer was delivered. Gateway's contract, contained in the box with the computer,

Continued on page 321

---

[53] *Uniform Commercial Code (UCC)* § 2–204(1): 'A contract for sale of goods may be made in any manner sufficient to show agreement, including conduct by both parties which recognises the existence of such a contract'.

[54] *Step-Saver Data Systems Inc. v Wyse Technology and Software Link Inc.* 939 F.2d91 (3rd Cir. 1991); *Arizona Retail Systems Inc. v Software Link Inc.* 831 F. Supp. 759 (D Ariz 1993).

[55] *Hill v Gateway 2000 Inc.* 105 F.3d 1147 (7th Cir. 1997); *Brower v Gateway 2000* Inc. 246 AD 2d 246, 37 UCC Rep Serv 2d 54 (NY 1998).

Continued from page 320

said that if they kept the computer for more than 30 days, they agreed to be bound by the terms and conditions of the contract.

The Hills family kept the computer for more than 30 days and then filed a class action, arguing, among other things, that the product's shortcomings made Gateway a 'racketeer'—the predicate offences being mail and wire fraud. Gateway's licensing terms required the Hills to settle the dispute by arbitration. A U.S. court held that they had to abide by the terms of the contract as set forth in the licensing agreement. This is because they had been given adequate notice and the Hills had accepted those terms by using and keeping the computer for over 30 days.

In these cases the goods were delivered on an 'accept or return' basis, with customers having the right to return the delivered goods within a set period of receipt if they did not wish to be bound by the pre-packaged terms and conditions. Such agreements are described as adhesion contracts. Retention of the goods beyond the return date constituted an acceptance of the terms and conditions contained in the packaging, such that those terms and conditions bound the customer. In a further U.S. case in *M A Mortenson Company Inc. v Timberline Software Corporation*,[56] this position was further strengthened by the fact that the terms and conditions were placed outside the goods, and a reference to the licence agreement appeared on the screen each time the software was accessed. The licence provided that the use of the program signified that the customer had consented to be bound by the terms of the licence.

Case Note: ProCD Inc. v Zeidenberg[57]

In *ProCD Inc. v Zeidenberg* the issue concerned a shrink-wrap licence terms and conditions. ProCD was a software vendor who compiled information from more than 3000 telephone directories and put these into an electronic database. It cost in excess of US$10 million to compile and was very expensive to keep up-to-date. ProCD charged different prices to different users of the database,

Continued on page 322

---

[56] 998 P.2d 305 (Wash 2000).
[57] 86 F.3d 1447 (7th Cir. 1996).

Continued from page 321

with an individual consumer having to pay less than an institutional user.

Matthew Zeidenberg bought the consumer version of the software that included a shrink-wrap licence stating that it could only be used for non-commercial purposes. Matthew ignored the term of the licence and began to re-sell the database on the Internet to anyone willing to pay a price that was less than that of ProCD's rates for institutional users. He argued that ProCD could not enforce the restrictions on his use of the software because those restrictive terms did not appear on the face of the packaging. He further argued that he had no opportunity to reject the terms and conditions offered by ProCD.

The Court ruled that it is not unfair or commercially destructive to allow the taking of data assembled with a significant investment of time, effort, and money, albeit in an uncopyrightable form, and to use it for commercial purposes without paying any compensation. Furthermore, federal copyright law pre-empted the manufacturer's claims of misappropriation and unfair competition, as well as violation of a state computer crimes act.

The Court held that as the terms of a licence agreement were found not to place an unreasonable burden on the licensee the shrink-wrap licence was deemed enforceable. The Court considered the sale of the software to be a sale of goods subject to the UCC. The Court then went on to rule that the parties had entered into a binding contract for that sale when it was purchased off-the-shelf by paying for it at a retail outlet. At that time, the court observed, the terms of the user agreement were not presented to the purchaser.

It was further observed that '[t]he sole reference to the user agreement was a disclosure in small print at the bottom of the package, stating that defendants were subject to the terms and conditions of the enclosed license agreement. The defendant did not receive the opportunity to inspect or consider those terms.' This 'mere reference' in the user agreement did not give buyers an adequate opportunity to decide, at the time of contract formation, whether the terms were acceptable. So viewed, the terms did not become a part of the contract.

Since this landmark decision, courts have tended to enforce shrink-wrap agreements that are 'commercially reasonable' and valid under contract law. The provisos being that customers are given prior notice of them, have the right to reject the terms and conditions, and have the prospect of receiving a refund after opening the product.

The terms 'layered contracting' and 'rolling contract' have been used to describe these sorts of agreements in which terms are presented only after the buyer has received the product.

Thus, licence provisions protect both customers as well as owners of IT products. Licence conditions may be included to give warranty protection to consumers; or it may limit liability of the owner and to control the use and distribution of the software. Further, under consumer protection laws, the terms under which the goods may be returned because of defects should be included. As a comparative note, it may be observed that U.S. courts are more accommodating than other jurisdictions elsewhere in the world in so far as shrink-wrap licences are concerned.

5.5.1 Why is Spatial Data Special?

Before leaving the substantive law of contract, the theme of why spatial data is special needs further reiteration. As foreshadowed in the introductory statements to this chapter, there may be occasions when a contract may not be the only way to proceed in GIS for either the supply of goods or services.

The very nature of spatial data suggests that contract terms must vary from data set to data set and the use that these data are put to. Hence, it may be necessary that for each and every sale there must be negotiated terms and conditions that relate to that particular sale. Sometimes, the same terms may be used in cases where the data are of the same kind and used for the same purposes. In other cases there may be variations, so that the contract terms must be varied in order to cover for these differences. As has been observed by Onsrud (1999) 'because geographic data sets may sometimes be acquired through a string of successive sources that may each alter the data set, a seller's ability to sell or licence full interests in a geographic data set may be an issue in some cases. Additional warranties may [need to] be express[d] or implied'.[58]

In the sale of GIS databases and data, it may be necessary to specify the nature of the data product as completely as possible. This will mean specifying as far as possible what the data were collected for, what the data have been used for, what the data may be unsuitable for, and the

---

[58] Onsurd, HJ 1999 'Liability in the use of GIS and geographical datasets' in Longley, P, Goodchild, M, Maguire, D and Rhind, D (eds) *Geographical Information Systems: Vol. 2, Management Issues and Applications*, London: John Wiley & Sons, Inc., pp. 643–652.

coverage of the data set. Specifications will be necessary in order to ensure that buyers know what they are buying and what types of warranties are being given. By doing so the vendor is also limiting liability should the data be found to be deficient when used in applications.

An Environmental Science Research Institute (ESRI) (1995) White Paper has given useful practical tips and guidelines for data publishing.[59] It suggests that the use of disclaimers on log-on screens and user manuals is a particularly useful way to remind users of the nature of the data being used. The disclaimer is to be followed by a statement giving a clear understanding of what is known about the true nature of the data, its strengths and weaknesses and how robust it may be in particular applications and transformations. There should also be a notice giving details of the lineage of the data and how and where it may be verified. Finally, to limit risks and minimise damage, vendors must take care in how they advertise and market the material that is being sold.

Practice Notes: Summary Recommendations to Purchasers of Data/Information

It is strongly recommended that you:

- Always check and verify the data/information before use.
- Are aware that the data/information are published as is and its accuracy, uncertainties, and possible dangers in its use for particular purposes have been given in a disclaimer. Read the disclaimers carefully.
- Seek expert help to verify usage of the data/information as well as testing the use of the data on different platforms. Test the data for sensitivity at various levels of details and at various scales of resolution. Evaluate its robustness under different circumstances.
- Ensure that the use of the data does not infringe third party proprietary rights, for example, data that are copyright to another party other than the vendor.
- Buy insurance against the risk of liability in the use of the data/information in the event of loss, litigation or an award of damages by the courts.

---

[59] ESRI 1995 Data Publishing Guidelines for ESRI Software, September, J-6040. Redlands, CA: ESRI.

To conclude this section on the requirements of the law of contract, the following Licence Agreement checkpoints must to be 'ticked off' in order to ensure that these are covered in the agreement to be signed between a vendor and customer. Readers may also wish to compare this checklist with the summary of legal issues on contracts found at the end of this chapter in order to consolidate what the law requires and what usually happens in practice.

Practice Notes: Licence Agreement Check Points

- What is being licensed? To avoid later disputes, the parties must be as specific as possible. For example, if small excerpts of a work are being licensed and not the whole work, the agreement should make this point clear and identify these excerpts carefully.
- Whether licence rights, including rights of production, distribution, manufacture, and sale and the right to transfer licences a third party are permitted.
- In what projects, products or publications can a customer be permitted to use the licensed material?
- What rights are being granted?
- Is the license exclusive or non-exclusive?
- Will the owner get credit and if so, how will this be shown?
- What intellectual property rights are retained by the licensor?
- What is the licence fee: a single one-time fee, an annual fee, or a royalty percentage of revenues? The fee need not be in monetary form, but could be in kind such as products, services, publicity or just a credit acknowledgment on a publication.
- What is the duration of the licence and can it be renewed?
- What warranties are being given for the use of this product by the licensor? Note that a warranty is a legally binding promise that certain assertions are true. For example, a licensor might need to warrant that it owns the material and all the intellectual property rights protecting them and that the use of the materials will not violate the rights of third parties. Thus, if material provided infringes a third

Continued on page 326

Continued from page 325

party's rights, that fact alone may also make the user liable for infringement of the third party's rights even if one was unaware of the infringement. Also note further that the warranties are limited by time and to specific rights only. Usually the licensor will add that the warranties stated are the sole and exclusive warranties under the agreement.

- What are the liabilities of the licensor? Normally this will include liability to repair and replace defects, provide advice as to how to achieve functionality and to return the purchase price or fees paid where there has been notice of a breach of the warranties given previously. This will also include any damages awarded for any inadvertent breaches to third parties, which may have been caused by the licensor.

- What remedies are available if the products and services are not warranted? An indemnification provision from the licensor will mean that the licensor agrees to defend and hold the licensee blameless against any claims resulting from the breach of the warranties.

- What obligations are there as to confidentiality of proprietary information?

- Check if licensors have excluded liabilities for any indirect, incidental, special or consequential damages.

- Depending on the jurisdiction, ensure that there is a statement as to whether the agreement is subject to export control laws, regulations and requirements.

- Under miscellaneous provisions check if there are provisions in the Agreement that may be severable and whether the invalidity or enforceability of one of the provisions affects any other.

- Ensure that the relationship between the parties in the licence is that between independent contractors.

- Check if all claims and disputes relating to the agreement are subject to final and binding arbitration, and under what jurisdiction.

- The Agreement should conclude by stating that it contains the entire agreement of the parties and that it supersedes all prior oral or written understandings or agreements between the parties with respect to the subject matter. Service of notices, contact officers, and postal address, etc. must be shown here as well.

5.6 Liability Implications and the Privity of Contract

Legal duties and responsibilities arise from contract, and these may overlap with consumer protection legislation. For example, a person suffering an injury may suggest a breach of strict product liability as well as a breakdown of the contract. Civil liability may arise from contractual, quasi-contractual or delictual (legal) responsibilities. Delictual responsibilities are owed to strangers and third parties that have no privity of contract, but nevertheless may attract protection because of a breach of statutory duty or negligence. The general principle on delictual liability is found in the *Consumer Protection Act* 1987 (UK).

The topic in this section is that of a privity of contract. A general rule in contract law is that only a party to a contract who has privity, and who has given consideration, can enforce it. That is, the contract cannot confer rights or impose obligations arising under it on any person except the parties to it. The rule also states that a non-party to the agreement cannot generally be burdened or benefited in law by the formation of a contract.

In Australia the principle of the privity of contract was exhaustively reviewed by the High Court in the leading case of *Trident General Insurance Co. Ltd v McNiece Bros Pty Ltd*[60] A majority of the court favoured certain special qualifications to the common law rule denying third party beneficiary rights. But the doctrine of privity itself was not explicitly rejected or abandoned.

The doctrine of privity has long been settled in England, but not Scotland, and is thought to be a doctrine of general application in most common law jurisdictions. Under Scottish law, third party rights can only be created where the doctrine of *jus quaesitum tertio* [literally a right acquired by or vested in a third party] applies. This doctrine requires that in order to establish a third party right it must be clear from the terms of the contract that the right is being created and the right must be irrevocable. The issue has since been clarified with the passage in the U.K. of the *Contracts (Rights of Third Parties) Act* 1999 (UK)[61] and clarified under Scottish law in *Beta Computers (Europe) Ltd v Adobe Systems (Europe) Ltd* in 1996.[62]

---

[60] (1988) 165 CLR 107; 80 ALR 574.

[61] At http://www.hmso.gov.uk/acts/acts1999/19990031.htm.

[62] FSR [1996] 367 and Bainbridge, D 1996 Beta Computers (Europe) Ltd v Adobe Systems (Europe) Ltd *Computer Law and Security Report*, v. 12(5), pp. 310–312.

The Supreme Court of Canada adopted this principle of third party rights being created in *Greenwood Shopping Plaza Ltd v Beattie.*[63] The N.Z. *Contracts (Privity) Act* 1982 (NZ) ss 4 and 8 allows a third party to enforce a contract made for that party's benefit. This is achieved by imposing an obligation on the promisor in favour of the third party and in cases where the parties intended that the third party should be able to enforce the provisions for benefit. As it now stands in most states in the U.S., third parties can sue directly upon contracts made for their benefit by others—a principle of law given in Chapter 14 of the *Second Restatement of Contracts* (1979).

5.6.1 Exclusion Clauses

In the U.K. and Australia, any standard form contracts use exclusion clauses to exempt or exclude liability. Some exclusion clauses fix a financial limit on any claim against the party seeking redress, while others insert a time limit in which claims must be made. In some consumer contracts, exclusion clauses may result in an unfair avoidance of liability by a corporation resulting in an otherwise valid consumer claim being defeated.

In order to overcome such anomalies, English and Australian courts have developed certain practice rules which protect the consumer from exclusion clauses by preventing contracts from unfairly extending these to situations that they were not intended to cover.

1. The principle for the communication of exclusion clauses is one where a party must do everything that is reasonably necessary to bring the exclusion conditions to the notice of the other party either before or contemporaneously with the making of the contract.[64]
2. Courts will not allow a party to rely on the protection of an exclusion clause where the party does not carry out the expected performance under contract.[65]

---

[63] (1980) 111 D.L.R. (3d) 257.

[64] In *Olley v Marlborough Court Ltd* [1949] 1 KB 532 a hotel guest could recover losses due to a theft from his room just as the contract was being completed at the reception desk. Similarly in *Chapelton v Barry Urban District Council* [1940] 1 KB 532 the injury to a deck chair hirer was deemed recoverable as the ticket of hire was considered only a voucher and the conditions printed on the back of the ticket did not form part of the contract.

[65] In *TNT (Melbourne) Pty Ltd v May and Baker (Australia) Pty Ltd* (1966) 40 ALJR 189 a driver for a transport company was given permission to take the truck home in the event the depot was closed. Fire destroyed the truck parked in the home garage and the High Court held that the exclusion clause did not apply as the driver, on instructions from the company, had not performed as expected (per WINDEYER J at 377).

3. Exclusion clauses are strictly construed against the party seeking protection from the benefit of the clause. Thus, any ambiguity will be resolved in favour of the complaining party.

Some Parliaments have passed legislation to limit the effectiveness of exclusion clauses. In the U.K., the *Unfair Contract Terms Act* 1977 (UK) has made all exclusion clauses subject to a test of reasonableness. In N.S.W. the *Contracts Review Act* 1980 (NSW) has brought in a test in which a contract deemed unjust at the time it is made, permits a court to declare it invalid, delete the terms or refuse to enforce it. Other statutory relief from harsh exclusion clauses can be found in the *Trade Practices Act* and various *Fair Trading Acts* adopted in each of the Australian states.[66]

Parties may expressly limit or exclude liability for the breach of the whole or part of a contract. Thus, where a certain function to be performed in a computer program is defined in a contract, whatever else it might connote that defined function is the one to which the parties will be bound.[67] However, whether one is bound or not may be subject to the ruling legislation which may deny such an exclusion or limitation. All exclusion clauses have to be clear, unequivocal and unambiguous. Any dispute and difficulty of interpretation will be read *against* the party claiming to benefit from it. An example of a disclaimer that limits liability with no warranties is given below.[68]

---

[66] *Trade Practices Act* 1974 (Cwlth), *Fair Trading Act* 1958 (Vic); *Fair Trading Act* 1987 (SA); *Fair Trading Act* 1987 (WA); *Fair Trading Act* 1989 (Qld); *Fair Trading Act* 1990 (Tas).

[67] In *Kennison v Davie* (1986) 162 CLR 126 the accused had closed a cheque account that was held at a bank. Aware that there was no money in it, he used his automatic teller machine (ATM) card to withdraw $200. As the ATM was disconnected from the central computer at the time, the accused was able to withdraw the money without triggering the fact that his cheque account had already been closed. The accused was convicted of obtaining money by false pretences. It was argued as a defence that the ATM was a machine and that a machine cannot be deceived because it is simply programmed to respond to a customer's requests. JACOBS J held that 'the machine disgorged the money in the circumstances in which it had been programmed and instructed by the bank to do so, and I do not see how the bank could be heard to say that it did not consent to part with the money'. The conviction, however, was upheld on appeal.

[68] Personal communication. Mineral Resources Tasmania, 1999 in response to an Australia Research Council-funded questionnaire survey on GIS and the Law conducted by the author.

Practice Notes: Illustration of a Warranty and Liability Disclaimer

While every care has been taken in the preparation of this report, no warranty is given as to the correctness of the information and no liability is accepted for any statement or opinion or for any error or omission. No reader should act or fail to act on the basis of any material contained herein. Readers should consult professional advisers.

As a result the Crown in Right of the State of Tasmania and its employees, contractors and agents expressly disclaim all and any liability (including all liability from or attributable to any negligent or wrongful act or omission) to any persons whatsoever in respect of anything done or omitted to be done by any such person in reliance whether in whole or in part upon any of the material in this report.

Source: Mineral Resources Tasmania, 1999.

5.7 Contract Execution: Discharged, Failed Contracts, and Remedies

A contract may be 'discharged' or 'terminated' by performance, by agreement, by frustration, by repudiation and by breach of contract. When both parties have carried out all their obligations under the contract, the contract will automatically come to an end. Parties may also agree that the contract is at an end at a particular time or because of the occurrence or non-occurrence of a particular event, such as the customer failing to obtain finance. Also parties may agree that a contract should cease at any time; and because this agreement will also be a contract all the elements discussed above will apply. In particular, 'consideration' will be required because where parties may still have obligations under the former contract, the contract will consist of promises by each party not to enforce the promises made by the other party under the old contract, and hence each party will receive something of value.

Sometimes unforeseen events occur which result in a situation that is fundamentally different from that which the parties had in mind when the contract was first made. For example, where the contract is one of personal service and the party to perform the service becomes seriously ill or dies; or where the item being developed is destroyed before performance of the contract can take place. When such an event occurs, the law regards the agreement between the parties having been 'frustrated' and at an end. In N.S.W. the position of parties to a frustrated contract is governed by the *Frustrated Contracts Act* 1978 (NSW).

Where one party by word or action indicates either before or during performance of its obligations under the contract that it is unwilling to perform or continue to perform those obligations, that party is said to have repudiated the contract. Repudiation by one party gives the other party the right to terminate the contract and sue for damages.

In the case in which one party fails to carry out obligations under the contract that party will be in breach of contract. The consequence of the breach will depend on the seriousness of the breach and the terms of the contract. Certain types of breach of contract will give an innocent party the right to terminate the contract immediately as well as a right to sue for damages suffered as a result of the other party's breach. In other cases, the innocent party will not have a right to terminate, but merely to claim damages. Termination of a contract implies that the parties are released from their future obligations under the contract.

If a contract makes no express provision for termination, only a serious breach of contract will give the other party a right to terminate. A serious breach of contract is either a breach of a condition or fundamental term of the contract. A fundamental term is one which the other party would not have entered the contract unless they had believed that the term would be fulfilled. A serious breach is also any breach which has the effect of substantially depriving the other party of what they had intended to obtain under the contract.

It is sometimes difficult to determine whether a breach has occurred and whether it gives rise to a right either at law or under the contract to terminate the contract. It may be unwise to terminate a contract unless one is certain of both the breach and the right to terminate.

Damages refer to the compensation for losses suffered by an innocent party because of a breach of contract. However, not every loss caused by the breach will be compensated. If it can be shown that, at the time of entering the contract, the party in breach should have realised that the sort of loss was reasonably likely to result from the breach of contract then compensation for the loss will be payable. On the other hand, unusual losses will be compensated only if it was made clear at the time of contracting that a special loss might occur in the circumstances of the particular case.

The general principle in the award of damages is to put the innocent party in the same position they would have been in had the contract been properly performed. Parties are free to agree in the contract on the amount or manner of calculating the amount of damages to be paid in the event of a breach of contract. The only restriction is that the agreed amount must be a genuine pre-estimate of the likely loss that is incurred as a result of the breach of contract. If this is the case, the pre-estimate will be the amount to be paid if there is a breach, irrespective of the actual loss suffered. If there is nothing in the contract about the amount of damages to be paid, the party claiming damages will have to establish the actual amount of the loss caused by the breach of contract.

5.8 Web-based Contracts

With the new generation of GIS products and enhanced services it is now feasible to deliver these in an on-line Web-based environment. A GIS organisation may decide to provide customer and/or technical support via a website. Customer service is an area that lends itself very well to delivery via the website to all users. Clients may view data sheets, specifications for products, catalogues and frequently asked questions (FAQs). The organisation may also be ready to sell products via the Internet. However, to do so the organisation may need to put in place several different kinds of agreements categorised by the different products and services to be delivered. These may include the following:

- access content of GIS databases only;
- access and serve interactive map content via the Web;
- serve static Web maps only;
- provide transaction services, including ordering and catalogue facilities;
- provide business-to-consumer (B2C) e-commerce facilities, including secure payment; and
- provide business-to-business (B2B) e-commerce facilities, including exchange and tracking of legal and financial instruments such as requests for proposals (RFPs), purchase orders, requisitions, shipping manifests and letters of credit for international trade.

Thus, given the diversity of services and products that may be delivered, it is imperative that legal mechanisms, such as contracts, are put in place to protect both the organisation as well as its customers from financial loss or harm. An understanding of the limits of such electronic transactions that are permissible and supported by the legal infrastructure is necessary. This section discusses the nature of electronic B2B and B2C transactions and what steps need to be taken to ensure legal compliance and to minimise legal risks.

5.8.1 Electronic Transaction Regulations

Increasingly, business is being undertaken by electronic means rather than by traditional face-to-face business and transactions carried out by the exchange of paper-based documents. In Australia, in order to encourage e-commerce the *Electronic Transactions Act* (ETA) 1999 (Cwlth) was passed on the recommendations of an expert group on e-commerce (ECEG).[69] In turn the ETA is based on the UN Commission on International Trade Law (UNCITRAL) Model Law on Electronic Commerce 1996.[70] The ETA received Royal Assent in December 1999 and was proclaimed on 15 March 2000. This is a significant enactment especially as it relates to e-commerce applications including GIS.

The UNCITRAL Model Law on Electronic Commerce 1996 provides that the legal requirements for a person's signature are met on two conditions. First, that the person uses a method that identifies that person and indicates his or her approval of the electronic text. Second, that the method is reliable as is appropriate for the purposes for which it is used. To further review this issue on electronic signatures, a Working Party was set up in 1999 to draft Uniform Rules on Electronic Signatures. The Rules provide for the development of cross-recognition frameworks for electronic signatures between different countries.[71]

The 39th session of the UNCITRAL Working Group on Electronic Commerce was held in New York in 2002. The Working Group has developed a draft Convention on Electronic Contracting—Working Paper 95. The draft Convention aims to eliminate legal obstacles to the use of modern means of communication in contract formation. The draft Convention addresses issues such as the location of parties, time of contract

---

[69] Electronic Commerce Expert Group (ECEG) 1998 *Electronic Commerce: Building the Legal Framework* (April) http://www.law.gov.au/aghome/eceg/ecegreport.html.

[70] UNCITRAL Model Law on Electronic Commerce with Guide to Enactment, 1996, General Assembly Resolution 51/162 at http://www.uncitral.org/en-index.htm.

[71] Draft Uniform Rules on Electronic Signatures, UNCITRAL Working Group on Economic Commerce available at http://www.uncitral.org/english/sessions/wg_ec/wp-84.pdf or at http://www.uncitral.org/en-index.htm. See also Rezende, PAD 2001 'The possible laws on Digital/Electronic signature on the proposed UNCITRAL Model', *SCI'2001 Proceedings*, v. X, pp. 87–92 and archived at http://www.cic.unb.br/docentes/pedro/segdadtop|UNCITRAL.

formation, distinction between offers and invitations to make offers, time and place of offer and acceptance, automated transactions, and the information to be made available to the parties. In general, Australian contract law has closely emulated the approach of the draft Convention.[72]

The ETA aims to prevent the invalidation of transactions merely because they have been conducted electronically. This Act is part of the Australian government's strategic framework for developing the information economy in Australia. The strategic framework reflects the Australian government's commitment to ensuring that all appropriate government services are available on-line by 2001.[73]

In general, the ETA allows e-commerce to fit into the same legal framework as traditional paper-based transactions. The goal is to ensure that the technological developments brought about by e-commerce can be dealt with within the existing legal system, while allowing business and consumers to determine the most effective technological choice for their purposes.[74] Various states in Australia have also passed legislation mirroring the Commonwealth Act.[75]

The ETA is built on three underlying principles, namely:

- *functional equivalence*, or the guarantee that paper and electronic documents will be on the same legal footing;
- *technological neutrality*, that is, the law will not discriminate between different forms off technology; and
- a principled approach to electronic signatures.[76]

The passage of the ETA is aimed at creating a 'light-handed regulatory regime for the use of e-commerce in transactions'. This Act may encourage the development of the information economy and the use of electronic communications in Australia. It provides a mechanism for

---

[72] For details of the draft Convention see http://www.ag.gov.au. See also the relevant reports at http://www.uncitral.org/english/sessions/unc/unc-35/509e.pdf and at http://www.uncitral.org/english/sessions/unc/unc-36/acn9-527-e.pdf.

[73] For further information about the strategic framework see the National Office of Information Economy's website at http://www.noie.gov.au.

[74] Giles, D 2000 'You've got mail . . . or have you?', *Internet Law Bulletin*, v. 3(1) pp. 12, 14.

[75] *Electronic Transactions Act* 2000 (NSW); *Electronic Transactions Act* 2000 (Vic); *Electronic Transactions Act* 2000 (SA); *Electronic Transactions Act* 2000 (WA); *Electronic Transactions Act* 2000 (NT); *Electronic Transactions Act* 2001 (Tas); *Electronic Transactions Act* 2001 (Qld); *Electronic Transactions Act* 2001 (ACT).

[76] See UNCITRAL Uniform Rules on Electronic Signatures—Australian Issues Paper http://law.gov.au/publications/ecommerce/welcome.html. See also Sneddon, M 1998 'Legislating to facilitate electronic signatures and records: Exceptions, standards and the impact of the statute book', *UNSW Law Journal* (May) http://www.law.unsw.edu.au/unswlj/ecommerce/sneddon.html.

business and commerce to voluntarily choose electronic communications when dealing with Government agencies and will remove a number of existing legal impediments in this area. It would put e-commerce and paper-based commerce on the same legal footing with each other, and not discriminate between different forms of technology. This is consistent with international developments.

The passage of the ETA also coincides with the presentation of similar legislation in the U.K., E.U., U.S. and elsewhere.[77]

5.8.2 Electronic Contracts

Internet Web-based contracts are no different from the paper-based contracts discussed in previous sections. Web-based contracts can be narrowly defined to encompass only those types of contracts that can be accomplished with the 'click' of a mouse on-line within an electronic business (e-business) environment. The legal and practical issues remain the same; only the delivery mechanism is different. Indeed, Katsh (1995) has observed that while paper contracts may bind parties to an act; an electronic contract binds parties to a process.[78] Arguably, while there may be a difference in form, the substance of a contract remains the same. Nimmer (1996) has said that with machine supplied information the nature of contracts based on human consent will change radically.[79]

Electronic Contract Formation

The formation of a contract in the electronic environment may begin with an exchange of e-mail messages, or through a page on the website, at an on-line auction or through an on-line portal of the firm. As with paper-based contracts, an offer which is accepted with valuable consideration for the promises made and a consensus has been reached between the parties creates a contract. Similarly, in an on-line context the bargain may be struck at arms length through the media of the Internet.

---

[77] *Electronic Communications Act* 2000 (UK); European Directive on Electronic Signatures approved on 30 November 1999; *Uniform Electronic Transactions Act* 1999 (US); *Electronic Signatures in Global and National Commerce Act (e-Sign)* 2000 (US); *Digital Signature Act* 1997 (Malaysia); *Electronic Signatures Act* 1998 (Singapore); *Electronic Transactions Ordinance* 2000 (HK SAR); and *Information Technology Act* 2000 (India).

[78] Katsh, ME 1995 *Law in a Digital World*, New York: Oxford University Press, p. 129.

[79] Nimmer, RT 1996 'Electronic contracting: Legal issues', 14 *John Marshall J of Computers and Information Law*, Winter, pp. 211–246.

Below is set out sample language for an electronic contract 'offer' in which human beings participate. The following may be presented on a computer screen by the computer system making the offer or soliciting offers.

Practice Notes: Draft Language for Electronic Offers of Contracts

You are about to enter into a contract. By typing the words '*I accept*' in the space provided at the bottom of your screen, you accept our offer to connect you to the database containing spatial data for your area of interest. You will be billed at a price of $20 per hour for your usage of this database, and by typing '*I accept*' at the bottom of this screen you agree to pay that charge and also acknowledge your use of the database according to records we will keep.

If you do not wish to enter into this contract at this time, simply press the Enter or Escape key and you will be returned to the preceding screen.

The on-line version of an invitation to treat may be posted on a website, on an electronic mailing list or sent by e-mail to a specific recipient. However, to avoid the possibility that the invitation may constitute a unilateral contract, that is, a contract taken up by a customer without reference to the offeror, websites now customarily include disclaimers. These disclaimers indicate that the statements made on the websites are not offers or unilateral contracts but are mere invitations to treat.[80] Websites and electronic mailing lists issue invitations to treat through the advertisements and act as shop windows that display goods and prices and give customers an opportunity to look at their goods before purchase.

Are Machines Capable of Accepting Offers?

In the U.K. and also in Australia the so-called ticket cases have suggested machines can accept offers. In *Thornton v Shoe Lane Parking Ltd* an English court has held that a car park machine had made an offer which was accepted when the customer put money into the slot in the machine.[81]

---

[80] See Chissick, M and Kelman, A 2000 *Electronic Commerce: Law and Practice* (2nd edn), London: Sweet & Maxwell Ltd.
[81] [1971] 2 QB 163; 1 All ER 686.

Similarly in the U.S. *State Farm Mutual Auto Insurance Co. v Brockhurst* it was held that a computer was capable of forming a contract for the renewal of an insurance policy as the computer operated to instructions in a program written by the insurance company.[82]

There is strong evidence to suggest that, while on-line businesses are bound by contracts 'accepted' by their computers, such businesses will also be bound if their computers and software systems behaved in a faulty manner (see Groebner 2004).[83] Computer generated contracts are only acceptable 'as a whole' as the machines are incapable of accepting contracts in which a customer may have altered some of its terms. Any changes to the standard computer-generated and accepted contract would require a face-to-face encounter in order to make the changes and additional terms and conditions valid.

Jurisdiction

A further issue with the acceptance of on-line contracts is the question of jurisdiction. By jurisdiction is meant the power of a particular government's laws and court system that binds parties or makes judgements in relation to a party on a particular matter. Jurisdiction may be limited by geography, by subject matter that a court may deal with, by the type of relief that is being sought, and the point of law or fact that is to be decided. The limits of a court's jurisdiction are usually defined by statute. The law governing the contract will usually be the jurisdiction in which the contract was concluded.

However, the jurisdiction where a contract is formed in an on-line environment can be problematic. Similarly whether the postal acceptance rule is applicable in such environments adds to the complexity. The problems arise in regard to the 'click-wrap' or 'click-through' acceptance of the contract. Is the contract formed in the country where the customer clicks the '*I accept*' icon or where the server for the website is located or when the acceptance is communicated to the vendor of the product wherever they may be located? Common law principles suggest that a contract will be formed where it is accepted and when acceptance is actually communicated to the offeror. This is known as the receipt rule.

---

[82] 453 F.2d 533 (10th Cir. 1972).

[83] Groebner, B 2004 'Oops! The legal consequences of and solutions to online pricing errors', 1 *Shidler J.L. Computers and Technology*, 2 and at http://www.lctjournal.washington.edu/vol1/a002Groebner.html.

An exception to the receipt rule is the 'postal acceptance rule' that deals with situations where acceptance is by post and a contract is formed when a properly pre-paid and addressed letter is posted. In England, whether the postal rule is applicable when methods other than by post are used has been examined in *Entores Ltd v Miles Far East Corporation*.[84] The case suggests that the postal acceptance rule will not apply to methods of communication which are 'virtually instantaneous'—phone, facsimile, telegram, but not by telex. The test is whether the communication is instantaneous; however, this is not decisive since cases involving 'click-wrap' agreements have yet to be litigated.

But the advice here and private customary practice suggests that it may be prudent to use the fax-then-mail procedure to ensure receipt, maintain a log of correspondence and keep a paper record of the agreement made. Also, some websites advise their customers to print out a copy of the agreement for reference purposes. This makes good business sense, as there will always be some documentary evidence of the transaction having taken place.

Arguably, a dishonest person may seek to withdraw from the agreement by claiming that the fax was not the final communication and that the deal was still in negotiation stage until the paper agreement was mailed. In such cases the parties may need to clarify the operative mode of communication at the outset.

Furthermore, it suggested that the postal acceptance rule may apply to e-mail messages and an acceptance becomes effective once the e-mail has been sent. The jurisdiction of the contract in such cases will be where the offeree is located when sending the e-mail. Downing and Harrington (2000) however, have argued to the contrary in saying that the rule should not apply to e-mail acceptances because it may place an unfair burden on offerors.[85]

In the case of *Cybersell Inc.* (AZ) *v Cybersell Inc.* (FL)[86] the U.S. Ninth Circuit Court of Appeals concluded that the mere fact that an Internet advertisement could be viewed in the plaintiff's jurisdiction was insufficient to make the defendant subject to that jurisdiction. The court found that 'something more' was needed, for example, an interactive website. But in Australia, it would be likely that where interactive websites are involved, or in situations where a consumer engages in a positive act in

---

[84] [1955] 2 QB 327.

[85] Downing, S and Harrington, J 2000 'The postal rule in electronic commerce: A re-consideration' (2000) 5(2) *Communications Law* 43 at 44.

[86] 130 F.3d 414 (9th Cir. 1997).

response to a website advertisement, the requirement that the defendant is 'engaged in conduct' will be met.[87]

Placement of Terms and Conditions on Websites

The issue of acceptance and agreement in on-line context has been particularly difficult despite its seeming simplicity. The problem is that a website may offer to supply goods and services on one page, but its terms and conditions are found on a different part of the website or in an obscure hypertext link. There is no certain way to ensure that the customer has accessed these terms, let alone read them. Therefore, the question arises as to whether the offer and terms and conditions have been adequately communicated to the offeree. To ensure that customers are made aware of terms and conditions, customers may be forced to go through a set of screens and click the '*I agree*' button before proceeding to the next stage of contract formation. This 'click-wrap' technique widely in use by many U.S. websites provides an effective way of ensuring that the terms and conditions of the proposed contract have been communicated to the customers. Thus, in *Hotmail Corp. v van Money Pie Inc.*[88] it was held that a 'click-wrap' was a valid means of acceptance.

However, there have been cases where visitors to a website could be 'deep-linked' into the pages of a different website and hence may miss reading the terms and conditions associated with that different website's offerings. This has happened in the case of *Ticketmaster Corp. v Tickets.com Inc.*[89] where a court concluded that a contract had not been created between the plaintiff and defendant simply because the plaintiff had displayed terms and conditions at the bottom of its website. However, in *Specht v Netscape Communications Corp.* a U.S. Court of Appeals for the Second Circuit has ruled that the downloading of software did not constitute an adequate manifestation of assent.[90] It therefore meant that an arbitration clause would not be enforceable where the text is located behind a link off the bottom of the download screen. In this case the terms were behind a link that could be seen only by scrolling down on a Web page, a 'reasonably prudent' user would not have learnt of the existence of the terms before responding to an invitation to download free software.

---

[87] Finch, B 1999 'Consumer protection on the Internet', in Fitzgerald, A *et al.* (eds) *Going Digital 2000: Legal Issues for E-Commerce, Software and the Internet*, St Leonards: Prospect Media Pty Ltd, p. 265.
[88] 1988 WL 388389, 47 USPQ 2d 1020 (BNA) (ND Cal. 16 April 1998).
[89] 54 USPQ 2d (BNA) (1344 US D Cal. 27 March 2000); 2000 WL 525390 (C.D.Cal.2000).
[90] 150 F.Supp.2d 585 (S.D.N.Y. 2001); *Electronic Commerce and Law Report*, v. 7 n. 39, p. 999.

Valuable Consideration

Valuable consideration is the price for a promise indicating that more than good intentions are involved. On-line, valuable consideration may simply involve the giving of a credit card number, even though no funds are transferred; the information is given merely as security to further progress negotiations. A problem that may arise is when erroneous or stolen credit card numbers are proffered and it is here that the offeror bears the risks of non-payment for goods and services rendered.

On the Internet today there are various security devices that are used to ensure that financial and credit card information given electronically are encrypted to protect both the sender and the receiver. One hint that a user is entering a secure website may be found in the universal resource locator (URL) address itself. The usual form of an URL is given as http:// followed by the address, where the prefix refers to the hypertext transport protocol—the method this is used to access a Web page. However, secure websites use the following form https:// followed by the address. The extra 's' in the protocol indicates that it is a secure site and that the site uses secure socket layer (SSL) encryption of all data and information within the site.[91] Hence, one can reliably send credit card details that are encrypted in 'packets' which only the merchant website is able to decrypt.

Writing and Digital Signatures

The formal requirements of a contract in writing and that of a signature have now been overcome in an electronic environment. In Australia, the requirement for a written contract in electronic form is provided for by s 9 *Electronic Transactions Act* 1999 (Cwlth) and corresponding State and Territory legislation. Also a digital signature under s 10 of the same Act may now be used.[92] Digital signatures refer to the data included within a digital document that identifies the person who produced it. A digital signature can also be used to detect and track any changes that may have been made to the document in the interim between its sending and its receipt.

---

[91] Secure socket layer (SSL) is a general-purpose encryption system developed at Netscape Communications. Version 2.0 was released in 1994 as a feature bundled into the Netscape Navigator browser. SSL is designed to secure the transmission of messages in a network. Thus, an SSL transaction is referenced via https:// protocol rather than http:// protocol. See Brinson, J *et al.* 2001 *Analyzing E-Commerce & Internet Law*, Upper Saddle River, NJ: Prentice-Hall, pp. 173–174.

[92] The corresponding U.S. version is the federal *Electronic Signatures in Global and National Commerce Act (e-Sign)* 2000.

In Practice—Keep a Paper Copy

In e-commerce, there has emerged a practice in contract formation where one party will require its customers to download terms and conditions from a website and then to accept these by signing a hard copy of the contract and returning this to the other party by post. This simple act thus creates a higher degree of certainty in the transactions as not only is the notice for such terms given but also an agreement to particular terms of the contract. This will avoid misinterpreting representations made either in writing or during negotiations.

Terms and Conditions on Websites

Representations made during early negotiations, but not repeated later, are unlikely to form a part of the contract. Similarly, representations made after the formation of the contract will not be included in an agreement. On-line vendors must make their customers aware of relevant documents or further Web pages that contain contractual terms and conditions, or else these might not form part of the contract. In the event that such terms and conditions are unread, when a dispute arises, a court would examine the transaction and ask where a reasonable person would expect to find such terms and conditions on the website. On commercial websites, a reasonable person should be able to easily discover such terms and conditions.

Sometimes consumers are asked to waive their rights to certain privileges; for example, intellectual property rights. This is again possible using the click of a mouse to consent to the waiver as shown in the illustration below. The following is a form of waiver of intellectual property rights and demonstrates a slightly different approach to the acceptance of terms.

Practice Notes: Waiver of Intellectual Property Rights

You are about to enter a discussion area in which you are allowed to post messages and comments. By participating in this discussion group, you are entering into a contractual waiver of any intellectual property rights you may have in anything you post or up-load or type. In other words, you are giving up any claim of copyright to your messages that you post on this service.

Continued on page 342

Continued from page 341

There are two buttons shown at the bottom of your screen. If you agree to waive your intellectual property rights and to continue in the discussion group, select the button labelled '*I agree to waive intellectual property rights*'. If you do not agree, select the button '*I do not agree*' or hit any key on your keyboard, and you will be returned to the preceding screen.

Forum Selection Clauses

In the U.S. forum selection clauses in 'click-wrap' contracts have received favourable treatment in the courts. In *Hughes v McMenamon*[93] a Magistrate Judge of the U.S. District Court for the District of Massachusetts held that the forum selection clause in a click-wrap agreement between an Internet service provider and a subscriber was enforceable. In *Forrest v Verizon Communications Inc.*[94] the District of Columbia Court of Appeal held that a forum selection clause included in a click-wrap agreement for digital subscriber line service was reasonable and enforceable against a subscriber. The court was not moved by arguments that the agreement was unenforceable, either because it was contained in a lengthy on-line text message, was viewable only in a small scroll box or because the forum selected by the agreement did not provide a remedy in a class action. The court used the 'modern rule' granting presumptions of validity since the clauses were found not to be unfair or unreasonable.

The cases now suggest that consumer protection in cyberspace is returning one full circle to that of *caveat emptor* [buyer beware]. Legislation can protect the consumer only up to a point, for other eventualities, consumers must have their wits about them when contracting on-line (Schu 1997).[95]

5.8.3 Summary of Legal Issues

The development of a GIS contract is rarely simple. The drafting of a good contract is comparable to writing good software. Generally it is

---

[93] Civil Action No. 2001-10891-RBC, 28 May 2002 (2002) *Electronic Commerce and Law Report*, v. 7 n. 23, p. 568.

[94] D.C. No. 01-CV-1101, 28 August (2002) *Electronic Commerce and Law Report*, v. 7 n. 35, p. 905.

[95] Schu, R 1997 'The applicable law to consumer contracts made over the Internet: Consumer protection through private international law', 5 *International Journal of Law and Information Technology*, Summer, pp. 192–299.

a multistage process, beginning with the vendor proposing a design schedule and then creating a beta (working) version of the product, testing it and delivering it to the client for trials. Then the vendor creates the final version and delivers it to the client for final acceptance-testing including 'scripts' for software control and development.

Minimum Provisions

At a minimum such an agreement may need to include the following provisions:

- *Duties and obligations*: provide a detailed description of the duties and obligations of the parties and the deadlines for performance. Include detailed specifications here.
- *Testing and acceptance*: a GIS development agreement should include 'testing and acceptance' clause stating the number of days a client has to test and accept or reject the work. Such clauses generally provide that, if the client has not given notice of rejection within the stated period, the work will be deemed accepted. Set out a project management time-line Gantt chart.
- *Payment*: GIS development agreements often provide for several payments over the course of the development process. Typically, the first payment is due when the contract is signed. The next payment is due when the completed project is accepted by the client. Additional payments are due when the client accepts the beta version and the final version. Sometimes the agreement provides that the developer will receive a bonus for completing the entire project by a certain date. Set out a schedule of payments.
- *Proprietary rights*: address copyright and other intellectual property ownership and licensing issues here. Include also royalty payments where the products are licensed for use only.
- *Warranties and conditions of use*: as a warranty is a legal promise that a statement or representation is true, this clause must be included to state what the promises have been. List the promises made.
- *Other uses*: what use may the system be put to, for example, commercial, consultancy services, and academic research and teaching purposes? What has been agreed to in regard to value-adding and resale of data and information? State permissible and impermissible uses.

- *Developer's credit*: what form will the Developer's 'credit' take? Credits usually consist of the Developer's logo and a few lines of text and a Web address or other contact details. Give an example of how the credits will look.

- *Termination*: termination clauses provide that either or both parties have the right to terminate the contract under certain circumstances. Generally, termination clauses describe breach of contract events that trigger the right to terminate the contract. Some contracts provide that one or both parties may terminate the contract for convenience, on notice to the other party and on payment of any fees. Termination clauses also describe the methods of giving notice of the exercise of a termination right and whether the breaching party must be given an opportunity to mitigate the breach before the other party may terminate the contract. Give examples of circumstances under which a contract may be terminated.

- *Arbitration clause*: this clause will state that disputes arising under the contract must be settled through arbitration rather than through court litigation. State the form of alternative dispute resolution: conciliation, arbitration, and mediation, and how the forum and arbitrator are to be selected.

- *Merger clauses*: merger clauses state that the written document contains the entire understanding of the parties. The purpose of merger clauses is to ensure that evidence outside the written document will not be admissible in court to contradict or supplement the terms of the written agreement. State if there are any such clauses to either be included or excluded in the contract.

- *Exclusion of liability*: state here what is being excluded and what may be included.

- *Replacement, refunds and upgrades*: state what replacement, refund and upgrade agreements are in place. Identify the escrow agent if one is to be used.

A facsimile of common agreement provisions used by Geoscience Australia is given below, including its policies on licensing and copyright taken from their *Product Catalogue 2002*.[96]

---

[96] Geoscience Australia 2002 *Product Catalogue and Sampler 2002 CD*, Canberra: Department of Industry, Tourism and Resources.

Practice Notes: Geoscience Australia Copyright and Licensing Policy

Copyright and Ownership of Data

In general terms, copyright is the exclusive right given to an author by the *Copyright Act* 1968 (as amended in 1980 and 1984) to reproduce, publish, perform, broadcast and adapt a work. Geoscience Australia is the custodian of the copyright of the products it produces on behalf of the Australian Government. Geoscience Australia does not give away copyright. Rather, it grants a right to use its products as specified in any application, i.e. a licence.

In the case of all satellite image products, with the exception of Landsat products, the Satellite Operator holds the copyright of the imagery. However, Geoscience Australia is authorised to licence the data on behalf of the Operator.

Licensing

Rather than selling the actual data, Geoscience Australia sells a licence to use the data. For most digital products this requires a signed licence agreement to be completed by the customer before a licence is granted. For some products, a licence is granted upon a gesture signifying agreement to the licence conditions, for example, breaking the seal on a product or registering on the Internet.

Conditions of Use of Data and Imagery

Geoscience Australia supplies data products to customers under a licence. Copyright is retained in any data made available to another party, even if no licence fees or royalties are sought, and even if the other party is extended very liberal rights in the use of the data.

For example, there is no licence fee for most products downloadable from the Geoscience Australia Internet site. However, the Commonwealth still retains copyright to the data and users need to abide by the conditions of the license which is provided as part of the download.

A copyright notice must accompany the distributed product. Full details of the conditions of use of Geoscience Australia data products are contained in the licence agreement.

Continued on page 346

Continued from page 345

Exclusion of Liability

Geoscience Australia gives no warranty regarding a product's accuracy, completeness, currency or suitability for any particular purpose. Geoscience Australia's liability for breach of any statutory warranty is limited to replacement of the product or refund for the purchase price. Geoscience Australia disclaims all other liability for all loss, damages, expense and costs incurred by any person as a result of relying on the information and data in the product.

Commercial Applications

Geoscience Australia encourages the widest possible use of any material over which it administers copyright within the provisions of the *Copyright Act.*

Minimal restrictions apply to the value-adding or commercial use of fundamental spatial data. ... The restrictions include completion of a licence and inclusion of a copyright notice with the distributed product as well as maintenance of distribution records that can be accessed by the Commonwealth.

Commercial use of data and map products *not* included in the Spatial Data Access and Pricing Policy will require a special licence.

Specific conditions apply to a commercial licence and they include payment of royalties. The royalty required will depend upon the recommended retail price of the publication or product, the size of the publication, the proportion of Geoscience Australia material in the product compared to its total size, the proposed production or print run, type of product and method of distribution or access to the value- added product and other information which will enable the licence to be determined.

Geoscience Australia undertakes to give prompt consideration to all requests for use of Geoscience Australia material. All requests are treated strictly Commercial-in-Confidence.

Value-added Reselling of Data

Value-added reselling of data and map products not included in the Spatial Data Access and Pricing Policy is defined as the use of a Geoscience Australia product in the provision of another product or service to a third party whether or not a fee is charged for that service or product.

Continued on page 347

Continued from page 346

Value-added resellers (VARs) are required to purchase a licence (at the normal list price) to use the data for a particular VAR application plus pay a royalty fee for each use of that data in the application.

Royalty payments are usually negotiated on a case by case basis.

Replacement and Refund Guarantee

Geoscience Australia is committed to providing quality products and services. In the unlikely event you are provided with a defective product, we guarantee to promptly provide either:

- a replacement or
- a full refund.

A replacement or refund will be provided only where:

1. The product is returned within 30 days of purchase and includes the complete product (i.e. all data, software, media, documentation and packaging) to Geoscience Australia or the place of purchase where the product has been purchased through Geoscience Australia's registered network of retailers. You may retain no copy of the product or anything derived from the product (other conditions pertaining to termination, which are specified in the Licence Agreement, also apply).
2. Proof of purchase is provided within the returned product.
3. Geoscience Australia agrees the product is defective and has not been purposefully or wilfully damaged.
4. The product has been damaged during shipment.

Geoscience Australia will not replace or refund products where the product cannot be proven to be defective.

Source: Geoscience Australia 2002 *Product Catalogue and Sampler 2002 CD*, Canberra: Department of Industry, Tourism and Resources.

5.8.4 Contract Precedents

Contract precedents provide a template for those who are either contemplating developing a contract for their own use or who wish merely to use those that have already been developed. A sample list of contracts and precedents from Knight and FitzSimons (1990)[97] and Morgan and Stedman

---

[97] Knight, P and FitzSimons, J 1990 *op. cit.* pp. 211–340.

$(1995)^{98}$ is given below. The precedents serve to assist in their rewriting and reworking for particular purposes and readers are advised to consult these before venturing out and writing their own contracts. None of the precedents from the two references have been reproduced, as it would have unnecessarily increased the length of this book.

Practice Notes: Sample List of Contracts and Precedents

From: *Knight and FitzSimons* (1990: 211–340)

A   Boiler plate clauses
B   End-user single software licences
C   Software maintenance agreements
D   Hardware purchase agreements
E   Equipment loan agreements
F   Hardware maintenance agreements
G   Installation and implementation agreements
H   Software development agreements
I   General contractor agreements
J   Employment agreements
K   Assignment of copyright
L   Confidentiality and technology disclosure agreements
M   Escrow agreements
N   Reciprocal back-up site agreements
O   Database/network access agreements
P   Software distribution agreements
Q   Bureau processing agreements
R   Facilities agreements

From: *Morgan and Stedman* (1995: 251–612)

A   Hardware sale and installation agreement
B   Rental maintenance agreement
C   Software licence agreement
D   Commissioned software agreement

Continued on page 349

---

[98] Morgan, R and Stedman, G 1995 *Computer Contracts*, London: FT Law & Tax, pp. 251–262.

Continued from page 348

E   Shrink-wrap licence agreement
F   Multiple copy software licence agreement
G   Exclusive distributorship agreement
H   Software marketing agreement
I   Software distribution agreement
J   Beta test agreement
K   Hardware maintenance agreement
L   Software maintenance agreement: licence
M   Software maintenance agreement: commissioned
N   Software code deposit agreement
O   Turnkey agreement
P   Bureau service: general on-line
Q   Bureau service: batch
R   Network service level agreement
S   Consultancy agreement
T   Master agreement for programming services
U   Non-disclosure agreement
V   Unsolicited disclosure agreement

Summary

Contractual rights and obligations cannot take place in isolation. This is because there are interactions between various theories of law, including tort, intellectual property, and statutory provisions. In many instances common law contract rights and obligations have been superseded by legislative provisions which have been written to formalise the rules, especially those relating to personal liability, assessments of loss, damage, injury and to define clearly the respective roles of all parties. In part these developments have been in response to the rise of consumer protection concerns, especially where the two parties are contracting at arm's length, for example, for Internet services and products, and in some cases where the parties are in unequal bargaining positions.

The question of ownership, copyright rights and patent is an area both of contract law and of intellectual property rights. Most software development contracts do not assign copyright, but only a licence to the end-user to use the computer programs. There is an expectation on the part of the customer to have 'rights' to use as well as claims to 'ownership' of the software products as a result of the contract. However, industry practice seems to suggest that there are no such 'rights' or claims to 'ownership' unless the developer has explicitly 'assigned' ownership to the customer. In the absence of such an assignment the customer may only have 'moral' rights to the computer program.

Some consideration should also be given to software maintenance after delivery of both the computer hardware and software. In the case of a contractor/consultant brought in to develop software programs the practice seems to be that the contractor is entitled to keep working documentation with no obligation to deliver the source code. The argument is that the situation is analogous to architects and lawyers who keep and retain documents relating to their clients. However, the contractor may agree to assign these to the customer.

Chapter 6
Geographic Information and Liability Standards

Learning Objectives

After reading this chapter you will:

- Understand the context of legal liability when it arises in geographic information (GI) systems and services and how the legal theories are slowly evolving.

- Be able to outline legal standards of liability in contract, tort or statute impinging on geographic information products, and services or a combination of both.

- Know how to avoid legal exposure and to minimise legal risks in the provision of geographic information systems and services.

- Attain a mastery of the application legal risk management strategies in order to minimise liability risks.

- Grasp the different strategies for use in global positioning systems and cartographic products together with data quality assurance seals and accuracy standards.

6.1 Introduction

Liability in law is a broad term that includes almost every type of duty, obligation, debt, responsibility or risk arising by way of a contract, a tort

Geographic Information Science: Mastering the Legal Issues   George Cho
© 2005 John Wiley & Sons, Ltd   ISBNs: 0-470-85009-4 (HB); 0-470-85010-8 (PB)

or a statute. While a contract may regulate the extent of liability, common law tort also governs who should be legally responsible for an act or activity that has caused harm to a person. The scope, degree, and extent of liability may also be determined by reference to a statute. These 'liabilities' under the different theories of law such as common law contract and tort, and statute are measured against established legal standards. While seemingly objective and clear-cut in law, the liability in the use of GI systems and services is not fully understood and is in the process of evolving, but slowly. This uncertainty is because of the wide array of current applications of geospatial technology, but also the future applications of GI science in ways that have yet to be contemplated. But, the 'source of liability would be an important and possibly controversial issue in the determination of which entities might be liable for damage resulting from the use of geographic information outputs or geographic information systems' (Johnson and Dansby 1995).[1] This complexity and diversity of environments in which GI is used presents broad, undefined liability concerns, and legal standards in use by the industry are only now being developed.

This chapter explores what those legal standards are, whether in contract, tort or statute and evaluates each of these theories in so far as it impinges on the provision of geographic information (GI) either as a product, service or a combination of product and service. These various legal theories are explored in order to understand the basis of litigation, such as the provision of erroneous information and services deemed unprofessional, and to examine the defences that may be raised against those claims. The examination itself may suggest a framework for limiting liability and better practice in the provision of information and services.

The first part of this chapter establishes the importance of legal standards and guidelines. These legal standards have been evolving and changing over time, becoming more refined and stringent in the face of technological progress. The second part interweaves an examination of contract, tort and statute liability exposure and risks in the context of the provision of GI. In the main, liability is imposed principally on the provider of GI systems, databases, and reports, but this liability is also moderated in part by the user's role in contributing to the fault-based equation. This section is extended to include a sub-section on other

---

[1] Johnson, JP and Dansby, HB 1995 'Liability in private sector geographic information systems', *Proceedings of the Conference on Law and Information Policy for Spatial Databases*: Tempe, AZ (NCGIA, University of Maine, Orono, 1995) and at http://www.spatial.maine.edu/tempe/johnson.html.

potential liability risks. In the third part there is an evaluation of various liability risk management strategies both in risk minimisation as well as adopting tactics to eliminate liability altogether. Finally this chapter suggests recommendations and an agenda for users and providers of GI to raise standards that avoid legal liability. These suggestions are made within the parameters of recent developments in global spatial data infrastructures and distributed GI.

6.2 Legal Standards and Guidelines

A discussion of the legal standards should begin with first principles and proceed towards practice and application. Such a discussion may provide a review of how well principles accord with practice. However, the discussion may also proceed from an observation of real-world incidents and trace the cause of loss and damage. It seems that the latter path should be taken to better appreciate the legal dilemmas without an assessment of the standards that may be produced by the context itself. The following describes five cases that have common and abiding themes—error in maps, navigation charts, and/or instruments that have caused accidents and loss of life.

Case 1: A Coal Mine Disaster

At 05:30 hrs on 14 November 1996 four men of the night shift at Gretley Colliery, Newcastle Wallsend Coal Company Pty Ltd in New South Wales (N.S.W.), Australia were engulfed by water, swept away and drowned. The four men were in a team of eight who were in the process of developing a roadway in an area of the mine called the 50/51 panel, using a continuous mining machine. The water came from a long abandoned old working of the Young Wallsend Colliery. The mine was using a plan approved by the N.S.W. Department of Mineral Resources (DMR). The plan showed that the Young Wallsend Colliery was more than 100 m away from the point of holing-in. Investigations after the event have since shown that the plan was wrong. At the commencement of the night shift at 23:00 hrs on 13 November 1996 the Young Wallsend Colliery was only 7–8 m away from the point of holing-in. Also the workings of the old mine were full of water. As a result of this accident the owners of the mine and other employees have been found guilty of breaches of the *Occupational, Health and Safety Act* 1983 (NSW).

A formal investigation of the causes and circumstances of the accident was undertaken by JUDGE JAMES STAUNTON who issued a two-volume report in 1998.[2] The inquiry found that the department had made the original mistake of creating the incorrect maps—the 'fatal error' that sparked the tragic chain of events. Key personnel of the DMR had shown 'an absence of reasonable care' and the maps, which the department had incorrectly created 'sat like a loaded gun in the archives of the department'.[3] The DMR has since spent about AUD$0.5 million on reviewing and digitising all records of mine workings it holds with a view of improving safety.

Case 2: Air Safety Corridors

Marrickville City Council, in Sydney, N.S.W. raised serious safety concerns with the Civil Aviation Authority (CAA) about flight paths into and out of Sydney's Kingsford Smith Airport. The CAA published maps in 1995 of new flight paths at the airport and also issued pilots with new latitude and longitude coordinates. These flight paths define the routes with navigational allowances indicating how far the pilots could deviate from them. Marrickville Council used their own surveyors to independently chart the coordinates of the flight and landing paths. The Council's maps show that the flight paths for planes landing from the north onto the parallel runways 16L and 16R covered wide areas of the east and west of the inner western suburbs and intersect closer to the airports than are shown on the official CAA maps. Pilots flying into Sydney on unofficial routes thus affect more residents than is officially acknowledged. Moreover, using the path for the new runway 16L, aircraft will be flying at a higher altitude than those landing at the old runway 16R since the new runway begins 2 km south of the old one. This difference in distance would mean that there is a vertical separation in altitude of about 1000 feet, leaving very little room for error. All landings on parallel runways are staggered and the installation of new precision radar also permits simultaneous landings. This means that landing planes could be occupying the same airspace separated vertically by only about 1000 feet of altitude.[4]

---

[2] Staunton, JH 1998 *Report of a Formal Investigation under Section 98 of the Coal Mines Regulation Act, 1982 (NSW)*, 2 vols. Sydney: NSW Government.

[3] See Connolly, E 2000a 'Department of disaster escapes Gretley charges', *The Sydney Morning*, 17 April 2000, p. 5; and Connolly, E 2000b 'Gretley: Mine and men face charges', *The Sydney Morning Herald*, 18 April 2000, p. 6.

[4] Whelan, J 1995 'Planes straying off the official flight paths, new maps reveal', *The Sydney Morning Herald*, 18 April, p. 2.

Case 3: National Park Bush Fire

Four N.S.W. National Parks and Wildlife Service (NPWS) officers were killed from smoke suffocation in a burn-off operation. The officers were given maps that showed two possible escape routes, but which ended in impenetrable bush or a line of cliffs. In a Coronial Inquiry, the Senior Deputy State Coroner informed the court that the hazard reduction burn by fire fighters at Ku-ring-gai Chase National Park on the 8 June 2000 on behalf of the NPWS did not follow proper management procedures. There was insufficient planning and preparation for the two-day burn-off. In addition, there were deficiencies in the maps used in the operations. The information contained in a map issued by the NPWS showed a cleared hilltop, which potentially could have provided shelter from the fire. But the map did not show a 30 m cliff which stood between anyone trying to escape the fire and the cleared area. The map also showed a path known as Wallaby Track running directly towards a local motorway. In reality this path, after the junction with a spur track, twists into impenetrable bush. A conclusion of the Deputy State Coroner was that the original botanical map had not been ground-truthed to include specific details and did not mark areas with safe refuges to retreat to as required in the fire management procedures guidelines.[5]

Case 4: Grounding of Cruise Ship

The 15 December 1998 grounding of the Royal Caribbean cruise ship *Monarch of the Seas* on a coral reef off St Maarten, Netherlands Antilles was the result of operational errors on the part of a bridge crew. This was the conclusion of a team of investigators from the US Coast Guard Office of Investigations and Analysis and the Norwegian Maritime Directorate.[6] Most significantly, the grounding was the direct result of the crew's failure to update the ship's charts to include a change in the location of a buoy marking a reef and the subsequent reliance on this buoy as a sole means of fixing the ship's position. The incident could have resulted in the total loss of the ship, passengers and crew. The *Monarch of the Seas* displaces 74 000 tonnes, is 880 feet long and is capable of carrying 2557 passengers and crew.[7]

---

[5] Brown, M 2001 'Maps led firefighters up paths with no exits', *The Sydney Morning Herald*, 17 July, pp. 1, 2.
[6] See Report 2003 at http://www.uscg.mil/hq/g-m/moa/casualty.htm.
[7] Braden, T 2003 'Crew of cruise ship that hit reef faulted for failing to update charts', *Professional Mariner*, v. 75 October/November.

Case 5: Sinking of a Yacht

The sinking of the Sydney, Australia yacht *Redrock Communications* was blamed on a faulty global positioning system (GPS) receiver. The yacht sank after it ran onto chartered rocks northeast of Coffs Harbour near South Solitary Island ten minutes after a GPS reading on the yacht showed its path was clear. No lives were lost and a boat from the Gold Coast rescued the crew. The GPS receiver was reported to have had poor coverage and failed to give an accurate position on previous occasions.[8]

Map Quality and Liability

The Gretley mine disaster raises the issue of map quality and who has the responsibility of checking the data. The maps used by the miners were copied from previous plans that in hindsight are known to have been wrong. The errors in those map sheets have been perpetuated and indeed reinforced by its incorporation into the maps that were in use. Apart from the unusual and disturbing lack of a legend, the original plan carries an inscription that reads 'Copied from the Colliery plans at the Coalfield Office by Herbert Winchester 21st March 1892'. In the Sydney airport flight path case there are apparent misstatements given by the authorities, especially where the public are given one set of 'facts' while the pilots have been given a different set in the air navigation charts. Only when these charts are compared, recalculated and calibrated to a common coordinate referencing system that the differences show up. The question here is to ask whether there has been negligence. Has there been a negligent misstatement or is there a mistake of judgement in the release of two sets of flight paths to different sets of people? Only through due diligence and the result of a civic protest against the new flight paths by the Council have these concerns been brought to the forefront.

The death of the firefighters in Ku-ring-gai National Park identified the lack of 'ground-truthing' of the maps given to the fire fighters. Here one may observe the classic dilemma of the very nature of a map—the problem of attempting to depict a three-dimensional world onto a two-dimensional surface. Whether a set of contours depict a hill or a valley can be ascertained only if one reads the values of the contour intervals carefully. In an emergency, fine details such as contour values can easily be missed by inexpert users with disastrous consequences. In the incident

---

[8] Gibson, J 2003 'Satellite's gone, we're going to hit: Skipper', *The Sydney Morning Herald*, 29 July, p. 3.

involving the cruise ship the breakdown in the chain of responsibilities and failure to follow set procedures led to its grounding. A failure to update the ship's chart was an act of omission of a statutory requirement. The lack of experience on the part of the Second Officer, the strained relations among the bridge crew, fatigue, and the vicious bout of diarrhoea that impaired the Master's ability to remain on the bridge the moment of the accident were identified as contributory factors. There was no equipment failure of any kind, unlike the loss of the yacht off the NSW coast. In the sinking of the yacht, the apparent error in the GPS readings precipitated the disaster. It has been observed by the Australian Maritime Safety Authority (AMSA) that while as a rule GPS were 'extremely reliable', glitches of the kind reported were not unheard of.

These five cases demonstrate the classic issues of standards to be attained if liability is to be avoided. The duty to take care, the responsibility of due diligence to those who may be affected by a lack of care, the reliance on information to one's detriment, and the subsequent injury, damage and loss that occurs, are established legal standards. Such standards have been set either through the common law or through statutory mandates in order to ensure compliance. While there appears to have been due diligence in all cases, the losses suggest that these standards of care have not been high enough. Noteworthy too is the fact that, while blame has been laid on the maps and charts, the makers of these navigation instruments have escaped sanction. It poses the question: should they be responsible?

The incidents indicate that there has been human error, omissions, misjudgment as well as equipment failure, but no one factor was identified as totally responsible for the losses. It is also possible to argue that the incidents were in the nature of things and occurred as an accident and thus it is difficult to lay blame on any one or any thing. In defence, an underlying message is that the mapmaker alone may not be wholly responsible for either errors or omissions and that the map user must also take due caution and care. The reality is that maps contain errors and may already be out-of-date even before the maps leave the shelves of the shop.[9]

In an increasingly digital world, the map of the future will be embedded in IT systems of various kinds. Such systems may include the digital terrain models (DTM) used in training airline pilots to the location-based services (LBS) that rely on geo-codes and unique street addresses through to road centre-lines as reference points in emergency dispatch

---

[9] Fairall, J 2002 'Death in a back paddock. Liability for bad data rests with the user', *GISUser*, Issue 50, February–March, p. 7.

systems. The embedded data will be invisible to the user and error detection will become even more difficult. Data can be easily manipulated and mixed with data from other sources, sometimes of unknown lineage and perhaps at inappropriate scales. The easy means with which to access these 'user-friendly' systems exacerbate the situation when used in inappropriate circumstances. While the authorship and ownership of the data and information is becoming more uncertain, the quality of the data may have degraded to the point that it cannot be guaranteed. The first indications that there are serious problems with the data are when accidents take place. By then it has become too late to redress the situation other than make corrections and error checks to mitigate faults.[10]

Liability has been shown to possibly arise from inaccurate, incomplete, and misleading information or data as well as incorrect decisions. The following table summarises liability risks, exposures, and defences in the provision of GI products and services. Table 6.1 captures some of the legal liability theories that are described in the next section.

Table 6.1 Liability risks, exposure and defences: provision of GI products and services

Activity	Liability	Exposure to	Defences
Sale of GI	Yes	Negligence Misrepresentation	Contractual Requisite duty of care
Gratis distribution of GI data	Yes	Negligence Failure to secure intellectual property	Due diligence
Compilation of GI products and sale to third party	Depends on use by third party and whether resale is contemplated	Strict liability for products Negligence for services Intellectual property rights	Professional— industrial standards Metadata
Distribution of GI data under public duty, community obligation—FOI request, open records, redaction	Possible exposure to liability	Statutory liability Malpractice Negligent misrepresentation	Security Official secrets Privacy Sovereign immunity First Amendment

---

[10] Fairall, J 2000 'Accuracy and liability. A looming nightmare for GIS Officers', *GISUser*, Issue 42, October–November, p. 9.

6.3 Legal Liability Theories

The principal legal liability theories are found in contract, tort and legislation. The focus here is on liability for GI data and information as well as for products from an information system. What is liability is explored with a view to avoiding or minimising one's exposure to it by taking care and by using liability risk management techniques. In information systems the potential liability from harm may result as much from use, misuse and inappropriate uses of the data and information as well as from the decisions that are made using inaccurate data or in the misuse of accurate information. Advertently, users may be unaware of data errors and may not appreciate what the information represents. Also, users of GI products may not appreciate how the information has been derived from the data or what computations have been performed to the raw data (Epstein 1991: 499).[11] Inadvertently, users may rely on seemingly correct information, but which may turn out to be wrong and to their detriment.

Legal liability arises from a breach of standards, these standards have been set either by statute or common law, and the breach of these standards suggests some form of negligence. Figure 6.1 attempts a simple encapsulation of a causal negligence–liability chain, identifying an issue, its cause leading to an injury and loss and the question of who pays.

To further explore the negligence–liability chain, the *first* subsection below examines contract and strict product liability issues because GI products are featured prominently. After discussing the goods–service distinction from the viewpoint of consumer protection and sale of goods legislation, strict product liability imposed on products with GI elements are then analysed. The implications of some exclusions and limitations arising from contractual liabilities are explored as possible solutions to the duties imposed by the cases. In the *second* subsection tortious liability is described in regards to its scope, professional standards, and acts and omissions causing pure economic loss. Also discussed here are the roles of negligent misstatements and misrepresentation in bringing about liability. The relevance of these themes is given against the background of GI services and products. Some defences against liability are discussed such

---

[11] Epstein, EF 1991 'Legal aspects of geographic information systems' in Maguire, DJ, Goodchild, MF and Rhind, DW (eds) *Geographic Information Systems: Principles and Applications*, 2 vols. London: Longman, pp. 489–502.

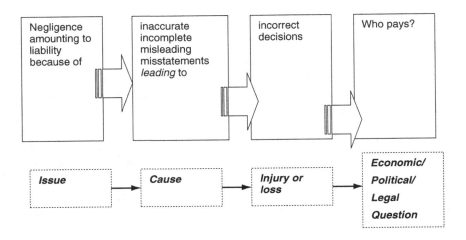

Figure 6.1 A causal negligence–liability chain

as contributory negligence and vicarious liability, and in an on-line setting, the use in the U.S. of the so-called First Amendment defence to stave off liability. The *third* subsection on statutory liability examines those duties imposed by legislation and the *final* subsection briefly reviews other liability theories that might be looming on the horizon that may have an impact on the GI industry including privacy, and defamation.

6.3.1 Contract and Strict Products Liability

Good–Service Distinction

The good–service distinction is especially problematic in GI because the standards applied may vary, depending on the end result. In the U.S. if the end result is a product or good then the Uniform Commercial Code (UCC) standards are invoked. If this is the case there may be implications arising from strict product liability laws. However, if the end result is the provision of a professional service then issues of various types of negligence may be implicated. Where the end result is both a product and a service then an arbiter of fact may have to weigh up the respective contributions to the loss suffered and apportion the blame accordingly in the circumstances of the case (Table 6.2).

In Australia, contractual liability arises from the implied terms of merchantability, fitness for purpose and due care and skill in consumer agreements. In addition, these obligations may be governed by the *Trade*

Table 6.2 GI product–service exposure to liability scenarios

Supply of	Exposure to strict liability	Exposure to negligence
GI product	Yes	No
GI service	No	Yes
GI product and service	Yes	Yes

Practices Act 1974 (Cwlth) (TPA) as well as the various state and territory consumer protection acts.[12] In Australia, for constitutional reasons, such consumer protection laws are based on Commonwealth, State and Territory consumer protection legislation. At the federal level the TPA applies generally to corporations rather than individuals, other than those individuals engaged in inter-state trade and commerce. Australian consumer protection laws imply non-excludable terms in certain contracts in order to prevent unfair conduct in the allocation of liabilities in one-sided agreements.

Under Part V Division 2 of the TPA, a number of non-excludable conditions and warranties are implied into all consumer contracts. Any term of a contract that excludes, restricts or modifies rights or liabilities under implied terms will be void.[13] For the purposes of the TPA, a consumer would have acquired goods or services where the price was less than AUD$40 000 or if the amount was more than AUD$40 000 the goods or services were of a kind ordinarily acquired for personal, domestic or household use.[14] Under the TPA warranties and conditions for *goods* include:

- a condition that the goods supplied by description will comply with that description (s 70);
- a condition that the goods will be of merchantable quality (s 66(2));
- a condition where the purpose is made known that the goods are fit for the purpose (s 71(2)).

Warranties and conditions for *services* include:

- a warranty that the service will be rendered with due care and skill (s 74(1));

---

[12] *Fair Trading Act* 1992 (ACT), *Consumer Affairs and Fair Trading Act* 1990 (NT); *Fair Trading Act* 1987 (NSW); *Fair Trading Act* 1989 (Qld); *Fair Trading Act* 1987 (SA); *Fair Trading Act* 1990 (Tas); *Fair Trading Act* 1985 (Vic); and *Fair Trading Act* 1987 (WA).
[13] TPA s 68.
[14] TPA s 48.

- a warranty that any material supplied in connection with the service will be reasonably fit for the purpose for which they are supplied (s 74(1));
- where the consumer makes known any particular purpose for which the services are required, a warranty that the services and supply of materials will be reasonably fit for that purpose (s 74(2)).

Where there is a breach on any condition or warranty set out above, a consumer has the right to take action for breach of contract rather than for a breach of the TPA.[15] Under the TPA, it is therefore not possible to exclude conditions and warranties. On the other hand, the TPA permits suppliers to limit their liabilities to the replacement or repair of goods or the cost of their replacement and in the case of services to resupply the service or the cost of the resupply. A breach of contract by a supplier will normally mean potential liability in damages to the customer. The latter may seek damages to cover loss suffered as a result of the breach. In certain circumstances the customer may treat the contract as at an end.

In the U.K. a company supplying goods and services to another for a fee will normally be bound by a contract that exists between the two. Ideally the terms of the contract will record in writing the terms that have been agreed between the parties. These express terms may also be supplemented by implied terms, in one of several ways, the chief of which is that implied by statute. For software and data the *Sales of Goods Act* 1979 (UK) and the *Supply of Goods and Services Act* 1982 (UK) will govern most contracts. For goods, the terms implied include that the goods are of satisfactory quality, be fit for purpose for which they are sold or supplied and conform to any description applying to them. However, for the supply of services, the implied term is that the service must be supplied with reasonable care and skill.

In the U.K. as well as in Australia, the quality of the software or data poses considerable difficulties with the interpretation of the terms used since the legislation has not been drafted specifically to cover the supply of software or data. *Satisfactory quality* is a term that has been substituted for merchantable quality, but with no precise definition other than meaning that there is a 'freedom from minor defects'. With physical objects the issues are not problematic, but with intangible objects such as digital data and information the problems can be greater. Providing the data and information on a disk may amount to the supply of a product—the disk

---

[15] *E v Australian Red Cross Society* (1991) 27 FCR 310.

itself. However, in cases where it involves simply the electronic transfer over the Internet, the customer may never receive a physical object.[16]

In relation to the supply of software and data, the reliance is normally on express terms. A detailed document will specify the operational and functional environment for the supply and use of the data and software. A fixed-term warranty will also accompany this agreement which may state that the software will operate as described in the specifications. The express terms may sometimes be written in such a way to exclude the effects of implied terms. In the U.K. the *Unfair Contract Terms Act* 1977 (UK) addresses issues of exclusions and limitation clauses in contracts. This piece of legislation also permits implied terms to be excluded in the contract on the proviso that it is reasonable to so exclude those terms.

Under English law there is an implication in all contracts for the supply of services that it be performed with reasonable skill and care. This will also include representations given for professional advice. Thus, in *Greaves & Co. (Contractors) Ltd v Baynham, Meikle & Partners*[17] a court invoked such an implied warranty. The case involves the design of the first floor of a building expressly stipulated to withstand the weight of loaded fork-lift trucks, but which was later found to be inadequate. The Court of Appeal held that even with the absence of an express warranty of suitability of a product given by the designer, such a warranty might be implied from the surrounding circumstances. By analogy, the message for GI professionals is that they should be wary of giving assurances and/or representations that a system or software application will produce a given result.[18]

Strict Product Liability

The issue of a strict product liability in the supply of goods with elements of GI has been debated in the courts in the U.S. and the extant litigated cases there provide good examples of the inclination of judicial opinion. It is to that jurisdiction that we turn in order to focus on developments in the law and policy directions that have been taken.

The law on strict liability and product liability in the U.S. is founded on § 402A of the *Restatement (Second) of Torts* 1965 and the *Restatement*

---

[16] See *St Albans City and District Council v International Computers Ltd* [1995] FSR 686 (QBD), unrept. 26 June 1996 (Ct of Appeal).

[17] [1975] 3 All ER 99.

[18] See *Ross v Caunters* [1980] 1 Ch. 27; *Independent Broadcasting Authority v EMI (Electronics) Ltd* (1980) 14 Build LR 1.

(Third) Torts: Products Liability 1997.[19] The principle of strict liability in the U.S. is that a seller is strictly liable, even though the seller has exercised all possible care and attention in the preparation and sale of the product. The seller is strictly liable to the buyer of the product if the latter suffered an injury as a result of using it. This principle of strict liability for producers of defective products has also been adopted by statute in the E.U., in the U.K. and Australia.[20] The rationale for such developments in strict liability stems from concerns for consumer protection as well as accident deterrence by placing responsibility for defective products on producers and manufacturers. This regime relieves consumers of the burden of proof and imposes on manufacturers a duty to keep unsafe defective products off the market.

Failure to warn of risks may make a product defective. Warnings may either reduce risks or help potential users to make an 'informed choice'.[21] Moreover, strict liability does not require proof of failure to use reasonable care. So long as the defect was discoverable in the light of scientific and technical knowledge there is a duty to warn. This duty may also arise after the product is marketed. In the context of geospatial products the issue is whether such warnings accompanying products containing elements of GI, such as maps and other information incorporated in a physical product, addressed to the user, would absolve producers of all responsibility.

Justices in some U.S. courts have indicated that they are prepared to impose strict liability in the case of electronic maps and charts when decisions hinged on the condition of the product and not the conduct of the user. However, the product–service distinction debate to sheet home liability has been an unfruitful one. It is settled that physical, tangible items such as cars, toasters and other consumer items, included within the product definition, and if defective and cause harm, would render the manufacturer strictly liable. On the other hand, services are not considered suitable grounds for a liability action and the courts are reluctant, for example, to apply strict liability principles to engineers and architects for rendered services.[22] One

---

[19] *Restatement (Third) Torts: Products Liability* 1997, St Paul, MN: American Law Institute.

[20] E.U.: *Product Liability Directive* 1985; E.U. 1999c *Green Paper on Product Liability*, COM (1999) 396 Final and 2nd Report, January COM (2000) 893 Final and at http://www.europa.eu.int/comm/dg15/en/index.htm [30 June 2004]. U.K.: *Consumer Protection Act* 1987 (UK); and Australia: *Trade Practices Act* 1992 (Cwlth) Part V A.

[21] Fleming, JG 1998 *The Law of Torts*, Sydney: LBC Information Services. Chapter 23 'Product Liability', pp. 530–558, at p. 554. E.U. 1999b Directive 1999/34/*EEC on the approximation of the laws, regulations and administrative provisions of the Member States concerning liability for defective products*, Official Journal L 141, 04/06/1999, pp. 0020–0021.

[22] Genignoni, MC 1981 'Product liability and software', 8 *Rutgers Computers and Technology Law Journal* 173, at 197.

reason may be that such services are furnished for one client at a time and so the cost of liability cannot be spread as widely compared with a whole line of tangible products. Moreover, the services rendered by professionals may cause economic loss rather than physical injury, and are thus excluded in any event in the liability balance sheet.

A series of cases in the U.S. have had a profound impact on the GI industry on two counts. First, the courts in these cases have held that a chart or an aid to navigation is a 'product'. Second, one may be strictly liable if information, although correct, but its design in regard to scale, graphics, symbology, may be misleading. The courts have viewed both maps and charts as 'practical' rather than 'creative' products whose purpose is to guide. In *Brocklesby v United States*[23] navigational charts were considered products. These charts were distinguished from other publications such as books in that the former were to be used by pilots who made decisions based on information contained in them. As a result, the duty of care is pivotal because of the element of reliance. First, one will be liable when it may be established that an injury resulted from the use of the data. Secondly, liability may also arise from negligence in that the creators of the work have failed to perform in accordance to a standard of care reasonably required of them when undertaking professional work.

In *Brocklesby v Jeppesen*[24] a pilot using a chart produced by Jeppesen to make a landing crashed into a mountain, killing the entire crew. Action brought against Jeppesen and the U.S. government claimed that the accident was caused by defects in the instrument approach procedure developed by the government and published by Jeppesen. The company was held liable as the information provided involved the gathering, repackaging and selling of federal aviation data. The Federal Aviation Administration (FAA) 'instrument approach procedure' was considered a compilation of data. Jeppesen had taken these data from the FAA and portrayed the instrument approach procedures in a graphical form. The court's final judgement against Jeppesen was an award of damages in excess of US$11 million for negligence, breach of warranty and strict liability on the basis that they were under a duty of care to test their products and to warn of any inherent dangers.

Sometimes the information provided may be correct, but has been misapplied, causing loss. In *Aetna Casualty & Security Co. v Jeppesen & Co.*[25] the issue of representation of error-free data was raised. Here, liability

---

[23] 753 F.2d. (9th Cir. 1985).

[24] 767 F.2d. 1288 (9th Cir. 1985) *cert. den.* 474 US 1101 (1986).

[25] 642 F.2d. 339 (9th Cir. 1981).

was found because the form of the representation was either confusing or inappropriate for the user's purposes. In this case a fatal airplane crash was attributed to a defective aeronautical chart published by Jeppesen. The chart in question depicted an instrument approach to an airport. The FAA supplied the information described in tabular form. The chart derived from the information was displayed in graphical form, giving two perspective views—one from above (a plan view) and the other giving a side or profile view. The court found that the chart was defective not in the accuracy of the representation, but in the graphical manner in which the data were shown. The side or profile view appeared to be on the same scale as the plan view when in fact there was a difference by a factor of five. The plane crashed on approach and the cause of the accident was attributed to the pilot's mistaken reliance on the chart feature which appeared to be 24 km (15 miles) from the airport, but which in fact was only 4.8 km (3 miles) from the airport. The crew was misled in part by the graphical representation that may have contributed to the crash. The publishers of the chart were found to be at fault since any 'professional must be able to rely on the accuracy of this information if he or she is to benefit from the mechanisation or special compilation of the data'. The crew were also partly at fault because a 'professional will be expected to use his or her professional judgement in evaluating the information and not to rely blindly on what he or she is told by the computer or the charts'.[26]

While cases such as these support the notion of a duty of care owed by the creator of a navigation chart to the user, the court in *Aetna Casualty* noted that a professional must also play a part in exercising certain judgements in evaluating the information. The requirement is that the professional must not rely unquestioningly on what is shown by the computer or the charts. In addition, a professional must question any obvious anomalies and exercise professional judgement at all times. A professional is generally held to exercise a higher standard of care than a non-professional on the basis that he or she has a minimum standard of specialised knowledge, training, and ability not shared by the public at large (Dragich 1989).[27]

In *Saloomey v Jeppesen & Co.*[28] a U.S. Court of Appeal found that the manufacturer of a mass-produced navigation chart could be sued in tort. The navigation chart showed the location of Martinsburg, West Virginia airport with the letters ILS printed beside it, indicating that a 'full instrument

---

[26] *ibid.*

[27] Dragich, MJ 1989 'Information malpractice: Some thoughts on the potential liability of information professionals' 8 *Information Technology and Libraries*, p. 266.

[28] 707 F.2d 671 (2d Cir. 1983).

landing system' was available for use. In fact Martinsburg airport does not possess such a system. An experienced airline pilot relied upon the chart to set his instrument landing approach and was killed along with his father and son when the straight line vector led his airplane into a ridge.

Californian law applies a product classification to airline charts. In *Fluor Corp. v Jeppesen & Co.*[29] an airplane crash in New York's Adirondack region on a snowy, visually obscured evening was blamed on an air navigation chart. The chart at issue failed to indicate a 2257 feet hill into which the aircraft crashed, leaving no survivors. The Johnson hill, tallest in the area, was not marked on the chart although another nearby hill of 1991 feet was shown. Here the court did not rely on the *Saloomey* mass-produced rationale, but rather declared that an important policy consideration demanded that airline charts be considered as 'products'. 'A piece of paper would not be lethal in and of itself, but the altering of that paper into a map with latent defects opened the manufacturer to strict liability.'

The decisions from these cases should be contrasted to *Winter v P.G. Putnam & Sons.*[30] Here a U.S. Federal court in discussing the danger of consuming wild mushrooms commented on the 'product' status of charts and software as against the content of a book that portrayed that certain poisonous mushrooms were edible. The court held that the book was not a 'product' as it contained 'ideas and expressions' and that the comparison between the mushroom guide and aeronautical charts were not analogous. The book was simply an expression of ideas and the intended use was for it to 'being read' and the act of reading is not dangerous.

Given these cases Perrit (1996) has observed that the explanations of the distinctions without a difference given by the courts is unsatisfactory. The doctrinal caprice of U.S. courts does not explain why the shifting of the cost burden from the user to the seller justifies strict liability on producers of charts while not applying the same to publishers of how-to books and other practical publications.[31] The cases also raise interesting questions about the treatment of GI products under the strict liability regime where scale, graphics and symbology can be treated very differently from textual forms of expressions (as in mere instructions in a book) and lead to different conclusions. Judicial opinion from other jurisdictions is unavailable to confirm or contradict U.S. justices on this matter.

---

[29] 216 Cal. Reptr. 68 (1985).
[30] 983 F.2d 1033 (9th Cir. 1991).
[31] Perritt, JJ 1996 *Law and the Information Superhighway*, New York: John Wiley & Sons, Inc. pp. 182–183.

Exclusions and Limitations of Contract Liability

Most commercial contracts for the supply of software or data include clauses that attempt to limit or exclude liability. However, suppliers are not permitted to avoid liability simply by using such clauses. In the U.K. exclusions are regulated by the *Unfair Contract Terms Act* 1997 (UK), Part I of the *Consumer Protection Act* 1987 (UK) as well as by the *Unfair Terms in Consumer Contracts Regulations* 1994 (UK). It is thus impermissible to exclude or limit liability for death or personal injury caused by negligence, certain implied terms dealing with ownership of the goods in question, and fraudulent misrepresentation. Attempts to exclude or limit liability in the contract will render the relevant clause unlawful and unenforceable.[32]

Close observation will show that it is a common practice for suppliers to limit their liabilities under contract to the value of the contract, that is, the amount a customer will pay under the contract. Sometimes exclusions of liability for consequential and indirect losses are written into contracts. However, this is a question of whether it is reasonable to do so in the light of the circumstances surrounding the contract.

Practice Notes: Contract Liability Drafting Tips

User: ensure the contract provides adequate protection in relation to the quality of what is being supplied and the level of liability reflects the commercial reality of the transaction.

Supplier: ensure that liabilities are limited to a reasonable extent within the contract and that it is permissible within the regulations.

An additional tip on contract management:

Staff: when choosing a licence or systems administrator it is important to consider a person's psychological profile. Many GIS users are scientists or planners who tend to lack the aptitudes and personality traits required for effective policing of a contract.

---

[32] See Westell, S 1999c 'Potential liability for defective software or data. Part I', *GeoPlace.com Mapping Awareness* and at http://www.geoplace.com/ma/current/1199law.asp.

6.3.2 Tortious Liability

A *tort* is a wrongful act and sometimes also describes crooked conduct. The act causes harm to an identified person, whether intentionally or not, being the breach of a duty arising out of a personal relationship, and which is contrary to law. This conduct is said to fall below a standard established by the law for the protection of others against unreasonable risk of harm.[33] This rule is the same in all common law jurisdictions—Australia, New Zealand, U.K., Canada and the U.S. A tort may also be a failure to act in carrying out a specific legal duty. The focus here is on fault. Another way of describing the wrong is the principle of liability in negligence described as the failure to take reasonable precautions to avoid foreseeable risk of injury to an identified person.[34] To find negligence a plaintiff, as the identified person, must prove that the:

- defendant owed the plaintiff a duty to take reasonable care;
- defendant breached that duty by failing to take reasonable care;
- defendant's breach of duty caused the injury or damage suffered by the plaintiff; and
- defendant's injury or damage was not too remote a consequence of the breach of duty.

Scope

The scope of the *duty of care* determines the class of persons to whom a defendant will be liable in negligence. The two elements to the duty of care are: (a) *reasonable foreseeability*—an assessment of whether a reasonable person in the shoes of the defendant would have foreseen that the act (or omission) would cause injury to the plaintiff or to a class of persons of whom the plaintiff is a member; and, (b) *proximity* (which may be physical, circumstantial or causal) between the defendant and the plaintiff.[35] A court holding that the requisite degree of proximity exists to find a duty of care will often involve considerations of public policy which 'underlay and enlighten the concept [of proximity]'.[36]

These two elements indicate that negligence is defined by an objective standard rather than a subjective one. That is to say, what a reasonable

---

[33] Prosser *et al.* 1982 *Cases and Materials on Torts*, 3rd edn § 31.

[34] *Halsbury's Laws of England*, 415–470.

[35] *Sutherland Shire Council v Heyman* (1985) 157 CLR 424, per DEANE J at 510–511.

[36] *Hill v Van Erp* (1997) 188 CLR 159, per DAWSON J.

person would believe to be the case and what would be reasonable in the circumstances.

Professional Standards

Liability may also exist for a failure to act. In the case of GI professionals, and those engaged in any work that requires specialised skills, the standard of care may be at a higher, professional standard rather than that of an ordinary person. Thus, a GI consultant may be liable for failing to act reasonably in light of superior knowledge of GI systems. As there is no directly relevant case law for GI professionals, a close analogy would be that of computer consultants. In the U.S. case of *Diversified Graphics Ltd v Groves*[37] a computer consultant was held to a professional standard of care. However, this 'standard of the profession' has not been universally accepted, even in the U.S. For example, in *Hospital Computer Systems Inc. v Staten Island Hospital*[38] the cause of action for professional negligence against a computer consultant was not recognised.

Under English and Australian common law, the standard required of a professional person has been described by SCOTT LJ in *Mahon v Osborne*[39] 'If he [a professional person] professes an art, he must be reasonably skilled at it...[and the degree of care is that]...as a normally skilful member of the profession may reasonably be expected to exercise in the actual circumstances of the cased in question'. In *Bolam v Friern Hospital Management Committee*[40] McNair J described the standard of care as follows: 'Where you get a situation which involves the use of some special skill or competence... The test is the standard of the ordinary skilled man exercising and professing to have that special skill... A man need not possess the highest expert skill'. The Privy Council, the highest Court of Appeal in the U.K., approved this *Bolam* standard in the case of *Chin Keow v Government of Malaysia.*[41] Thus, for the purposes of ascertaining the standard of the ordinary skilled person in GI exercising and professing to have a special skill in GI systems, services, and science, the standards established by relevant professional GI Associations and Institutes should be taken into consideration.

---

[37] 868 F.2d 293 (8th Cir. 1989).
[38] 788 F.Supp. 1351 (D.N.J. 1992).
[39] [1939] 2 KB 14.
[40] [1957] 1 WLR 582.
[41] [1967] 1 WLR 813.

In the U.S. liability may also be said to exist where there is a fore-seeable risk of equipment malfunction and the failure to provide appropriate back-up to computer systems that may lead to loss of data and productivity. This failure to anticipate a system malfunction has been deemed to be negligence.[42]

Act or Omission Causing Purely Economic Loss

There is no general duty of care in Australia to avoid causing economic loss to another. However, such a duty may be owed to a specific class of persons who may be identifiable or ascertainable at the time of the act of negligence. It is not necessary to identify the class of persons accurately as the imposition of such a duty is more likely where the plaintiff(s) is unable to avoid the loss by taking reasonable steps to protect his or her own interests.[43] As acknowledged by the High Court of Australia in *Perre v Apand*,[44] the law relating to compensation for negligence causing purely economic loss is uncertain, unsatisfactory, and still nascent. There are no governing principles that may be applied which can provide general guidance to possible outcomes on particular facts. However, two previous cases have provided damages as compensation for loss suffered owing to the negligence of some other person. The principle involved is that the injured party should be put as nearly as possible in the same position, so far as money can do it, as if he or she had not been injured.

In the first case the Australian High Court decision of *Caltex Oil (Australia) Pty Ltd v The Dredge 'Willemstad'* and *Caltex Oil (Australia) Pty Ltd v Decca Survey Australia Ltd*[45] dealt with the issue of whether damages should not only recognise personal injury, but also include pure economic loss. MASON J at 593 summarised the majority decision by stating that 'a defendant will then be liable for economic damage due to his negligent conduct when he can reasonably foresee that a specified individual, as distinct from a general class of persons, will suffer financial loss as a consequence of his conduct'.

In a second case the English decision of the House of Lords in *Junior Books Ltd v Veitchi Co. Ltd*[46] held that where the proximity between a person who produced faulty work or a faulty article and the user was sufficiently close, the duty of care owed by the producer to the user

---

[42] *Blake v Woodford Bank & Trust Co.* 555 S.W.2d 589 (Ky. Ct. App. 1977).
[43] *Perre v Apand Pty Ltd* (1999) 164 ALR 606.
[44] *ibid.*
[45] (1976) 136 CLR 529; 51 ALJR 270.
[46] [1982] 3 All ER 201.

extended beyond a duty merely to prevent harm being caused by the faulty work or article. The duty is also to avoid faults being present in the work so that the producer was liable for the cost of remedying defects in the work and for any consequential economic or financial loss. This was so, even when there was no contractual relationship between the parties.

In the U.S. an opposite viewpoint reigns. According to Schmitthoff (1990: 176) purely economic damage law in the U.S. suggests that while 'the decisions are in disarray', the preponderance of judicial opinion holds that purely economic loss is not recoverable without privity of contract.[47]

In general, in most common law jurisdictions where negligence has been established, a claimant may recover money damage for physical injuries or property damage. In some U.S. jurisdictions whether economic loss, for example, loss of earnings or profits may be recovered in a negligence suit will depend on existing law of the state.[48] In Australia, whether omissions on a map or chart could amount to a liability suit will have to depend to a large extent on how serious the omissions are. For example, the judge investigating the Gretley coal mine disaster found it 'unusual and disturbing' that there was no legend in the mine plan that was traced from another mine plan that had a 1892 date on it.[49]

Negligent Misstatement and Misrepresentation

The cause of action for negligent misstatement and misrepresentation may be taken in both tort and contract under English, Australian and U.S. common law. Where someone has suffered pure economic loss in reliance on an alleged negligent misstatement, the elements of negligence to be established is found in the following landmark case. The House of Lords in *Hedley Byrne & Co. Ltd v Heller & Partners Ltd*[50] established the principle that '...a negligent, though honest, misrepresentation, spoken or written, may give rise to an action for damages for financial loss caused thereby, apart from any contract or fiduciary relationship'.[51] This principle does

---

[47] Schmitthoff, CM 1990 *Schmitthoff's Export Trade. The Law and Practice of International Trade* (9th edn), London: Stevens & Sons. p. 176.

[48] In the State of Illinois a court dismissed an action to recover economic loss as the claims were not ones permitted by Illinois law whereas in California a court has protected economic loss. See *Black, Jackson & Simmons Insurance Brokerage Inc. v International Business Machines Corporation* 109 111. App.3d 132; 440 N.E.2d 282 (1982) and *J'Aire Corp. v Gregory* 24 Cal.3d 799; 598 P.2d 60 (1979).

[49] Staunton *op. cit.* p. 5.

[50] [1964] AC 465.

[51] *ibid.*

not require that the statement made is a factual representation that induced the making of the contract. The term used in the U.S. is 'fraud in inducement' in securing a contract. The objective of the plaintiff here is to focus the court's attention solely on the events that occurred prior to and separate from the contract. Fabricated testimonials or test results may justifiably be considered as tortious conduct.[52] Moreover, the principle encompasses the giving of advice as well as the provision of factual information. Thus, a statement of opinion or an undertaking may give rise to tortious liability for negligent misrepresentation or misstatement, as much as the giving of careless information or advice.[53]

In the converse, a person is under no duty to take reasonable care that the advice or information given to another is correct. However, if the maker of the statement knows, or ought to know, that the other person would rely on and act in reliance on the advice or information which is given, then the maker of the statement should take reasonable care. Reliance is thus the key determinant of the existence of a requisite proximity between the two people. This was emphasised by the High Court of Australia in *San Sebastian Pty Ltd v Minister Administering Environmental Planning and Assessment Act 1979 (NSW)*.[54] This case stands for the proposition that:

> ...a relationship of sufficient proximity to raise a duty of care may well exist where a provider of advice or information directs his statement to a class of persons with the intention of inducing members of the class to act on the statement in circumstances where he ought to realise that they may suffer economic loss if the statement is not true. This means that a proximate relationship sufficient for a duty of care could exist, not only as between a speaker and an intended recipient and user of advice, but also as between the speaker and members of a class where such members were the intended recipients and users of that advice.[55]

There may also be circumstances where neither misstatements nor misrepresentations are involved. Commercial data providers may be concerned with keeping their own data to themselves, given its value, but willingly distribute 'degraded' data. This does not mean that the data are bad, but that the data are not as detailed as the original—the distributed

---

[52] See Onsrud, HJ 1999 'Liability in the use of GIS and geographical datasets' in Longley, P *et al.* (eds) *Geographical Information Systems. Vol. 2. Management Issues and Applications*, Chichester: John Wiley & Sons, Inc. pp. 643–652.

[53] In the U.S. negligent misrepresentation is found in § 552 and fraudulent misrepresentation in §§ 525–526 of the *Restatement (2d.) Torts op. cit.*

[54] (1986) 162 CLR 340 at 355 joint judgement.

[55] Katter, NA 1999 *Duty of Care in Australia*, Sydney: LBC Information Services, p. 161.

data set may be described as having been electronically sampled. Thus, a civil engineering firm may sell a version of the orthophoto with a greatly reduced resolution, for example 5 m instead of 5 cm. A casual user may find the product gives a pretty picture of the suburb in which they live and it may be suited to many research and local government applications at this scale. However, this product may not be sufficiently accurate for the firm's competitors to use in more detailed applications. The difference here is one of scale and resolution of the orthophoto and such information once clearly communicated to users by any means may place the onus of responsibility back onto the user to take care.

On-line Settings and First Amendment Defences in United States

In on-line settings the risk of false and misleading information may be higher. Analogous cases in GI are unavailable, but parallels may be drawn. The so-called First Amendment defence has been made in which the U.S. Supreme Court held in *Time Inc. v Hill*[56] that the press would be unconstitutionally 'chilled' in violation of the First Amendment if it were subject to sanctions for either innocent or negligent misstatements. The extent of this defence in the GI industry is unknown and untested and may be a possible defence especially in the on-line provision of databases and Web-based GI services. It is also arguable that the customer to an on-line service is merely a subscriber. In *Daniel v Dow Jones & Co.*[57] a New York court has ruled that there was no special relationship between an on-line news service and a subscriber to support a negligent misstatement claim. In this case the special relationship between the investor and Dow Jones was absent and the investor was likened to be functionally identical to that of a purchaser of a newspaper and the latter being treated as a news service provider.

Contributory Negligence

In Australia, at common law, contributory negligence is a complete defence to a claim of negligence.[58] However, in the case of *Astley v Austrust*[59] the High Court held that legislative provisions apportioning liability where

---

[56] 385 US 374, 389; 17 L.Ed.2s 456; 87 S.Ct. 534 (1967).

[57] 137 Misc.2d 94; 520 N.Y.S.2d 334 (Cir.Ct. 1987).

[58] The common law position has been altered in all Australian jurisdictions by apportionment legislation such as the *Wrongs Act* 1958 (Vic), s 26(1).

[59] (1999) 161 ALR 155.

there has been contributory negligence do not apply to actions for breach of contract. As a result, in the absence of an express contractual provision allowing for apportionment, a plaintiff will not be able to reduce an award of damages to the extent that the plaintiff has contributed to its own loss.

Vicarious Liability

In common law jurisdictions, vicarious liability, also known as *respondeat superior* [let the principal answer], holds that an employer is responsible for the torts of the employee. Those torts must occur in the course of that person's employment. It may be important to recognise that there may be duties that might have to be performed personally, that is, by the person subject to the duty, and those duties which a person may delegate to others, but not the legal responsibility arising out of the performance. Thus, the law can hold a person to be responsible in tort on the basis of negligence of another in breach of a personal duty of care, for example, as a supervisor. The other person as employee may also have a separate standard of duty. As a result the two persons involved have different sets of duties and standards of care.

A corporation or an individual may be liable for a tort, either directly or vicariously. A corporation is directly liable for torts committed by its managing or directing bodies. It is vicariously liable for acts of its employees within the course or scope of their employment.

A corporation may be liable notwithstanding that the tort involves wilful wrongdoing, malice or fraud. However, this might not be so where the tortious act or omission is totally in fraud of the corporation, and the corporation derives no benefit from it. The fact that the corporation derives no benefit from the tortious act does not of itself mean that it is not liable.

Except in the case of the tort of deceit or the use of a motor vehicle, a principal will not be vicariously liable for the acts of his or her agent unless the agent is also the 'servant'.

In an interesting New Zealand case of *Bevan Investments Ltd v Blackhall and Struthers (No. 2)*[60] BEATTIE J, in applying the neighbour principle, held that a consulting engineer owed a duty of care to the 'indirect' employer, despite there being no contract between them. However, in a GI context, one could consider the project manager to take on all responsibilities when performing both the administrative as well as the operational tasks of completing a GI project. Thus, the person doing the editing and

---

[60] [1973] 2 NZLR 45.

digitising of map features and entering these onto attribute files and so forth, will be responsible to the manager who in turn will be directly responsible for errors and omissions arising in the final map product. There may be no possibility of absolving oneself of the duties and responsibilities. For this and other reasons there should be clear lines of responsibilities demarcated in large GI projects and in many instances the 'chain of command' should be clearly hierarchical in order to remove doubts as to who is to do specific tasks and who is to be responsible to whom. The responsibilities are both to give as well as to receive instructions.

Concurrent Liability in Contract and Tort

A contract between two parties can also include a cause of action in damages arising from pre-contractual negligence. Thus, in contract negotiations, the giving of negligent information or advice may constitute a breach of an express or implied term of the contract. That contract may state that competent advice or requiring competent advice has been given. In such cases, causes of action in both contract and tort may be raised in situations where:

- a pre-contractual statement involved the giving of negligent information or advice is embodied in an express term. This pre-contractual negligence in the provision of advice may constitute a breach of contract giving rise to a claim for damages in both tort and contract;
- the information or advice is provided in the performance of a contract may also constitute the breach of a duty of care.

In general, Australian courts have been willing to recognise a choice in some cases between damages in tort or contract. While the High Court has not decided on this issue, the approach in such cases has been to treat the rights and obligations created by any resultant contract as not superseding and replacing the tortious duty of care.

In England the concept of remoteness in contract is more expansive and recovery consequently more limited than is the case in tort.[61] However, a plaintiff can generally recover for loss of bargain in contract—that is, the thing or consideration acquired, but may not recover the same in tort. Thus, choice of law under which to frame one's claim is an important consideration if more favourable results are sought. In the U.S. on the other hand, damages for the loss of bargain can be recovered in tort, and because the rules

---

[61] See *Kufos v C. Czarnikow Ltd (The Heron II)*, [1969] 1 AC 350.

dealing with remoteness are no more favourable in contract than in tort, it is generally desirable to use tort.[62] Also in the U.S. punitive damages are generally not permitted in an action in contract in the absence of an independent tort.[63]

In the U.S. the remedies for those who suffer loss, either in contract or tort, are quite similar. On the other hand, in England where a duty is owed independent of contract, a breach of that duty in a contractual context gives the plaintiff a choice of remedies. The plaintiff may sue, either in tort or in contract, according to what is most appropriate. *In tort*, damages may be recoverable because of either the fraudulent misrepresentation or negligent misstatement. In each case the plaintiff is restored to an original position but for the deceit. Damages are assessed in terms of the difference between the price paid and the value received. *In contract*, a plaintiff may be entitled to damages assessed as the value represented less the value already received.

Contractual limitation of liability may also be effective where there is a limitation of liability clause in the agreement to undertake the work. Such clauses are enforceable by the courts.

6.3.3 Statutory Liability

In some circumstances the appropriate standard is defined by legislature instead of being left to the courts. There are a myriad of governmental regulations that require the observance of fixed and specific rules of conduct for the safety of industrial operations, building construction and road travel. These statutes lay down a standard of conduct with the aim at preventing injury or loss, and non-compliance is indicative of negligence, whether it is a breach of a common law duty of care or negligence as a matter of law *per se*. While there are slight differences between jurisdictions, the intention is the same—that of finding someone responsible for the loss or injury incurred by another. The U.S. and Canadian approaches integrate statutory liability within the negligence framework while the Anglo-Australasian approach recognises statutory liability as a separate tort and the application of general rules of liability for a breach of statutory duty.[64]

---

[62] See *Sandy Koplowitz v Andre Girard*, Fla. DCA 4th District, No. 93–1781, 9 August 1995 (CCH ¶47,310); *Advanced Micro Devices Inc. v Intel Corp.* Arbitration Award 24 February 1992, Palo Alto, CA. (CCH ¶60,368).

[63] *Uniform Commercial Code* Art. 2.14. See also CCH ¶8400.

[64] See Fleming 1988 *op. cit.* 'Statutory standards', pp. 137–148. Linden, AM 1997 *Canadian Tort Law* (6th edn), Toronto: Butterworths.

From the discussion of the maps and chart cases earlier under strict liability it is now possible to understand why there are statutes and codes governing their use. The need for accurate products to ensure safety of navigation has now been codified under U.S. law.[65] A violation of these standards would confer a private remedy to those who have suffered loss and injury.

Government immunity adds to the confusion when dealing with the dissemination of information, and especially under a duty to do so. For example, a California statute exempts a government entity from liability in regard to the dissemination of information under duty, even of incorrect information.[66] Here it appears that the open information policies and laws compelling disclosure have promoted the confused state of affairs. Courts in the U.S. have also used the 'floodgates' argument. On this argument governments would have to defend innumerable suits if the government were to be held accountable for errors in information which under public records laws, the government was obliged to provide. Hence it may be no wonder that the provision of information under the *Freedom of Information Act* (FOIA) 1966 (US)[67] are often accompanied by disclaimers. Such disclaimers may state that the information is being provided only for the purposes of complying with the FOIA and the accuracy of the contents cannot be presumed.

6.3.4 Other Liability Theories

Privacy

Chapter 4 on privacy suggests that one could be liable for breaches under common law or under statute for the disclosure of personal information. While it is acknowledged that there are privacy interests involved, one should be prudent when using geospatial technologies, whether in the form of GIS, geodemographics or location-based services in order to avoid exposure to litigation. The liability could result from the release of personally 'sensitive' information.

---

[65] See 10 U.S.C. § 167; 1791–96 [Armed Forces]; 44 U.S.C. § 1336 [Nautical charts]; and 49 U.S.C. App. § 1301–1348(b)(3) [Aeronautical charts].
[66] 'A public entity is not liable for an injury caused by misrepresentation by an employee of the public entity, whether such misrepresentation be negligent or intentional', W. Ca. Gov. Code Ann. § 818.8.
[67] *Freedom of Information Act* 5 U.S.C. 552 [Government Organization and Employees].

Intellectual Property Rights

Liability theory under the guise of intellectual property rights discussed in a previous chapter may give rise to certain duties. However, copyright law provides only 'thin' protection for factual works and those consisting of compendia of public domain information such poisonous mushrooms as in the *Winter v Putnam* case discussed previously. Under the copyright regime, the policy is for an uninhibited use and wide dissemination of the publication subject to certain conditions of use. However, under tort law, publication of practical works such as maps and charts is inhibited because of the possibility of liability to the producers.

Criminal Liability

Recent developments in information management also suggest that information brokers can be held responsible when data they sell puts people at risk involving criminal misconduct. In the U.S. the case of *Helen Remsburg* involved the fatal shooting of 20 year old Amy Boyer of Nashua, NH by Liam Youens.[68] Youens had stalked Boyer after obtaining her date of birth, social security number, home and work address from an on-line information provided by Docusearch.com. Youens suicided after killing Boyer. A suit was filed by Boyer's mother against Docusearch claimed that it had violated Boyer's privacy and broke other laws. The State court held that information brokers and private investigators could be held liable for 'foreseeable harm' as a result of the disclosure of personal information to a client. This is an important privacy and data distribution decision and is the first time anywhere that a court has found information brokers to be responsible to people about whom they sell information.

In the U.S. the government may control the export of defence articles and services to foreign countries. For example, the *Arms Export Control Act* (AECA) 1976 lists articles and services subject to export controls on a U.S. Munitions List.[69] To export an item from that list a person must be registered with the State Department prior to submitting an export licence application. Unclassified technical data not in the 'public domain' may be subject to export licence requirements. In the context of GI the concern is

---

[68] *Helen Remsburg, Administrator of the Estate of Amy Lynn Boyer v Docusearch Inc*. U.S. District Court, SC New Hampshire, No.2002-255 at http://www.courts.state.nh/supreme/opinions/2003/remsb017.htm and also Hovanyetz, S 2003 'Court: Information brokers can be liable', *DMNews. The Online Newspaper of Record for Direct Marketers*, March 11. See http://www.dmnews.com/cgi-bin/artprevbot.egi?article_id = 23093.

[69] 22 U.S.C. Chapter 39 § 2778(a)(1) [Foreign Relations and Intercourse].

with geographic data, software and cryptographic equipment that may be embedded in GI systems. Even up until 1992 there were strict controls in the export of computer equipment to Socialist countries. However, since then the U.S. government has relaxed and streamlined export controls on mass-market software with encryption capabilities.

Sovereign Immunity

In the common law world, for example, in Australia and the U.K., for a long time government authorities and the Crown were sheltered from tort liability under the old rule that 'the king can do no wrong'. While Crown officers were liable for torts committed in the performance of duties, the state was not vicariously implicated, nor did it owe any direct duties to members of the public. The rationale behind Crown immunity lay in the government's discretionary power, in that policy formulation should be held separate from executive functions. Hence, policy formulation should not be interfered with by the threat of liability. In Australia, the Crown immunity shield has, however, been abrogated and the public is able to seek redress for wrongs by a review of bureaucratic decisions under various enactments by administrative tribunals such as the Administrative Appeals Tribunal (AAT).[70]

Two cases, each from different jurisdictions, illustrate the basis of Crown immunity. In England in the case of *Warewick Shipping Ltd v Her Majesty the Queen*[71] although a navigation chart was found to be misleading, the Crown was held not to be liable for the stranding of the *Golden Robin* because no duty of care was owed to individual mariners nor did their master or pilot rely on the chart. In the U.S., a court ruled that the then Defense Mapping Agency (DMA) was not liable for the stranding of the *Pioneer Commander*.[72] The court held that the U.S. government was immune from suits because the *Suits in Admiralty Act* 1948.[73] In the U.S. government immunity is provided for under Titles 28 U.S.C. §§ 1346(b) [Judiciary and Judicial Proceedings] and 2671 and 49 U.S.C. §§ 741–752 [Transport] while in the U.K. the *Crown Proceedings Act* 1947 (UK) and in Canada the *Crown Liability Act* 1952–53 (Can.) provide similar immunity.

---

[70] The *Administrative Decisions (Judicial Review) (ADJR) Act* 1977 (Cwlth) and the *Administrative Appeals Tribunal (AAT) Act* 1975 (Cwlth).

[71] (1982) 2 F.C. 147 and affirmed in (1983) 48 NMR 378.

[72] *United States Lines Inc. v United States*, No. 79 Civ. 4209 (S.D.N.Y. 1983). DMA is now known as the National Geospatial-Intelligence Agency (NGA).

[73] 46 App. U.S.C. 741 *et seq.* [Shipping].

6.4 Liability Risk Management

The successful management of liability risks comes from the identification of the problem and how to use the law to find a solution for it. However, there are uncertainties in parts of the law that cloud solutions to the problem. The following suggests some pro-active strategies that may assist in managing the liability risks such as disclaimers, data quality checks, and other practical advice of 'things you can do'.

There is a view among data managers that if the release of data causes liability problems, then they may simply have no hesitation to restrict access to that data. Including the *caveat emptor* principle of buyer beware, the growing ubiquity of GI services and products may prompt governments and agencies to revisit the issue of managing legal risks by undertaking cost-benefit studies to discover the balance between costs of litigation and costs of legal reform (Fairall 2002).[74] In the interim, one pro-active strategy would be to avoid legal liability by 'getting the data right' in the first place. Technology today enables high accuracy to be obtained using GPS in conjunction with aerial photography. The commercialisation of GI also means that customers are demanding greater use of the data that they purchase. Also, since spatial data are now embedded deep in GI systems the legal imperative is that the 'data better be right'. This is because of the possible adverse consequences arising from its use as well as the difficulty for customers to tunnel deep into the database in order to correct errors (Fairall 2000).[75]

6.4.1 Disclaimers

The use of disclaimers is a useful strategy. Such disclaimers put users on notice about the data itself and what it may or may not be used for. Also, while such disclaimers may not protect against mistakes, the protection is to limit economic loss arising from use of the data. As reported in Mason (2002: 129) and borrowing from the surveying industry, Harrison (1998: 16) has given several examples of disclaimers employed by surveying firms.[76]

---

[74] Fairall 2002 *op. cit.*

[75] Fairall 2000 *op. cit.*

[76] Mason, R 2002 *Developing Australian Spatial Data Policies—Existing Practice and Future Strategies*, Unisurv s 70 2002, Reports from the School of Surveying and Spatial Information Systems, Sydney: University of New South Wales; and Harrison, B 1998 'Disclaimers', *Azimuth*, February, pp. 16–19.

Better still, Elgin (1995) suggests putting a statement on the surveyed plans and maps of what the survey comprised of and what the survey did not cover.[77] A disclaimer statement suggested by Elgin (1995: 17) is given below:

> The location of underground utilities as shown hereon are based on above ground structures and record drawings provided to the surveyor. Location of underground utilities/structures may vary from locations shown here. Additional buried utilities/structures may be encountered. No evacuations were made during the progress of this survey to locate buried utilities/structures. Before evacuations are begun, the following officials should be contacted for verification of utility type and for field location...

6.4.2 Data Quality Issues

Data quality is a persistent theme in GI services and data and can be grounds for liability if 'bad' data causes personal injury and death. Users frequently ask what kinds of accuracy one can expect from the data and the limits to which the data may be stretched before it breaks down and becomes unsuitable for the purposes at hand. Data quality standards and national standards used in national mapping agencies should be a good starting point for a discussion of quality issues and liability.

In an early paper, and coincidentally with experiences from the surveying industry, Onsrud and Hintz (1989) contemplated the challenges surrounding the upgrading of boundary information in a GIS using an automated survey measurement management system.[78] One of the difficulties facing GIS developers is the current liability issue that hinders the creation of a spatially accurate, legally supportive and operationally efficient land ownership database. There are no methods that provide a reliable link between legal ownership interests in the land and the physical location of these interests. While tax assessor maps are available for use as a cadastral base for local GIS use, these are hardly suitable to provide the need for highly accurate, legally defensible descriptions of parcels of land. Aside from giving the false impression of high accuracy, it does not appear that GI systems products are sufficiently robust to produce legal documents of title and all the other attributes contained in title deeds, despite the characteristic inadequacies of deeds.

---

[77] Elgin, RL 1995 'Surveyor's notes for plats', *Point of Beginning* (POB), August–September, p. 17.
[78] Onsrud, HJ and Hintz, RJ 1989 'Upgrading boundary information in a GIS using an automated survey measurement management system', *Technical Papers: 1989 ACSM-ASPRS Annual Convention*, v. 4, pp. 275–284.

The robustness of the GI systems products raise concerns for debate, namely that of standards attached to each of the data elements, products and services as well as the work of the GI professional. This latter challenge is one occupying the minds of professional societies, including practising certificates, accreditation of the GI curriculum, codes of conduct, and professional standards of care. In regard to data systems, there is also doubt about the legal status of measurements made using electronic equipment such as GPS.

6.4.3 Global Positioning Systems (GPS) and Map Quality Standards

Of relevance here is the use of GPS for measuring distances, finding positions in geographical space and providing the data in cadastral surveys. In Australia, the status of GPS readings has yet to be legally defined. The idea of legal traceability as applied to GPS measurements is still in doubt. Legal traceability is taken to mean the ability to reduce any measurement to a legally defined standard. For example, 'A is 20 mm from B' becomes legally traceable if the distance is measured with a ruler and that ruler is compared against a standard length that has been legally defined by an entity such as that held by the National Measurement Laboratory. The issue with measurements from GPS receivers is that it is not possible to compare these with known standards because the GPS measurement results are only known after processing by a computer and more often than not when the data are 'downloaded' in a laboratory.

Secondly, GPS devices are subject to variable errors. With 'selective availability' turned off, one may be able to achieve an accuracy of 5–20 m, depending on the quality of the GPS receiver. Selective availability is the deliberate degradation of accuracy that is introduced by system operators. GPS technology was initially constructed to meet defence needs of the U.S. military and only recently permitted for civilian use. When GPS was given to civilian use system operators deliberately introduced an error into GPS readings as 'security safeguards' in case such instruments fell into the wrong hands and were used for illegitimate purposes. The 5–20 m level of accuracy may be adequate for the bushwalker, recreational yachtsman or general vehicle navigation uses. However, for the purpose of cadastral surveys, even a 10 m level of accuracy may be inadequate. Differential GPS (dGPS) may improve accuracy levels. dGPS is used to negate the effects of selective availability and other systemic errors. This method requires two GPS receivers to be run simultaneously, one known as the *base* being located at a known point while the other(s) as the *rove* between unknown points. A differential may be calculated to locate a position precisely on the ground. To obtain the highest possible levels of accuracy it may be necessary to log GPS readings over a period of time at each site. But this may prove tedious, inefficient and time consuming for surveying purposes when results are required more or less instantaneously *in situ* and on the surveyed site itself.

Finally, GPS readings are reliant on satellites and errors may be the result of a number of factors. Other than selective availability discussed previously, satellites can malfunction and give erroneous results as a consequence. Data may be affected by variations in transmissions over which the user may have little or no control, for example, weather conditions, land cover and terrain conditions.

In Australian surveying and geodetic circles, there is no general agreement over the need for legal traceability to a national standard for GPS. Some have expressed doubts as to whether it is either achievable or necessary and favour the principle of best practice, as is the case in most other countries (Baker 1996).[79]

Map Data Quality Elements

In terms of map data quality, the consensus is that up to seven components may be pertinent to its assessment. According to Graham (1997) and corroborated by Joffe (2002) the components include:[80]

1. *Positional accuracy*: the depiction of map objects from true ground positions, that is, how closely the mapped coordinates correspond with actual coordinates on the ground. These are often regarded as a primary measure of the quality for maps and GI databases.
2. *Attribute accuracy*: the accuracy of various descriptions of map features, whether the map objects are connected, the symbology that has been used and the annotations that portray them to make them easily understood to users.
3. *Logical consistency and identification accuracy*: the sound logical relationships among data elements and how reliable the mapped objects are portrayed as the actual object they represent and the

---

[79] Baker, H 1996 'Measuring the GPS', *GIS User*, No. 17, June–July, pp. 28–29.
[80] Graham, S 1997 'Products liability in GIS: Present complexions and future directions', *GIS Law* v. 4(1), pp. 12–16; Joffe, B 2002 'How good are your maps', *GeoWorld*, February. Also at http:// www.geoplace.com/gw/2002/00202/0202dac.asp.

number of relevant objects that were omitted from the set of mapped objects.

4. *Resolution accuracy*: the smallest discernible unit or object represented on the map, sometimes also known as *referential accuracy*, that is, how close the portrayed distance between the map object compared with field measurements is.
5. *Completeness*: in coverage, classification or verification.
6. *Time*: whether mapped information is up-to-date and how frequently these data are updated given that data degrades over time.
7. *Lineage*: the history of the dataset, including sources and processing steps. Lineage is the identification and quality assessment of source documents as well as the methods of data collection used and whether multiple data sources have been employed in the compilation of an integrated mapping database.

Further, the assessment of the quality levels in any given mapping database requires different standards and methods for each quality factor—map currency, map representation, map lineage and map accuracy (see Practice Notes below). 'It becomes imperative, from a liability perspective to check the positional and attribute accuracy, the GIS logical consistency and resolution as well as the completeness, timeliness, and especially the lineage of the data. Such efforts can redound to the advantage of both the GIS provider and ultimately to the consumer' (Graham 1997: 14).

Practice Notes: Geographic Information Quality Questions

Map currency

- Is a date of data collection annotated in the descriptive attributes for each mapped object?
- When were the original observations made?
- What is the date of compilation of this database?

Map representation

- What statistical sampling has been undertaken to assess elements of the database in terms of each of the accuracy measurements for positions, references, identification, and attributes?

Continued on page 386

Continued from page 385

Map lineage

- What standards and classification systems have been used for map compilation techniques?
- What assessments have been made for the following lineage factors: direct observations, use of legal source documents, constructions from derived maps, trace digitisation, measurement and adjustment factors in use, object measurements recorded as text and stored as attributes, and measures shown as text only with no measurements shown?

Map accuracy

- What is the statistical variance of a set of sample mapped points from a set of independently referenced measures for those points?
- What confidence levels have been achieved using statistical tests to measure locations for positions, for referential accuracy as well as for identification and attribute accuracy?

The suggestions made in compiling these GI quality questions resemble safety management systems imposed on the mining industry in Australia. Such a system consists of a planned organisational framework which has the capacity to adequately address the hazards associated with mining as well as a primary mechanism for employers to meet their basic duty of care obligations. The Standards Australia Quality System AS/NZS ISO9002-1994 program chart includes four work groupings divided into ten programs for minimal compliance to best practice (Figure 6.2). A process of self-auditing is established as an integral part of the safety management system.

Practice Notes: A Data Quality Improvement Strategy Checklist

When a data quality issue is identified:

- plan how you are going to address the issue;
- do something to fix both the existing data and problem causing it;
- check that what you did has been successful;

Continued on page 387

Continued from page 386

- if it has been a success, act on the information and start planning for the next data quality issue;
- if it has been a failure, take further remedial action and check it again;
- continue quality tasks always in the fashion of 'plan, do, check, act' in a reiterative manner.

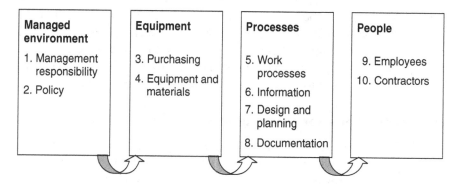

Managed environment	Equipment	Processes	People
1. Management responsibility 2. Policy	3. Purchasing 4. Equipment and materials	5. Work processes 6. Information 7. Design and planning 8. Documentation	9. Employees 10. Contractors

Figure 6.2 Australia: a standards quality system. Source: AS/NZS ISO9002-1994 Quality Systems

Map Accuracy Standards

In the U.S. since 1947 the National Map Accuracy Standard (NMAS) has governed locational accuracy characterised according to map scale. The standard prescribes that a map scale larger than 1:20 000 must locate at least 90% of objects correctly within 1/30th of an inch at scale. This means that 90% of the objects mapped at a scale of 1:1200 must be located within 3.33 feet. Objects mapped at scales smaller than 1:20 000 must be correctly located to within 1/50th of an inch at scale. The map accuracy standards have been redefined in 1998 by the Federal Geographic Data Committee (FGDC) and are known as the National Standard for Spatial Data Accuracy (NSSDA). This standard offers a method for comparing a sample of mapped points with independent measures of locations to derive statistical assessments valid for 95% of the points. As a result it would be possible to include a map accuracy statement of the following form: *The tested horizontal accuracy at a 95% confidence level is x feet.* With this statement in the metadata, that is, data describing the data, a user can expect locational errors of no worse than: *x feet for 95% of the time*, regardless of the scale at which the map is displayed.

Joffe (2002) has further suggested that there is no reason why such accuracy statements cannot be made, as a matter of course, a part of any GI map presentation.[81] In addition, there is every reason for incorporating several accuracy statements in a database, given that the data may have been compiled from various sources and at different times. To increase their usefulness, accuracy statements could also be colour coded or graphically indicated so that the user may make an informed decision as to the accuracy of that particular layer of the database. This is in recognition of the fact that different map layers may have different levels of accuracy. As with links in a chain, 'the accuracy of the compilation is no better than the accuracy of the worst layer'. But more than this, such a dataset may lend itself to a forensic audit trail as well as legal evidence presented to a court of law. The errant layer or data may be more easily identified and perhaps liability sheeted home to the agency responsible for that layer.

Electronic Charts for Navigation

In the U.S. the migration from the paper map to the electronic map has already taken place with the use of the electronic chart for digital navigation. Electronic charts now in use plan a track, determine a ship's position and plot the track made good. The U.S. Navy procured the Navigational Sensor System Interface (NAVSSI) which incorporates the National Geospatial-Intelligence Agency (NGA) (previously Defense Mapping Agency) digital nautical chart to process and distribute navigational data. The electronic chart can show selected features, simplify a cluttered chart, and provide navigational aids. When integrated with GIS it may assist navigation by providing dynamic, real-time capabilities. Also, potentially it may be possible to update charts using digital navigational information on a regular, automated and accurate basis. Such capabilities will raise the standards of liability with regard to negligent charting, safe navigation and vessel seaworthiness.

While the basic legal standards may remain unchanged the nature of the technology has changed and this development alone may force a re-thinking of liability-related questions. For example, the paper chart is fixed as to its contents whereas with electronic charts and databases there can be constant changes and updates in real-time. To take advantage of this, the International Hydrographic Organisation's (IHO) Legal Advisory Committee has proposed that certain minimum information is required in liability litigation. The minimum information with regard to the storage and reproduction of a ship's

---

[81] Joffe, B 2002 *ibid.*

past track should include the time, position, heading, and speed of the vessel. In addition, a record of the official data used such as chart source, edit date, cell, and update history in the past 12 hours, should be readily available. Further it should not be possible to change the recorded data for the previous 12 hours' data and voyage track should always be preserved. This is similar to a 'black box' used in the airline industry for maintenance purposes as well as forensic evidence in the case of an aircraft crash.

The adoption of the electronic chart will not eliminate the basic legal standards concerning liability for negligent charting. The supplier of the database is under the same duties to ensure timeliness and accuracy for the charts. What may be interesting is when and at what point such duties are said to have been discharged especially where there may be a 'moving target' in the form of dynamic update of the data and the chart itself in real-time (Obloy and Sharetts-Sullivan 1995).[82]

Practice Notes: Other Solutions Checklist

- In your contracts seek as much indemnification as you can get.
- Seek appropriate insurance for as many different liability scenarios.
- Ensure that the intellectual property rights are protected, but at the same do not infringe those rights of others.
- Display due diligence at all times in words, products, practices and processes.
- As a supplier of GI data, services and products, ensure that liability is limited to a reasonable and legally permissible extent.
- As a customer, ensure that you are adequately protected in relation to the quality of what is being supplied and the level of liability protection reflects the commercial reality of the transaction.
- As current and future applications of GI technology are constantly evolving, each raising peculiar liability issues, whether as a user or supplier, make a note of current litigation and case law to ensure against being ambushed by 'surprise' liabilities.

---

[82] Obloy, EJ and Sharetts-Sullivan, BH 1995 'Exploitation of intellectual property by electronic chartmakers: Liability, retrenchment and a proposal for change', *Proceedings of the Conference on Law and Information Policy for Spatial Databases*: Tempe, AZ (NCGIA, University of Maine, Orono, 1995) and at http://www.spatial.maine.edu/tempe/obloy.html.

6.4.4 Legal Risk Management Strategies

Legal risk management processes therefore identify and measure legal risks in relation to GI services and products. Risk management is also about developing a strategy to minimise, transfer, insure against, and if possible, eliminate those risks. The strategy is both pro-active and reactive. Regulatory compliance is to be factored into the strategy as well as the avoidance of loss and civil liability in contractual and non-contractual contexts.[83]

Risk management in general has been recognised as an important management tool for businesses. The advantages of using this tool are manifold and may help counter negatives such as those contained in the following list:

- the excessive time spent dealing with unanticipated losses;
- adverse effects on credit rating and costs of capital;
- reduction in cash flow, growth and profitability;
- deterioration of public image and loss of customers;
- difficulty in finding top management and qualified staff;
- abandonment of certain strategically desirable projects because of inadequate ability to manage loss exposures.

In the context of the commercial exploitation of GIS the focus on legal risks may provide the guidelines for formulating a risk management strategy. The Practice Notes below offer a starting point.

Practice Notes: Four Steps in Legal Risk Management

Step 1: identify the inherent legal risks
Step 2: develop a risk management system to quantify the risks
Step 3: develop strategies to reduce, transfer, manage or accept identifiable risks
Step 4: adopt procedures that monitor, audit and put in place training programmes for staff to ensure compliance with the risk management strategy

---

[83] See Reid, K, Clark, E and Cho, G 1996 'Legal risk management for Geographic Information Systems', *Journal of Law and Information Science*, v. 7(2), pp. 169–207.

6.4.5 Discussion

Emergent GI Science

At the beginning of the new millennium, GI science may not have matured sufficiently to have developed standards to the extent which could determine whether a particular spatial data system has been properly designed and developed. Even when GI products use metadata to ensure that map products meet cadastral data content standards, and display national map accuracy standards on the face of the maps, there is uncertainty as to whether these are sufficiently robust to defend a suit in negligence. These safeguards have yet to be tested in litigation in courts.

Contributory Negligence

Whether the provider of a spatial database or GI service has acted in a reasonable, competent, and professional manner may be clouded in part by the user. For instance, a user may request certain kinds of data or services and the GI provider has endeavoured to satisfy these requests. Whether these meet the needs of the user will be a matter of opinion and conjecture. In a legal sense it is important to note that the product and service being supplied is shaped by the request of a user. There thus arises an issue of contributing to the wrong in the form of contributory or comparative negligence if something should go wrong.

Consortia

A major strength of many GI systems (and perhaps also a source of error) is the facility to combine data from various sources. Hence, each 'layer' of a map display and data set may be combined in such a way as to produce a composite map showing a particular feature or item of interest to the user. Each of the layers in the data set may have been obtained from different providers and some may have been the result of work of multi-participant consortia comprising of government, public utility, private corporations and academic bodies. This kind of data integration is commonly used in the GI industry. While there may have been technical specifications provided to each of the data gathering parties, there may still be differences in the various modes of data gathering, data conversion, database design, query design, formatting and content of the final product. These differences may bring about compatibility problems as well as errors, which may have crept into some or all of the procedures. To identify the cause of the error may become a very tedious process, even if strict editorial guidelines have been followed. It would not be difficult to imagine that a court of law would try to unravel the data and get to the source of the offending data set in order to find the culprit and then lay the blame on to that source. Such a task would be both tedious and difficult indeed.

Subjective–Objective Criterion

The subjective–objective criterion of the law will also have a large bearing on whether the conduct of a GI professional has been reasonable or unreasonable. In general, the issue may be settled on whether the professional information provider has adhered to rudimentary standards of performance in order to have blame laid on the professional. Here again the final outcome will hinge on the circumstances and facts of each case.

Fair and Honest Communication

A final issue is that information providers have to be fair and honest in communicating the limitations of the data, information, and GI services delivered to customers. This is to protect the provider as much as it is to protect the customer. A full, frank and complete statement and explanation must be made in the GI report. Such a statement will spell out the limitations of the data as well as its shortcomings and how these may affect the results in the products and services given to the customer. This has been recognised previously by Blakemore's (1991) reminder that 'contractors and other information providers have an ethical, if not legal, responsibility to fully inform users including their government clients and the general public of the effects of aggregation and scale dependence [in their data and information]'.[84] To fail to do so would be to fall below a legal standard of care and duty that might be expected of GI professional.

Reports of unprofessional conduct have also emerged in other contexts. In a discussion on the siting of waste sites, cartographic power and public access, Monmonier (1995) quoted an attorney's anger and sense of betrayal thus: 'The siting Commission's staff told my clients and others...that the computer program the contractor used for the GIS is a commercial program called Arc/Info...but I discovered [from the

---

[84] Blakemore, M 1991 'Access and security issues in the provision of geographic information', in Medyckyj-Scott, D *et al.* (eds) *Metadata in the Geosciences*, Loughborough: Group D Publications Ltd, pp. 55–68 at p. 55.

Commission's attorney] that there is no Arc/info component to those tapes. They run solely on [a] trade secret program.'[85]

6.4.6 Minimising Liability and Damage Claims

The law of negligence attempts to find a cure for something that has gone wrong. For GI professionals simply having the thought of a liability claim against oneself or the organisation would make one take extra care at all times. As a minimum strategic position, the following reminders may assist in minimising legal liability and damage bills.

Care

The taking of care is inextricably linked with time. 'If only there was more time' is a lament often heard when deadlines approach and the pressure of getting the job done on time. Care at all stages of carrying out the professional task at hand is an important attribute and this includes the need to ensure that the work tasks, processes, and procedures are properly followed. Care and time are both critical factors and pre-planning work is vital for successful completion.

Competence

As most GI systems are complex in planning, implementation, and subsequent maintenance, there is a greater division of labour among personnel as to who is responsible for specific tasks. As a consequence, competencies should reside in several persons within an organisation. The GI system 'team' may often be identified as the experts and clients may come to rely on their skill and advice on particular matters. In the event of a failure to deliver the client may litigate on the basis of the different torts described earlier in this chapter and seek recovery for damages. Equally critical is the 'currency' of skills given that the field of GI technology is rapidly changing in light of the needs and the demands of clients. Professional bodies conduct continuing professional development programs and it behoves GI professionals to undertake such training to enhance, hone, and extend the skill base of everyone involved to ensure currency of skills as well as higher levels of competency in particular areas.

---

[85] Monmonier, M 1995 *Drawing the Line. Tales of Maps and Cartocontroversy*, New York: Henry Holt & Co., p. 252.

Communication

This is an important element to ensure that the GI professional is able to deliver what the consumer requires in order to complete the project at hand. Communication is also important because the GI professional will find out what skills might be required to undertake the project. It appears that communication is often overlooked and under-valued and the form it takes may have a bearing on what has been agreed. In addition to communication it may also be prudent to keep contemporaneous notes, even of phone conversations and oral discussions, so that these notes record what has actually be agreed upon. While the use of e-mail is now commonplace the habit of confirming what was actually said cannot be overemphasised.

Caveats

Do not be afraid to use caveats—that is, adopt a precautionary approach by stating that the task can be completed subject to several other events taking place, such as availability of relevant data and information. While this caveat might seem to be wholly inappropriate lest it be seen as an admission of the lack of skills and confidence it may be a necessary step if one wishes to limit one's liabilities.

Clients

Clients and customers are the *raisons d'être* of the whole enterprise. Identify and know your clients and build close relationships with them because a happy customer is a repeat customer and more than this, since they are the ones who would give unsolicited testimonials to your future customers. But every client is different and with every new client there begins yet another project from scratch.

Professional Indemnity Insurance

A professional indemnity insurance policy is now a necessity, given the potential exposure to liability. This exposure arises as soon as one holds out to be a professional person and begins to practise as such. The liability remains on going for the life of the practice. Choose a tailored policy that will provide assurance against an omission, mistake or misstatement to avoid the possibility that the insurer will not pay out. In insurance circles this is known as the 'innocent overrider clause'. A tailored policy should also include fraudulent and dishonest acts, not only of the firm's employees, but also of partners in the firm. The policy should not be limited solely to claims on a legal liability. Consideration must also be given to inserting a clause whereby the policy excludes any requirement to be compulsorily complied with before seeking a claim to be met. Also a decision needs to be made as to whether a 'claims made' policy is to be used rather than a 'retroactive limitation dating' policy. A *claims made policy* is one in which claims are made against the insured during the currency of the policy, irrespective of when the negligent act, error or omission took place. *Retroactive limitation dating* is where claims are met only if arising out of business transacted on or after a certain date.

Practice Notes: Minimising Liability and Damage Claims

Care: take care and make time available to complete tasks.

Competency: ensure competency of self and others in the team and keep skills current.

Communication: ask clients what they really want, not wish. Tell clients what is feasible given the competencies available. Keep notes.

Caveats: inform clients of what is not feasible and whether the project is subject to other circumstances. Spell out these circumstances simply.

Professional indemnity insurance: buy insurance, spread the risks. Choose policies carefully, better still, devise tailored insurance policies to suit your particular needs.

Summary

This chapter has demonstrated that liability issues are but one aspect of the legal framework under which GI professionals operate. There might be need for further policy and structural changes to foster a dynamic and more robust GI industry and commercial enterprise. Such changes may come about in one of two ways. First, through a better understanding and appreciation of the capabilities and economic benefits of the GI industry. Secondly, it may be the result of a public censure, for example, in a tragedy that caused loss of life and damage to property which in some way has been blamed on a product of the geospatial industry.

As noted in the first section the legal standards and guidelines have been illustrated by the five case studies that point to different aspects of the liability equation. Legal standards and guidelines were elucidated in four subsections, namely, contract and strict product liability, tortious liability, statutory liability and a collection of other cognate liability theories. The focus has been wholly on common law approaches to the question with some discussion of legislative provisions. Liability risk management was then discussed with the prime focus on data quality issues of GI products and services.

Even with the advent of the electronic chart, its producers would have to abide by the same strict rules and duties as imposed by legislation. As noted the production of navigational charts has been codified under U.S. law. The duties of these manufacturers are the same in regard to timeliness and accuracy. However, what might be doubtful is how are users to know that these duties have been discharged and who has the responsibility of ensuring that these have been satisfied. Would it take another tragedy or a court case to determine that the standards have not been satisfied? In the alternate, the question is whether the users of such products have assumed the risk and accept responsibility by assuming that the products are reliable?

In a concluding section it was postulated that the GI science industry is still nascent and a developing one. As a result all parties are taking small and cautious steps in unchartered waters with regard to not only of the technology and its applications, but also the extent to which the law may be stretched before it is infringed. To date, the law in its application to the GI industry has been by way of implication and analogy. In the future, when more litigation has taken place, some case law precedents might provide better guides to and solutions for the legal problems encountered in the GI industry specifically. This concluding section also provided some pointers to minimising liability and damage claims by way of a series of practice notes.

In summary, societal, economic, political, and legal issues interact with the conceptions and assumptions with which the various parties involved, that is, vendors, service providers and users, approach distributed GI. This becomes even more apposite with the proliferation of systems that may be delivered via the Internet.

References

Ackerman, L, Kempf, J and Miki, T 2003 'Wireless location privacy: Law and policy in the U.S., E.U., and Japan' at http://www.isoc.org/briefings/015/index.shtml [2 February 2004].

Agence France Presse (AFP) 2004 'Court finds against tabloid. Campbell privacy case win', *The Canberra Times*, 8 May, p. 20.

Agre, PE and Harbs, CA 1994 'Social choice about privacy: Intelligent Vehicle Highway Systems in the United States', *Information Technology and People*, v. 7(4), pp. 63–90.

Alpert, S and Haynes, KE 1994 'Privacy and the intersection of geographical information and intelligent transportation systems' at http://www.spatial.maine. edu/tempe/alpert.html [30 June 2004].

Alpert, S 1994 'Privacy on Intelligent Highway: Finding the right of way', *Santa Clara Computer & High Technology Law Journal*, v. 11, p. 1.

ANZLIC 1996 *Guidelines: Core Metadata Elements, Version 1*, Canberra: ANZLIC.

ANZLIC 2004 ANZLIC Position Paper. Engagement with the Spatial Information Industry, February, Canberra: ANZLIC.

ANZLIC 2004 Guiding Principles for Spatial Data Access and Pricing Policy, Canberra: ANZLIC and at http://www.anzlic.org.au [19 April 2004].

Appel, R 1995 'Copyright in a digital age. Chaos in the debate', *ANU Reporter*, 13 December, p. 7.

Attorney General's Department 2003 A Short Guide to Copyright at http:// www.law.gov.au/[20 February 2004].

AUSLIG Metadata Tools and Guidelines at http://www.auslig.gov.au/asdd/tech/ tools.html [7 April 2004].

Australia 1996 *Commonwealth Competitive Neutrality Policy Statement*, Canberra: AGPS.

Australia House of Representatives 2003 *A Nation Charred: Report on the Inquiry into Bushfires*, Canberra: AGPS. Also at http://www.aph.gov/house/ committee/bushfires/inquiry/report.htm [1 October 2003].

References

Australia House of Representatives Standing Committee on Legal and Constitutional Affairs 1999 *Advisory Report on the Copyright Amendment (Digital Agenda) Bill 1999*, Canberra: Australian Parliament House.

Australia Copyright Convergence Group 1994 *Highways to Change*, Canberra: AGPS.

Australia Department of Industry, Science and Resources (DISR) 2001 *Positioning for Growth. Spatial Information Industry Action Agenda*, September, Canberra: InfoProducts.

Australia Electronic Commerce Expert Group (ECEG) 1998 *Electronic Commerce: Building the Legal Framework* (April) http://www.law.gov.au/aghome/eceg/ecegreport.html [30 June 2004] and at http://152.91.15.15/aghome/advisory/eceg/ecegreport.html [30 June 2004].

Australia/New Zealand Standards 1994 AS/NZS ISO9002-1994 *Quality Systems*.

Australian Advisory Council on Intellectual Property (ACIP) 2004 'Report on a Review of the Patenting of Business Systems' at http://www.acip.gov.au/library/bsreport.pdf [30 June 2004].

Australian Communications Authority (ACA) 2004 *Location, Location, Location*, January, Melbourne: ACA.

Australian Communications Industry Forum 1999 'Mobile location indicator for emergency services', DR ACIF G530, Canberra: ACIF.

Australian Law Reform Commission (ALRC) 1983 *Report No. 23 Privacy*, Canberra: AGPS.

Australian Vice Chancellors Committee (AVCC) 2002 *Ownership of Intellectual Property in Universities: Policy and Good Practice Guide* 2002.

Bainbridge, D 1996 Beta Computers (Europe) Ltd v Adobe Systems (Europe) Ltd. *Computer Law and Security Report*, v. 12(5), pp. 310–312.

Baker, H 1996 'Measuring the GPS', *GIS User*, No. 17, June–July, pp. 28–29.

Ball, M 2003 'Concerning ourselves with privacy', *GIS World*, February and at http://www.geoplace.com/gw/2003/0302/0302ed.asp [24 February 2003].

Bangemann Report 1994 *Europe and the Global Information Society: Recommendations to the European Council*, Brussels: European Council.

Barr, R and Masser, I 1997 'Geographic information: a resource, a commodity, an asset or an infrastructure?' in Kemp, Z (ed.) *Innovations in GIS 4. Selected Papers from the Fourth National Conference on GIS Research UK (GISRUK)*, London: Taylor & Francis Ltd, pp. 234–248.

Barr, R 2002 'Choosing the best route for Ordnance Survey. An old friend looks to a new life – the Ordnance Survey Quinquennial Review' at http://www.ginews.co.uk/0402_35.htm [10 June 2002].

Bartlett, D 2001 *A Practical Guide to GPS-UTM* at http://members.rogers.com/don.bartlett/gpsutm.htm [30 June 2004].

Barton, C and Liberman, A 2004 'Who owns employee produced inventions?', *FindLaw Australia,* February at http://www.findlaw.com.au/articles/[3 March 2004].

Bennett, CJ and Raab, C 2003 *The Governance of Privacy: Policy Instruments in Global Perspective*, London: Ashgate Press.

Blainey, G 1967 *The Tyranny of Distance*, Melbourne: Pan Macmillan Australia.

Blakemore, M 1991 'Access and security issues in the provision of geographic information', in Medyckyj-Scott, D *et al.* (eds) *Metadata in the Geosciences*, Loughborough: Group D Publications Ltd, pp. 55–68.

Blakemore, M 2001 *Financing the NGDI* at http://www.gisdevelopment.net/ application/gii/planning/giip10009pf.htm [7 April 2001].

Blakemore, M and Singh, G 1992 *Cost Recovery Charging for Government Information: A False Economy?* London: GSA Ltd, pp. 30–34.

Blakeney, M and Kretschmer, M 2004 *Intellectual Property and Geographical Indications: A Legal and Economic Analysis*, London: Edward Elgar Publishing.

Borgman, CL 2000 The premise and promise of a global information infrastructure, *First Monday*, Issue 5(8) at http://www.firstmonday.org/issues/issue5_8/borgman/ index.html [19 May 2004].

Bradbrook, AJ 1988 'The relevance of the *cujus est solum* doctrine to the surface landowner's claims to natural resources located above and beneath the land' (1988) 1 *Adel LR* 462.

Braden, T 2003 'Crew of cruise ship that hit reef faulted for failing to update charts', *Professional Mariner*, v. 75, October/November.

Branscomb, AW 1994 *Who Owns Information? From Privacy to Public Access*, New York: Basic Books.

Brinson, JD, Dara-Abrams, B, Dara-Abrams, D, Masek, J, McDunn, R and White, B 2001 *Analyzing E-Commerce & Internet Law*, Upper Saddle River, NJ: Prentice Hall.

Broekhuyse, P 2000 'Copyrights and wrongs: Lawyer's view of the net', *The Australian IT, The Australian*, 25 July, p. 55.

Brown, M 2001 'Maps led firefighters up paths with no exits', *The Sydney Morning Herald*, 17 July, pp. 1, 2.

Brown, P 2002 'Khmer genocide in total detail', *The Australian*, 16 January, p. 20.

Brown, RH, Irving, L, Prabhakar, A and Katzen, S 1995 *The Global Information Infrastructure: Agenda for Cooperation* (March) at http://www.iitf.nist.gov/ documents/docs/gii/giiagenda.html [23 June 1999].

Calvert, M 1995 *Technology Contracts*, Sydney: Butterworths.

Canada 1997 *Cost Recovery and Charging Policy*, Ottawa: Treasury Board Canada Secretariat.

Canada, Treasury Board of Canada 1997 *Cost Recovery and Charging Policy* at http://www.tbs-sct.gc.ca/pubs_pol/opepubs/tb_h/crp_e.html [30 June 2004].

Canadian Geospatial Data Infrastructure 2001 'Canadian Geospatial Data Infra-structure Target Vision', *Geomatica*, 55, pp. 181–185.

Canadian Standards Association 1996 *Model Code for the Protection of Personal Information: A National Standard of Canada* Q830, Mississauga, Ontario: CSA.

Castagna, RG, Thornton, LL and Tyrawski, JM 1999 'Where's Ellis Island? GIS and Coastal Boundary Disputes', *ArcUser*, v. 2(4), October–December, pp. 67–69.

References

Chance, C 1996 (ed.) *European Computer Law: An Introductory Guide*, Current Issues Publication Series, London: The Computer Law Association.

Charlesworth, A 2000a 'Clash of the data titans: US and EU Data Privacy Regulation', *European Public Law* 6, pp. 253–274.

Charlesworth, A 2000b 'Data privacy in cyberspace: Not national vs. international but commercial vs. individual', in Edwards, L and Waelde, C *Law and the Internet: A Framework for Electronic Commerce*, Oxford: Hart Publishing, pp. 79–124.

Cheung, C 2004 'Feta cheese—geographic indication or generic term?', *Intellectual Property Law Bulletin*, v. 16(9), pp. 133–135.

Chissick, M and Kelman, A 2000 *Electronic Commerce: Law and Practice*, 2nd edn, London: Sweet & Maxwell Ltd.

Cho, G 1998 *Geographic Information Systems and the Law*, Chichester: John Wiley & Sons.

Cho, G 2002 'Now u c it, now you don't' *E-law Practice*, Issue 10 (October), pp. 45–46.

Clarke, R 1999a 'Introduction to dataveillance and information privacy, and definition of terms' at http://www.anu.edu.au/people/Roger.Clarke/DV/Intro. html [18 January 2004].

Clarke, R 1999b 'The legal context of privacy-enhancing and privacy-sympathetic technologies' at http://www.anu.edu.au/people/Roger.Clarke/DV/ Florham. html [18 January 2004].

Clarke, R 1999c 'The Packer/PBL/Acxiom InfoBase' at http://www.anu.edu.au/ people/Roger.Clarke/DV/InfoBase99.html [18 January 2004].

Clarke, R 2000a 'How to ensure that privacy concerns don't undermine e-transport investments' at http://www.anu.edu.au/people/Roger.Clarke/EC/eTP.html [18 January 2004].

Clarke, R 2000b 'Person-location and person-tracking: Technologies, risks and policy implications', *Information Technology & People*, v. 14(2), Summer, pp. 206–231 and at http://www.anu.edu.au/people/Roger.Clarke/DV/PLT.html [18 January 2004].

Clarke, R 2001 'The end of privacy: While you were sleeping . . . surveillance technologies arrived', *AQ: Journal of Contemporary Analysis*, v. 73(1), January–February, pp. 9–14.

Clarke, R 2002 'ENUM' at http://www.anu.edu.au/people/Roger.Clarke/DV/ enum.html [18 January 2004].

Cleveland, H 1985 'The twilight of hierarchy: Speculations on the Global Information Society', *Public Administration Review*, January–February, pp. 185–195.

Clinton, W 1994 Executive Order 12906—Coordinating Geographic Data Acquisition and Access: The National Spatial Data Infrastructure, *Federal Register*, v. 59(71), April 11, pp. 17671–17674 also at http://www.fgdc.gov/ publications/documents/geninfo/execord.html [30 June 2004]. Since amended by President George Bush Executive Order 13286, *Federal Register* v. 68(43) pp. 10619–10633, 5 March 2003.

Commerce Clearing House Inc. (CCH) 1992 *Guide to Computer Law*, CCH ¶7120 'Computer hardware and software as goods', CCH ¶7140 'Hybrid transactions: Contracts including system support services', and CCH ¶7160 'Time-sharing and service bureau contracts', and CCH ¶8400.

Commerce Clearing House Inc. 1992 *Guide to Computer Law*, CCH ¶21,100 for 'Employment Contracts'.

Commission for the European Communities 2000 *Commercial Exploitation of Europe's Public Sector Information: Final Report*, PIRA International Ltd at ftp://ftp.cordis.lu/pub/econtent/docs/commercial_final_report.pdf [30 June 2004].

Commonwealth Spatial Data Committee (CSDC) 2001 A proposal for a Commonwealth Policy on Spatial Data Access and Pricing, Belconnen, ACT: CSDC, also at http://osdm.gov.au/osdm/docs/Commonwealth_Policy_on_Spatial_Data_Access_and_Pricing.pdf [21 April 2004].

Connolly, E 2000a 'Department of disaster escapes Gretley charges', *The Sydney Morning*, 17 April 2000, p. 5.

Connolly, E 2000b 'Gretley: Mine and men face charges', *The Sydney Morning Herald*, 18 April 2000, p. 6.

Cranor, LF 2001 'The role of privacy enhancing technologies' at http://www.cdt.org/privacy/ccp/roleoftechnology1.shtml [1 February 2004].

Cranor, LF, Byers, S and Kormann, D 2003 'An analysis of P3P deployment on commercial, government, and children's websites as of May 2003', Technical Report prepared for the Federal Trade Commission (FTC) at http://www.research.att.com/projects/p3p [31 January 2004].

Cross, JT 1994 'Protecting computer databases under the United States copyright laws: Implications of the *Feist* decision', in Carr, I and Williams, K (eds) *Computers and Law*, London: Intellect Books, pp. 113–127.

Curry, DJ 1992 *The New Marketing Research Systems: How to Use Strategic Database Information for Better Marketing Decisions*, New York, NY: John Wiley & Sons.

Curry, MR 1994 'In plain and open view: Geographic information systems and the problem of privacy' at http://www.spatial.maine.edu/tempe/curry.html [30 June 2004].

Dando, LP 1994 'A survey of open records laws in relation to recovery of database development costs: An end in search of a means' in *Marketing Government Geographic Information*, Washington, D.C.: URISA, pp. 5–22.

Dansby, H 1992 'Public records and government liability. Part II', *GIS Law*, 1(1), pp. 7–13.

Daratech 2003 Press Release 'Daratech: GIS Revenues forecast to grow 8% to $1.75 billion in 2003: Utilities and Government Increase Spending' at http://www.daratech.com.press/2003/030807/[23 May 2004].

Dearne, K 2004 'Black listed. A landmark decision has exposed databases to expensive litigation', *The Australian IT Business*, 27 April, pp. 1, 4.

DeLamarter, RT 1986 *Big Blue: IBM's Use and Abuse of Power*, New York: Dood, Mead.

References

Department of Industry, Science and Resources 1999 *Shaping Australia's Future. Innovation Framework Paper*, Canberra: AGPS and at http://industry.gov.au/library/content_library/shaping.pdf [30 June 2004].

Department of Industry, Science and Resources (DISR) 2001 *Positioning for Growth: Spatial Information Industry Action Agenda*, September, Canberra: InfoProducts, and at http://www.isr.gov.au/agendas/sectors/siiaa/aa_presentation.pdf [30 June 2004].

Department of the Environment, Transport and the Regions (DETR) 1999 'National Interest Mapping Service Agreement' at http://www.dtlr.gov.uk/annual99/index.html [30 June 2004].

Dobson, J 1998 'Is GIS a privacy threat?', *GIS World*, v. 11(7) July, pp. 34–35 and at http://www.geoplace.com/gw/1998/0798/798onln.asp [13 January 2004].

Dobson, J 2000 'What are the ethical limits of GIS?', *GeoWorld*, May p. 24 and at http://www.geoplace.com/gw/2000/0500/0500g.asp [22 May 2000].

Dobson, J 2002 'The "G" in GIS – What's New about GIS?' at http://www.geoplace.com/gw/2002/0203/0203gngs.asp [23 February 2002].

Dodge, M and Kitchen, R 2001 *Atlas of Cyberspace*, Reading, MA: Addison Wesley.

Doherty, M 2002 'Gallery reveals new entrance. Controversial plans dumped after protest', *The Canberra Times,* 24 February, p. 3

Donato, DI 1985 A review of pricing issues and alternatives for 1:100 000 scale digital cartographic data. Reston, VA: USGS.

DOT Plan 1992 'IVHS Strategic Plan—Report to Congress', 18 December 1992 cited in Alpert, S and Haynes, KE 1994 'Privacy and the intersection of geographical information and intelligent transportation systems' at http://www.spatial.maine.edu/tempe/alpert.html [30 June 2004].

Downing, S and Harrington, J 2000 'The postal rule in electronic commerce: A re-consideration' (2000) 5(2) *Communications Law* 43.

Draft Agreement between the United Nations and the Government of Cambodia concerning the prosecution under Cambodian law of crimes committed during the period of Democratic Kampuchea at http://www.yale.edu/cgp/Cambodia_Draft_Agreement_17-03-03.doc [23 May 2004].

Dragich, MJ 1989 'Information malpractice: Some thoughts on the potential liability of information professionals', 8 *Information Technology and Libraries*, p. 266.

E.U. 1986 *Council Directive on Self-Employed Commercial Agents* 86/653/EEC.

E.U. 1987 *Single European Act* OJ L 169, 29.6.1987 also at http://europa.eu.int/abc/obj/treaties/en/entr14a.htm#41 [8 May 2004].

E.U. 1995 Data Protection Directive of the European Union Parliament and of the Council of 24 October 1995 *Directive on the Protection of Individuals with regard to the Processing of Personal Data and on the Free Movement of such Data*, Brussels: European Commission; Directive 95/46/EC; *Official Journal of the European Commission* (L 281) 31 and at http://europa.eu.int/eur-lex/en/lif/dat/1995/en_395L0046.html.

E.U. 1999a *Green Paper on Public Sector Information in the Information Society*, COM (98) 585 Final, adopted on 20 January 1999 at http://europa.eu.int/ISPO/docs/policy/docs/COM(98)585/[1 May 2004].

E.U. 1999b Directive 1999/34/*EEC on the approximation of the laws, regulations and administrative provisions of the Member States concerning liability for defective products*, Official Journal L 141, 04/06/1999 pp. 0020–0021.

E.U. 1999c *Green Paper on Product Liability*, COM (1999) 396 Final and 2nd Report, January COM (2000) 893 Final and at http://www.europa.eu.int/comm/dg15/en/index.htm [30 June 2004].

E.U. 1999d *Directive on the harmonisation of certain aspects of copyright and related rights in the Information Society* (COM (1999) 250 Final (Directive)).

E.U. 2002 Report on Comparative Study Carried Out Under Trilateral Project B3B, 14–16 June 2002 at http://www.european-patent-office.org/tws/front_page.pdf [30 June 2004].

E.U. 2002 Directive on Privacy and Electronic Communications Directive of the European Parliament and of the Council of 12 July 2002 concerning the processing of personal data and the protection of privacy in the electronic communications sector 2002/58/EC at http://europa.eu.int/comm/internal_market/privacy/law_en.htm [30 June 2004].

Editorial 2002 'Location, location, location. Global positioning is too vital to be left in the hands of the Pentagon', *New Scientist*, 30 March, p. 5.

EEC 1957 *Treaty of Rome* at http://www.europa.eu.int/abc/treaties_en.htm [8 May 2004].

Eldred, P 1995 'Geographic information systems and copyright: Are we on the right road?', *WALIS News* (March), pp. 13–18.

Electronic Communication Privacy 1989 'ECPA and Online Computer Privacy', 4 *Federal Communications Law Journal* 17, 39 (1989).

Electronic Funds Transfer Code of Conduct 2002 reproduced in *CCH Australian Consumer Credit Law Reporter*, pp. 95–250.

Elgin, RL 1995 'Surveyor's notes for plats', *Point of Beginning* (POB), August–September, p. 17.

Environmental Systems Research Institute 1995 *Data Publishing Guidelines for ESRI Software*, White Paper September, J-6040, Redlands, CA: ESRI.

Epstein, EF 1991 'Legal aspects of geographic information systems' in Maguire, DJ, Goodchild, MF and Rhind, DW (eds) *Geographic Information Systems: Principles and Applications*, 2 vols. London: Longman, pp. 489–502.

Equifax and National Decision Systems 1993 *InfoMark-GIS: Tomorrow's Technology for Today's Business Success*, Atlanta, GA: Equifax, Inc.

ESRI 1993 'GIS System for Marketing Applications Introduced', *ARC News*, v. 15(3) Summer, p. 6.

EUROGI 2000 'Towards a Strategy for Geographic Information in Europe' ver. 1.0, October, Apeldoorn, Netherlands: EUROGI, also at http://www.eurogi.org/geoinfo/eurogiprojects/strategy.pdf [30 June 2004].

References

European Commission 1998 *GI2000: Towards a European Policy Framework for Geographic Information*, Luxembourg: DG Information Society.

European Communities—Commission 1989 *Guidelines for Improving the Synergy between the Public and Private Sectors in the Information Market*, Luxembourg: Office for Official Publications of the European Communities.

European Science Foundation (ESF) 2000 *European Science Foundation Policy Briefing: Good Scientific Practice in Research and Scholarship*, December at http://www.esf.org/sciencepolicy/170/ESPB10.pdf [30 June 2004].

Fairall, J 2000 'Accuracy and liability. A looming nightmare for GIS Officers', *GISUser*, Issue 42, October–November, p. 9.

Fairall, J 2002 'Death in a back paddock. Liability for bad data rests with the user', *GISUser*, Issue 50, February–March, p. 7.

Farmer, D 2002 'The red soils of Coonawarra, part of a unique *terroir*', *The Australian and New Zealand Wine Industry Journal*, v. 17(6), December, pp. 30–39. Also at http://www.winetitles.com.au [30 June 2004].

FGDC 1993 *Content Standards for Digital Geospatial Metadata* (CSDGM) at http://www.fgdc.gov/Metadata/metahome.html [30 June 2004] and also at http://www.fgdc.gov/metadata/constan.html [30 June 2004].

FGDC Clearinghouse at http://www.fgdc.gov/clearinghouse/clearinghouse.htm [30 June 2004].

FGDC I-Team at http://www.fgdc.gov/I-Team/index.html [30 June 2004].

Finch, B 1999 'Consumer protection on the Internet', in Fitzgerald, A *et al.* (eds) *Going Digital 2000: Legal Issues for E-Commerce, Software and the Internet*, St Leonards: Prospect Media Pty Ltd.

Findlaw 2004 'Tenancy Database Operator Breaches Privacy Act', 20 April at http://www.findlaw.com.au/news/default.asp?task=real&id=19703&newstype=L&site=NE [3 May 2004].

Fingar, A 1999 A CEO's Guide to e-Commerce using Intergalactic Object-Oriented Intelligent Agents at http://home1.gte.net/pfingar/eba.htm [30 June 2004].

Flaherty, DH 1994 'Privacy protection in Geographic Information Systems: Alternative protection scenarios' at http://www.spatial.maine.edu/tempe/flaherty.html [30 June 2004].

Fleming, JG 1998 *The Law of Torts*, Sydney: LBC Information Services. Chapter 23 'Product Liability', pp. 530–558.

Founds, G 1999 'Shrink-wrap and click-wrap agreements: 2B or not 2B?' 52(1) *Federal Communications Law Journal* 100.

France 1994 *Circulaire du 14 fevrier 1994 relative à la diffusion des données publiques*, Paris at http://www.environment.gov.fr/donnees/diffdon/balladur.htm.

Free Software and Open Source Foundation for Africa (FOSSFA) Action Plan 2003–2005 at http://www.admi.net/cgi-bin/wiki?FOSSFA [30 June 2004].

Fuller, L 1941 'Consideration and form', 41 *Columbia L. Rev.* 799 (1941).

G-7 Ministerial Conference on the Information Society 1995 at http://europa.eu.int/ISPO/intcoop/g8/i_g8conference.html [13 May 2004].

Gale Group 2001 *Gale Directory of Databases*, 1993–2001, Farmington Hills, MI: Gale Group.

Garretson, C 2002 'Whose data is it, anyway?' at http://www.itworld.com/Tech/ 2987/020409whosedata/pfindex.htm [12 April 2002].

Gaze, B 1989 *Copyright Protection of Computer Programs*, Sydney: Federation Press.

Genignoni, MC 1981 'Product liability and software', 8 *Rutgers Computers & Technology Law Journal* 173.

Geoscience Australia 2002 *Product Catalogue and Sampler 2002 CD*, Canberra: Department of Industry, Tourism and Resources.

Gibson, J 2003 'Satellite's gone, we're going to hit: Skipper', *The Sydney Morning Herald*, 29 July, p. 3.

Giles, D 2000 'You've got mail . . . or have you?', *Internet Law Bulletin*, v. 3(1) pp. 12, 14.

GINIE 2004 *GINIE Final Report*, D-1.5.1 (January), 22 pp. at http://wwwlmu. jrc.it/ginie/doc/GINIE_finalreport.pdf [16 April 2004].

GINIE 2004 Introduction to building registries or catalogue services at http:// wwwlmu.jrc.it/ginie [16 April 2004].

GINIE Geographic Information Network in Europe 2004 *A Compendium of European SDI Best Practice*, Ver. 1 (January) at http://wwwlmu.jrc.it/ginie/ doc/d511_Book_v1.pdf [16 April 2004].

Ginsburg, JC 1992 'No "sweat"?' Copyright and other protection of works of information after *Feist v Rural Telephone*, 92 *Columbia L. Rev.* 338 (1992).

GIS Monitor Press Release on Daratech 2003 at http://www.gismonitor.com/ news/pr/2003/080803_Daratech.php [8 May 2004].

Givoni, S 2003 'Pushing the boundaries of copyright: protection of databases', *Australian Intellectual Property Bulletin*, v. 15(8) January, pp. 113–118.

Gore, A 1994 'Al Gore Speech on a U.S. vision for the Global Information Infrastructure', World Telecommunication Development Conference, Buenos Aires (March) quoted in Brown, RH, Irving, L, Prabhakar, A and Katzen, S (1995) *The Global Information Infrastructure: Agenda for Cooperation* (March) at http://www.iitf.nist.gov/documents/docs/gii/giiagenda.html [23 June 1999].

Gorman, RA 1992 'The *Feist* case: Reflections on a path-breaking copyright decision', 18 *Rutgers Computer & Tech LJ* 731 (1992).

Goss, JD 1994 'Marketing the new marketing. The strategic discourse of Geodemographic Information Systems' in Pickles, J (ed.) *Ground Truth: The Social Implications of Geographic Information Systems*, New York: Guildford Press, pp. 130–170.

Goss, J 1995 'We know who you are and we know where you live: The instrumental rationality of Geodemographic Systems', *Economic Geography,* v. 71(2) April, pp. 171–198.

Graham, SJ 1997 'Products liability in GIS: Present complexions and future directions', *GIS Law,* v. 4(1), pp. 12–16.

References

Grant, DM and Krogh, B 1995 'Partners in the spatial information systems industry – an "open marriage" between the public and private sectors', *GIS Law*, 2(4), pp. 9–17.

Grayson, I 1999 'Packer sets up Big Brother data store', at http://technology.news.com.au/news/4277059.htm [18 January 2004].

Greguras, F, Egger, MR and Wong, SJ 1995 'Multimedia content and the Super Highway: Rapid acceleration or foot on the brake?' at http://www.batnet.com/oikoumene/mmcopyright.html [19 March 2004].

Groebner, B 2004 'Oops! The legal consequences of and solutions to online pricing errors', 1 *Shidler J.L. Computers and Technology*, 2 and at http:// www.lctjournal.washington.edu/vol1/a002Groebner.html [5 August 2004].

Groot, R and Georgiadou, Y 2001 'Advancing the concept National Geospatial Data Infrastructure: Reflections on the "bottom line"' at http://www.gisdevelopment.net/application/gii/planning/giip10013pf.htm [7 April 2001].

Groot, R 2001 'Economic Issues in the Evolution of National Geospatial Data Infrastructure' Background paper for the 2nd meeting of the Committee on Development Information (CODI-2), 4–7 September, Addis Ababa, Ethopia, at http://www.uneca.org/eca_programmes/it_for_development/geoinfo/doc8EN (Economic%20issues%20in%20the%20evolution%20of%20Geo-information).pdf [21 April 2004].

Groves, P 1991 *Copyright and Designs Law. A Question of Balance. The Copyright, Designs and Patents Act, 1988*, London: Graham & Tratman.

GSDI 1997 *Conference Findings and Resolutions*, 2nd GSDI Conference, Chapel Hill, North Carolina, USA 19–21 October 1997 at http://www.gsdi.org/docs/gsdi97r.html [16 April 2004].

GSDI 2004 *Global Spatial Data Infrastructure. Strategic Development Plan*, Ver. 0.9 (January) draft at http://www.gsdi.org/DSP/GSDI_PlanD_v_0-9.htm [16 April 2004].

GSDI Technical Working Group 2001 *Developing Spatial Data Infrastructures: SDI Cookbook* Ver. 1.1 (May) at http://www.gsdi.org/pubs/cookbook/index.htm [15 May 2001].

Guest, AG (ed.) 1994 *Chitty on Contracts* (27th edn), London: Sweet & Maxwell.

Handelsmann, A 2001 'Strategies for complying with Australia's Privacy Principles' at http://www/gigalaw.com/articles/2001/handelsmann-2001-11-p1.html [26 January 2004].

Harris, R 1997 *Earth Observation Data Policy*, Chichester: John Wiley & Sons.

Harrison, B 1998 'Disclaimers', *Azimuth*, February, pp. 16–19.

Harvey, JA and Verska, KA 2001 'What the European Data Privacy obligations mean for U.S. Businesses' at http://www.gigalaw.com/articles/harvey-2001-02-p1.html [27 September 2001].

Hedberg, O, Paull, D and Bower, M 2003 'Spatially enabling Australia through collaboration and innovation' paper presented to the Cambridge Conference 2003, Southampton: Ordnance Survey, paper 7.1.

Heidemann, MA 1992 'Copyright and Copy wrong', *Planning* (American Planning Association) v. 58(2), pp. 22–23.

Henry, N 1975 *Copyright Information Technology Public Policy*, New York, Basle: Marcel Dekker.

Ho, D 2001 'Cellular phone carriers buy more time to develop emergency location technology' at http://www.nando.net/technology/v-text/story/124854p-1312017c.html [30 June 2004].

Hobson, D 2001 NSDI Development in Asia and the Pacific. Report on Australian Activities, paper presented at the 7th PCGIAP Meeting, Tsukuba, Japan, 24–27 April at http://www.gsi.go.jp/PCGPIAP/tsukuba/seminar/paper_au.pdf [21 April 2004].

Hocker, PJ and Heffey, PG 1994 *Contract: Commentary and Materials* (7th edn), Sydney: The Law Book Company Ltd.

Hovanyetz, S 2003 'Court: Information brokers can be liable', *DMNews. The Online Newspaper of Record for Direct Marketers*, 11 March at http://www.dmnews.com/cgi-bin/artprevbot.egi?article_id=23093 [11 March 2003].

Hugenholtz, PB 2001 'The New Database Right: Early Case Law from Europe', paper 9th Annual Conference on Intellectual Property Law and Policy, Fordham University School of Law, New York, 19–20 April 2001. Also at http://www.ivir/publications/hugenholtz/fordham2001.html [1 April 2004].

Hugenholtz, PB 2003 'The Database Right File' at http://www.ivir.nl/files/database/index.html [29 April 2004].

Hughes, A 2001 'A question of adequacy?' The European Union's approach to assessing the *Privacy Amendment (Private Sector) Act 2000* (Cwlth) [2001] *University of NSW Law Journal* 5, also at http://www.austlii.edu.au/au/journals/UNSWLJ/2001/5.html [6 February 2004].

Hughes, G 1994 'Government data matching to continue', 68 *Law Institute Journal* 488.

Hull, C 1994 'Privacy question in phone-book CDs', *The Canberra Times*, August 1, p. 13.

Industry Canada 2001 'Geomatics Industry' at http://www.strategis.ic.gc.ca [19 April 2004].

Information Infrastructure Task Force (IITF) Information Policy Committee, Privacy Working Group 1995 *Privacy and the National Information Infrastructure: Principles for Providing and Using Personal Information* at http://www.ntia.doc.gov/ntiahome/privwhitepaper.html [30 June 2004].

International Telecommunications Union (ITU) 2001 'ENUM' at http://www.itu.int/osg/spu/enum/index.phtml [30 June 2004].

Internet Engineering Task Force (IETF) 2000 'E.164 Number and DNS' RFC 2916 September 2000 at http://www.ietf.org/rfc/rfc2916.txt [30 June 2004].

IP Australia 2004 What is Intellectual Property? at http://www.ipaustralia.gov.au/ip [29 March 2004].

Joffe, B 2002 'How good are your maps', *GeoWorld*, February. Also at http://www.geoplace.com/gw/2002/00202/0202dac.asp [25 February 2002].

References

Joffe, BA 2003 *Open Data Consortium Project: Model Data Distribution Policy*, Oakland, CA: GIS Consultants. Also see http://www.joffes.com/[21 April 2004].

Johnson, JP and Dansby, HB 1995 'Liability in private sector geographic information systems', *Proceedings of the Conference on Law and Information Policy for Spatial Databases*: Tempe, AZ (NCGIA, University of Maine, Orono, 1995) and at http://www.spatial.maine.edu/tempe/johnson.html [30 June 2004].

Karjala, DS 1994 'Copyright in electronic maps', *Proceedings of the Conference on Law and Information Policy for Spatial Databases*, Tempe AZ, October, NCGIA and Center for the Study of Law, Science and Technology, Arizona State University College of Law, 35 *Jurimetrics J.* 395–415 (1995) and at http://www.spatial.maine.edu.tempe/karjala.html [30 June 2004].

Katsh, ME 1995 *Law in a Digital World*, New York: Oxford University Press.

Katter, NA 1999 *Duty of Care in Australia*, Sydney: LBC Information Services.

Knight, P and FitzSimons, J 1990 *The Legal Environment of Computing*, Sydney: Addison-Wesley Pub. Coy.

Krause, B 2001 'An Overview of the Canadian Personal Information Protection and Electronic Documents Act' at http://www.gigalaw.com/articles/2001/krause2001-02.html [26 January 2004].

Krim, J 2003a '8 million credit accounts exposed. FBI to investigate hacking of database' February 18, p. E01 at http://www.washingtonpost.com/ac2/wp-dyn/A27334-2003Feb18?language=printer [20 February 2003].

Krim, J 2003b 'States seen as lax on database security. Study faults efforts to police insurers, 26 March, p. E05 at http://www.washingtonpost.com/wp-srv/resetcookie/front.htm [28 March 2003].

Latimer, P 1999 *Australian Business Law* (18th edn), Sydney: CCH Australia Ltd.

Lee, C 2003 'High Court hangs up on issue of copyright and compilations', *Australian Intellectual Property Law Bulletin*, v. 16(4) August, pp. 45–60.

Leipnik, M, Bottelli, J, von Essen, I, Schmidt, A, Anderson, L and Cooper, T 2001 'Coordinates of a killer' at http://www.geoinfosystems.com/1101/1101spokane.html [24 November 2001].

Lemmens, MJPM 2001 'An European Perspective on Geo-Information Infrastructure Issues', at http://www.gisdevelopment.net/application/gii/global/giigp0007pf.htm [7 April 2001].

Leonard, PG and Spender, PA 1989 'Intellectual property protection of databases', 9 *Information Services and Use* 33–43.

Lessig, L 2001 *The Future of Ideas. The Fate of the Commons in a Connected World*, New York: Random House.

Lickson, CP 1968 'Protection of the privacy of data communications by contract' (1968) 23 *Bus. L.* 971.

Lillesand, TM and Keifer, RW 1999 *Remote Sensing and Image Interpretation* (4th edn), New York: John Wiley & Sons.

Linden, AM 1997 *Canadian Tort Law* (6th edn), Toronto: Butterworths.

Lindley, M 1995 'GIS Customer Service vs Legal Status', AURISA/SIRC'95 – 7th Colloquium of the Spatial Information Research Centre, University of Otago, AURISA NZ and Massey University, 26–28 April.

Lindsay, P 1993 'Copyright protection of electronic databases', *Journal of Law and Information Science*, v. 4(1) pp. 287–292.

Litman, J 1992 'Copyright and information policy', 55 *Law and Contemporary Problems* 185, 1992.

Longhorn, RA 1998 Strategic Initiatives for the Global Spatial Data Infrastructure (GSDI), GSDI-3 Conference, 17–19 November, Canberra, Australia at http://www.gsdi.org/docs1998/canberra/ralstrat.html [20 April 2004].

Longhorn, RA 2001 'The impact of data access policies on regional spatial data infrastructure' available at http://wwwlmu.jrc.it/Workshops/7ec-gis/papers/html/longhorn/longhorn.htm [22 April 2004].

Longhorn, RA 2004 *GINIE: Geographic Information Network in Europe. Document Peer Review Report* (January), 6 pp. at http://wwwlmu.jrc.it/ginie/doc/ginie_peer_review.pdf [22 April 2004].

Longley, PA, Goodchild, MF, Maguire, DF and Rhind, DW 2001 *Geographic Information Systems and Science*, Chichester: John Wiley & Sons Ltd.

Lowe, S 1994 'Indecent disclosures', *The Sydney Morning Herald*, 21 March, p. 47.

Luckett, B 2004 'Case could open public access' at http://www.casperstartribune.net/articles/2004/04/11/news/wyoming/df0fc4b4ae49db6287256e73001aeff2.prt [16 April 2004].

Macey, R 2001 'Nightmare on bogus street: A work of fiction in your glovebox', *The Sydney Morning Herald*, July 7–8, p. 1.

Marx, GT 1994 'Some information-age techno-fallacies and some principles for protecting privacy' at http://www.spatial.maine.edu/tempe/marx.html [30 June 2004].

Mason, R 2002 *Developing Australian Spatial Data Policies—Existing Practice and Future Strategies*, Unisurv s 70 2002, Reports from the School of Surveying and Spatial Information Systems, Sydney: University of New South Wales.

Masser, I and Stevens, AR 2003 'Global Spatial Data Infrastructure (GSDI): At the Crossroads, Moving Forward', paper presented to the Cambridge Conference 2003, Southampton: Ordnance Survey, paper 6.2.

Maurer, SM, Hugenholtz, PB and Onsurd, HJ 2001 'Europe's Database Experiment', *Science*, v. 294 (26 October) pp. 789–790 and at http://www.sciencemag.org [18 January 2004].

McCoy, MD and Spence, AT 2001 'Lessons from the United States and Europe on Computer-related Patents' at http://www.gigalaw.com/articles/2001-all/mccoy-2001-08-all.html [30 April 2003].

McCullagh, D 2003 'Supreme Court nixes copyright challenge', 15 January at http://news.com.com/2100-1023-980792.html [18 January 2003].

McSherry, C 2001 *Who Owns Academic Work: Battling for Control of Intellectual Property*, Boston, MA: Harvard University Press.

References

Millar, P, Grey, S and Rufford, N 1997 'Prying eyes', *The Sunday Times* (U.K.), 15 June 1997, p. 16.

Miller, P and Greenstein, D 1997 *Discovering On-line Resources Across the Humanities: A Practical Implementation of the Dublin Core*, Bath: UKOLN.

Millhouse, D 1994 'A merchant banker's view of GIS', *AURISA News*, 55, pp. 1, 6–11.

Milrad, L 1994 'Database ownership, Canadian copyright principles and multi-participant GIS projects', *GIS Law*, v. 2(2), pp. 7–10.

Minchin, N 2001 Media Release Senator Nick Minchin 25 September 2001 at http://www.minister.industry.gov.au/minchin/releases/2001/September/cmr485-01.doc [25 September 2001].

Monmonier, M 1995 *Drawing the Line. Tales of Maps and Cartocontroversy*, New York: Henry Holt & Co.

Monmonier, M 2002 *Spying with Maps. Surveillance Technologies and the Future of Privacy*, Chicago and London: The University of Chicago Press.

Morgan, R and Stedman, G 1995 *Computer Contracts*, London: FT Law & Tax.

Multimap patent challenge at http://mapserver.gis.umn.edu/wilma/mapserver-users/0203/msg00495.html [1 April 2004].

Multimap.com U.S. Patent No. 6,240,360 see http://patft.uspto.gov and search under patent number 6240360 [1 April 2004].

Nairn, AD and Holland, P 2001 'The NGDI of Australia—Achievements and Challenges from a Federal Perspective', paper presented at a Workshop on NGDI—Towards a Road Map for India, 5–6 February, New Delhi, India at http://www.gisdevelopment.net/policy/international/interna007.htm [7 April 2001].

National Academy of Public Administration (NAPA) 1998 *Geographic Information for the 21st Century—Building a Strategy for the Nation*, January at http://www.napawash.org/[30 June 2004] and http://38.217.229.6/NAPA/NAPAPubs.nsf [30 June 2004] and at http://www.napawash.org/resources/testimony/testimony_06_09_99.html [30 June 2004].

National Archives Australia (NAA) 1999 *Recordkeeping Metadata Standard for Commonwealth Agencies*, Ver. 1.0, May, 136 pp. Canberra: NAA.

National Office for the Information Economy (NOIE) 2003 *The Interoperability Technical Framework for the Australian Government*, Canberra: NOIE now known as Australian Government Information Management Office (AGIMO) at http://www.agimo.gov.au/publications/2003/08/framework/overview [30 June 2004].

National Research Council 1999 *A Question of Balance: Private Rights and the Public Interest in Scientific and Technical Databases*, Washington, D.C.: National Academy Press at http://books.nap.edu/books/0309068258/html/index.html [13 December 2003].

National Research Council 1999 *Distributed Geolibraries: Spatial Information Resources*, Washington, D.C.: National Academies Press.

New South Wales Treasury 2001 *Guidelines for Pricing User Charges*, Sydney: NSW Treasury.

New Zealand Treasury 2002 *Guidelines for Setting Charges in the Public Sector*, Wellington: NZ Treasury and at http://www.treasury.govt.nz/publicsector/charges/setcharges.pdf [30 June 2004].

Nimmer, MB and Nimmer, D (eds) 1993 *Nimmer on Copyright* § 4.12(A) Bethesda, MD: Matthew Bender Publishers and Lexis Nexis.

Nimmer, RT 1996 'Electronic contracting: Legal issues', 14 *John Marshall J of Computers and Information Law*, Winter, pp. 211–246.

Obloy, EJ and Sharetts-Sullivan, BH 1995 'Exploitation of intellectual property by electronic chartmakers: Liability, retrenchment and a proposal for change', *Proceedings of the Conference on Law and Information Policy for Spatial Databases*: Tempe, AZ (NCGIA, University of Maine, Orono, 1995) and at http://www.spatial.maine.edu/tempe/obloy.html [30 June 2004].

Organisation for Economic Cooperation and Development (OECD) 1980 *Guidelines on the Protection of Privacy and Transborder Flows of Personal Data*, Paris: OECD. Recommendation by the OECD Council of 23 September 1980. See http://www.oecd.org/e/droit/doneperso//ocdeprive/priv-en.htm [7 February 2004] and at http://www.oecd.org/dsti/sti/it/secur/prod/PRIV-en.htm [7 February 2004] and at http://www.oecd.org/documentprint/0,2744,en_2649_201185_15589524_1_1_1_1,00.html [7 February 2004].

OECD 1998 *User Charging for Government Services*, Public Management Occasional papers no. 22, Paris: OECD.

OECD 1999 *Guidelines for Consumer Protection in the Context of Electronic Commerce*, adopted 9th December 1999 by 29 OECD member countries, at http://www.oecd.org/documents/51/0,2340,en_2649_34267_1824435_1_1_1_1,00.html [30 June 2004].

Office of the Federal Privacy Commissioner Australia 1994 *Fifth Annual Report, 1992–1993*, Canberra: Human Rights Australia.

Office of the Federal Privacy Commissioner Australia 1999 *National Principles for the Fair Handling of Personal Information* (Revised edn January and at http://www.privacy.gov.au/news/media/99_1.html [30 June 2004].

Office of the Federal Privacy Commissioner Australia 2000 *National Privacy Principles extracted from the Privacy Amendment (Private Sector) Act 2000 at* http://privacy.gov.au/publications/npps01_print.html [30 June 2004].

Office of the Federal Privacy Commissioner Australia 2001 'Information Sheet 14. Privacy obligations for Commonwealth contracts' at http://www.privacy.gov.au/publications/IS_14_01.html [30 June 2004].

Office of Management and Budget (OMB) 1993 *Circular A-25 Revised*. Washington, D.C.: OMB also at http://www.whitehouse.gov/omb/circulars/a025/a025.html [30 June 2004].

OMB 1996 Circular A-130 *Management of Federal Information Resources*. Washington, D.C.: OMB also at http://www.whitehouse.gov/omb/circulars/a130/a130.html [30 June 2004].

OMB 2001 Circular A-16 Revised at http://www.whitehouse.gov/omb/circulars/a016/a016_rev.html [30 June 2004].

References

Onsrud, HJ 1992 'In support of open access for publicly held geographic information', *GIS Law*, 1(1), pp. 3–6.

Onsrud, HJ 1993 'Law, Information Policy and Spatial Databases', *NCGIA Working Paper* (April), Orono, Maine: NCGIA, 18 pp.

Onsrud, HJ 1995 'Role of Law in impeding and facilitating the sharing of Geographic Information', in Onsrud, HJ and Rushton, G (eds) 1995 *Sharing Geographic Information*, News Brunswick, NJ: Rutgers, The State University of New Jersey, Center for Urban Policy Research, pp. 292–306.

Onsrud, HJ 1998 'Access to geographic information in the United States' in Free Accessibility of Geo-information in the Netherlands, the United States, and the European Community, *Proceedings*, Delft, The Netherlands, October 2, pp. 33–41.

Onsurd, HJ 1999 'Liability in the use of GIS and geographical datasets' in Longley, P, Goodchild, M, Maguire, D and Rhind, D (eds) *Geographical Information Systems: Vol. 2, Management Issues and Applications*, London: John Wiley & Sons Inc., pp. 643–652.

Onsrud, HJ 2001 SIE 525 *Information Systems Law. First Readings in GIS Law*, at http://www.maine.edu/~onsrud/GISlaw.htm [13 February 2003].

Onsrud, HJ and Hintz, RJ 1989 'Upgrading boundary information in a GIS using an automated survey measurement management system', *Technical Papers: 1989 ACSM-ASPRS Annual Convention*, v. 4, pp. 275–284.

Onsrud, HJ, Johnson, JP and Lopez, X 1994 'Protecting personal privacy in using geographic information systems', *Photogrammetric Engineering and Remote Sensing*, v. 60(9), September, pp. 1083–1095. Also at http://www.spatial.maine.edu/tempe/onsrud.html [30 June 2004].

Onsrud, HJ and Rushton, G (eds) 1995 *Sharing Geographic Information*, News Brunswick, NJ: Rutgers, The State University of New Jersey, Center for Urban Policy Research.

Openshaw, S and Goddard, JB 1987 'Some implications of the commodification of information and the emerging information economy for applied geographical analysis in the United Kingdom', *Environment and Planning A* 19: 1423–1439.

Ordnance Survey Media Release 22 July 2002 at http://www.ordnancesurvey.co.uk [5 May 2004].

Orwell, G 1990 (first published in 1949) *Nineteen-Eighty-Four*, New York: The New American Library Inc.

OXERA 1999 *The Economic Contribution of Ordnance Survey GB'*, Oxford Economic Research Associates Ltd at http://www.ordnancesurvey.co.uk/oswebsite/aboutus/reports/oxera/index.html [30 June 2004].

Pain, C and Nightingale, J 2001 'Geology proves fruitful for viticulture', *AusGeo News* 63 October/November, p. 3.

Pappas, C 2004 'Can copyright subsist in Chinese characters?', *Intellectual Property Law Bulletin*, v. 16(9) February/March, pp. 150.

Perritt, JJ 1996 *Law and the Information Superhighway*, New York: John Wiley & Sons, Inc.

PIRA International Ltd, University of East Anglia, and Knowledge Ltd 2000. *Commercial Exploitation of Europe's Public Sector Information*, Final Report for the European Commission, Directorate General for the Information Society, Luxembourg: EC DG INFSO. See also ftp://ftp.cordis.lu/pub/econtent/docs/commercial_final_report.pdf [30 June 2004].

Plewe, B 1997 *GIS Online: Information Retrieval, Mapping, and the Internet*, Santa Fe, NM: OnWord Press.

Pocket GPS World 2004 'US and EU agree to link Navstar and Galileo GPS Systems' at http://pocketgpsworld.com/modules.php?name=News&file=article&sid=370 [5 August 2004].

Powell, S 1995 'Melbourne changes tack on copyright', *The Australian*, 18 July.

Pratt, M and Donaldson, J 2003 'Drawing the Line: Mapmakers and International Boundaries', 2003 Cambridge Conference, Southampton: Ordnance Survey, Paper 4D.3.

Privacy Protection Study Commission (US) 1977 *Personal Privacy in an Information Society*, Washington, D.C.: OMB and at http://www.epic.org/privacy/ppsc1977report/[30 June 2004].

Productivity Commission 2001 *Cost Recovery by Government Agencies*. Inquiry Report No. 15, Canberra: AusInfo, also at http://www.pc.gov.au/inquiry/costrecovery/[18 April 2004].

Prosser, WL, Wade, JW and Schwartz, VE 1982 *Cases and Materials on Torts* (3rd edn), Brooklyn: Foundation Press University Casebook Series.

Queensland Treasury 1996 *Full Cost Pricing Policy*, Brisbane: Queensland Treasury.

Raab, CD 1994 'European perspectives on the protection of privacy' at http://www.spatial.maine.edu/tempe/raab.html [30 June 2004].

Radcliffe, J 2003 'Death of Copyright—Long Live Patents', 2003 Cambridge Conference, Ordnance Survey, Southampton, U.K. Paper 4D.2B.

Radin, MJ, Rothchild, JA and Silverman, GM 2002 *Internet Commerce: The Emerging Legal Framework, Cases and Materials*, New York: Foundation Press.

Rappa, M 2002 'Managing the Digital Enterprise' at http://www.digitalenterprise.org/models/models_text.html [30 June 2004].

Reichman, JH 1994 'Legal hybrids between the patent and copyright paradigms', 94 *Columbia L. Rev.* 2432 (1994).

Reid, K, Clark, E and Cho, G 1996 'Legal risk management for Geographic Information Systems', *Journal of Law and Information Science*, v. 7(2), pp. 169–207.

Restatement (Second) Contracts 1979, St Paul, MN: American Law Institute.

Restatement (Second) Torts 1965, St Paul, MN: American Law Institute.

Restatement (Third) Torts: Products Liability 1997, St Paul, MN: American Law Institute.

Rezende, PAD 2001 'The possible laws on Digital/Electronic signature on the proposed UNCITRAL Model', *SCI'2001 Proceedings*, v. X, pp. 87–92 and archived at http://www.cic.unb.br/docentes/pedro/segdadtop|UNCITRAL [8 January 2004].

References

Ricketson, S 1984 *The Law of Intellectual Property*, Sydney: Law Book Company.

Ricketson, S 1989 'The use of copyright works in electronic databases', 1989 *Law Inst. J.* 480–482.

Ricketson, S 1990 'Copyrights and databases' in Hughes, G (ed.) *Essays on Computer Law*, Melbourne: Longman Cheshire, pp. 67–98.

Risset, JC 1979 *Problems Arising from the use of Electronic Computers for the Creation of Works*, UNESCO/WIPO doc. GTO/3 1979, Paris and Geneva: UNESCO, WIPO.

Rose, M 1993 *Authors and Owners. The Invention of Copyright*, Cambridge: Harvard University Press.

Rose, N and Radcliffe, J 2003 'Death of copyright—Long live patents and database rights', 2003 Cambridge Conference, Ordnance Survey, Southampton, U.K. Paper 4D.2A.

Rowe, H and McGilligan, R 2001 'Data protection. Location technology and data protection', *Computer Law & Security Report*, v. 17(5), pp. 333–335.

Rushworth, D 1999 'Geographic support to courts and states involved in boundary dispute settlement' in Dahlitz, J (ed.) *Peaceful Resolution of Major International Disputes*, New York & Geneva: United Nations, pp. 172–173.

Samuelson, P 1992 'Computer programs, user interfaces and § 102(b) of the *Copyright Act* of 1976: A critique of *Lotus v Paperback*', 55(2) *Law and Contemporary Problems* 311.

Samuelson, P 1995 'A manifesto concerning the legal protection of computer programs: Why existing laws fail to provide adequate protection' in Brunnstein, K and Sint, PP (eds) *Intellectual Property Rights and New Technology*, Proceedings KnowRight 95 Conference, Wein, München: Österreichische Computer Gesellschaft and Oldenbourg, pp. 105–115.

Samuelson, P 2003 'Mapping the digital public domain: Threats and Opportunities', 66 *Law and Contemporary Problems*, 147 (Winter Spring).

Schilit, B, Hong, J and Gruteser, M 2003 'Wireless location privacy protection' at http://www.computer.org/computer/homepage/1203/invisible/rz135.pdf [1 December 2003]. See also http://www.allnetdevices.com/wireless/news/2001/ 07/13/location-privacy.html [1 December 2003] and http://www.techlawjournal. com/cong107/privacy/location/s1164is.asp [1 December 2003]

Schmitthoff, CM 1990 *Schmitthoff's Export Trade. The Law and Practice of International Trade* (9th edn), London: Stevens & Sons.

Schu, R 1997 'The applicable law to consumer contracts made over the Internet: Consumer protection through private international law', 5 *International Journal of Law and Information Technology,* Summer, pp. 192–299.

Scoglio, S 1998 *Transforming Privacy: A Transpersonal Philosophy of Rights*, Westport: Praeger.

Sears, G 2001 *Geospatial Data Policy Study*, Ottawa, Ontario, Canada: KPMG Consulting Inc., 28 March, p. 18.

Seddon, NC and Ellinghaus, MP 1997 *Cheshire and Fifoot's Law of Contract* (7th Australian edn), Sydney: Butterworths.

Segkar, A 2004 'The declining relevance of copyright law—how contract may become the key', *Internet Law Bulletin*, v. 6(10), pp. 121–125.

Sexton, J 2001 'Its his gallery and he'll sigh if he wants to', *The Australian* 29 June, p. 10.

Siebrasse, N 1994 'Copyright in facts and information: *Feist* is not, and should not be the law in Canada', 1 *Canadian Intellectual Property Review*, 191.

Smith, G 2003 'Chapter 1: Software contracts' in Reed, C and Angel, J (eds) *Computer Law* (5th edn), London: Oxford University Press.

Sneddon, M 1998 'Legislating to facilitate electronic signatures and records: Exceptions, standards and the impact of the statute book', *UNSW Law Journal* (May) v. 21(2) and at http://austlii.law.uts.edu.au/au/other/unswlj.OLD/thematic/1998/vol21no2/sneddon.html [30 June 2004].

South Australia Treasury and Finance 1998 *A Guide to the Implementation of Cost Reflective Pricing*, Adelaide: SA Treasury & Finance.

Staunton, JH 1998 *Report of a Formal Investigation under Section 98 of the Coal Mines Regulation Act, 1982 (NSW)*, 2 vols. Sydney: NSW Government.

Suppa, C 2004 'Credit agency reports security breach' at http://www.computerworld.com/printthis/2004/0,4814,91319,00.html [19 April 2004].

Tapper, C 1989 *Computer Law* (4th edn), London: Longman.

Tasmania Treasury and Finance 1998 *Costing Fees and Charges: Guidelines for Use by Agencies*, Hobart: Tasmania Treasury and Finance.

Taylor, G 2001 'Protect your online database', *International Internet Law Review, Issue 13* (April) pp. 33–36 also at http://www.netlawreview.com [5 April 2004].

The Birmingham News 2004 'To catch a thief, charter captains outlaw the GPS' April 25 at http://www.al.com/printer/printer.ssf?/base/sports/10828.html [1 May 2004].

The Economist 2003 'The revenge of geography' March 13. Also at http://www.economist.com/science/tq/PrinterFriendly.cfm?story_ID=1620794 [29 March 2003].

The Guardian 2004 'Tabloid's fear Naomi's court win will spell end for exposés', *The Sydney Morning Herald*, 8 May, p. 15.

The Privacy Bulletin 1990 Special Issue, August, v. 6(2), Sydney, Australia, p. 2.

Tobin, R 'Invasion of Privacy' (2000) *New Zealand Law Journal* 216.

U.K. Association of University Teachers (AUT) IPR code of conduct at http://www.aut.org.uk [11 April 2004].

U.K. Department of the Environment 1979 *Report of the Ordnance Survey Review Committee*, Chairman: Sir David Serpell, London: HMSO.

U.K. Department of Trade and Industry 1988 *Government-held Tradeable Information: Guidelines for Government Departments Dealing with the Private Sector* (2nd edn), London: HMSO, 35 pp.

U.K. HMSO 1981 *White Paper: Government Statistical Services*, Chairman: Sir Derek Rayner, London: HMSO.

U.K. Treasury 1992 *The Fees and Charges Guide*, London: HMSO.

References

U.S. 2002 *E-Government Act* 2002, H.R. 2458/S. 803 signed 17 December 2002 effective 17 April 2003, at http://www.whitehouse.gov/omb/egov/pres_state2. htm [10 May 2004].

U.S. Coast Guard Office of Investigations and Analysis and Norwegian Maritime Directorate 2003 Report on *Monarch of the Seas* at http://www.uscg.mil/hq/ gm/moa/boards/monarch.pdf [30 June 2004].

U.S. Department of Commerce 1995 *Privacy and the NII: Safeguarding Tele-communications-related Personal Information*, Washington D.C.: DoC.

U.S. Department of Health, Education and Welfare *1973 Records, Computers and the Rights of Citizens*, Washington, D.C.: HEW.

U.S. Federal Trade Commission (FTC) 1998, Privacy Online: A Report to Congress (June) at http://www.ftc.gov/reports/privacy3/toc.htm [30 June 2004].

U.S. Federal Trade Commission 2000 Privacy Online: Fair Information Practices in the Electronic Marketplace, May, Washington, D.C.: FTC at http:// www.ftc.gov/os/2000/05/index.htm#22 [30 June 2004].

U.S. Geological Survey (USGS) 2003 *The National Map*, US Geological Survey Report at http://nationalmap. usgs.gov/nmreports.html [30 June 2004].

UNCITRAL Uniform Rules on Electronic Signatures—Australian Issues Paper 1999 http://law.gov.au/agd/selaw/issuesPaper8-99.htm [30 June 2004].

United Nations 1990 Guidelines for the Regulation of Computerized Personal Data Files at http://www.datenschutz-berlin.de/gesetze/internat/aen.htm [7 February 2004].

University of Strathclyde 2000 'Policy Approaches to Copyright in Higher Education Institutions' at http://www.strath.ac.uk/ces/projects/jiscipr/report. html [30 June 2004].

Victoria Treasury and Finance 2000 *Guidelines for Setting Fees and Charges Imposed by Departments and Budget Sector Agencies*, Melbourne: Victoria Treasury and Finance.

Vijayan, J 2002 'Recent breaches raise spectre of liability risks' at http:// www.computerworld.com/printthis/2002/0,4814,71609,00.html [19 October 2002].

Warren, S and Brandeis, L 1890 'The right to privacy', 4 *Harvard L. Rev.* 193.

Weiss, P 2002 'Borders in Cyberspace: Conflicting Public Sector Information Policies and their Economic Impacts' p. 13 at ftp://ftp.cordis.lu/pub/content/ docs/peter_weiss.pdf [24 April 2004].

Westell, S 1999a 'Legal Protection of Databases' http://www.geoplace.com/ma/ 1999/0899/899law.asp [20 October 2001].

Westell, S 1999b 'New data protection legislation. How will this affect a geographic information business?' *Mapping Awareness*, April and at http://www.geoplace. com/ma/1999/0499/499gis.asp [20 October 2001].

Westell, S 1999c 'Potential liability for defective software or data. Part I', *GeoPlace. com Mapping Awareness* and at http://www.geoplace.com/ma/current/1199law. asp [20 October 2001].

Westermeier, JT 1994 ' "Sweat of the brow" protection survives', *GIS Law* v. 2(1), pp. 1–3.

Western Australia Treasury 1998 *Costing and Pricing Government Outputs*, Perth: WA Treasury.

Westin, AF 1967 *Privacy and Freedom*, New York, NY: Atheneum.

Westin, AF 1971 *Information Technology in a Democracy*, Cambridge, MA: Harvard University Press.

Whelan, J 1995 'Planes straying off the official flight paths, new maps reveal', *The Sydney Morning Herald*, 18 April, p. 2.

White House, Office of the Press Secretary 2000 'Statement by the President regarding the United States' decision to stop degrading global positioning system accuracy', May 1 at http://www.ostp.gov/html/0053_2.html [30 June 2004].

Wilkes, GA 1978 *A Dictionary of Australian Colloquialisms*, Sydney: Fontana Books, p. 150.

Winer, J 2001 'What the Canadian Privacy Act means for U.S. companies' at http://www.gigalaw.com/articles/2001/winer-2001-02.html [26 January 2004].

Wolf, DB 1992 'Is there copyright protection for maps after *Feist*?' 39 *J. Copyright Society* 224.

Wolf, DB 1993 'New landscape in the copyright protection for maps: *Mason v Montgomery Data Inc.*', *GIS Law* v. 1(4), pp. 14–17.

Wood, DJ 2002 'Best practices for avoiding linking and framing legal liability' at http://www.gigalaw.com/articles/2002-all/wood-2002-06-all.html [11 April 2004].

Wriston, WB 1992 *Twilight of Sovereignty: How the Information Revolution is Transforming our World*, New York: Scribner.

Yu, P 2001 'An introduction to the EU Directive on the Protection of Personal Data', http://www.gigalaw.com/articles/2001/yu-2001-07a-p1.html [27 September 2001].

Zhong-Ren Peng and Ming-Hsiang Tsou 2003 *Internet GIS: Distributed Geographic Information Services for the Internet and Wireless Networks*, Hoboken, N.J.: John Wiley & Sons, Inc.

Internet URL References

[Note: the square brackets and date after the universal resource locator (URL) denotes the date on which the web page was last viewed by the author.]

African SDI website at http://www.nsif.org.za/africasdi_main.htm [30 June 2004].
ANZLIC Council at http://www.anzlic.org.au [22 April 2002].
Asia Pacific Smart Card Forum 1997 Code of Conduct at http://www. smartcardforum.asn.au/ [28 January 2004].
Associated Credit Bureau (ACB) at http://www.acb-credit.com [30 June 2004].
Atlas of Canada at http://atlas.gc.ca/site/english/learning_resources/ccatlas/index. html [30 June 2004].
Australia Copyright Agency Ltd (CAL) at http://www.cal.com.au [20 February 2004].
Australia On Disc (AOD) http://www.bsgi.biz/aodspecial.htm [30 June 2004].
Australia Spatial Industry Action Agenda at http://www.isr.gov.au/agendas/ Sectors/siiaa/index.htm [13 May 2004].
Australia Spatial Sciences Institute (SSI) at http://www.spatialsciences.org.au [4 January 2004].
Australian Communications Industry Forum (ACIF) 1999 Telecommunications Code at http://www.acif.org.au/publications/codes [30 June 2004].
Australian Consortium for Social and Political Research Inc. (ACSPRI) access to Australian social science data at http://acspri.anu.edu.au/ [11 April 2004].
Australian Direct Marketing Association (ADMA) Code of Practice 2001 at http://www.adma.com.au/asp/index.asp?pgid=1985 [30 June 2004].
Australian Government Information Management Office (AGIMO) 2003 http:// www.agimo.gov.au/publications/2003/08/framework/overview [8 May 2004].
Australian National Privacy Principles at http://www.privacy.gov.au/act/index. html [30 June 2004].
Australian Spatial Data Directory (ASDD) at http://www.asdd.ga.gov/asdd//tech/ tools.html [30 June 2004].
Beowulf Project at http://www.beowulf.org [11 April 2004] and at http://www. canonical.org/~kragen/beowulf-faq.txt [11 April 2004].
California Coast Project at http://www.californiacoastline.org [17 May 2004].

Geographic Information Science: Mastering the Legal Issues   George Cho
© 2005 John Wiley & Sons, Ltd   ISBNs: 0-470-85009-4 (HB); 0-470-85010-8 (PB)

California Coast Streisand lawsuit at http://www.californiacoast.org/streisand/ lawsuit.html [17 May 2004].

Canada Oil and Gas Consortium at http://www.geoconnections.org/iacg/barrier1. htm [30 June 2004].

Candian Geospatial Data Infrastructure (CGDI) GeoConnections at http://www. geoconnections.org/cgdi.cfm [30 June 2004].

Choicepoint at http://www.choicepointinc.com [30 June 2004].

Copyright Basics U.S. at http://www.copyright.gov/circs/circ1.htm#cr [5 August 2004].

Creative Commons Founders Copyright at http://www.creativecommons.org/ projects/founderscopyright [18 January 2003].

Creative Commons Open Content at http://www.opencontent.org/openpub/ [18 January 2003].

Deep linking definition at http://www.selu.edu/Academics/FacultyExcellence/ Pattie/DeepLinking/intro.html [11 April 2004].

Department of Industry, Science and Resources 1999 *Shaping Australia's Future. Innovation Framework Paper*, Canberra: AGPS and at http://industry. gov.au/library/content_library/shaping.pdf [30 June 2004].

Dublin Core metadata standard at http://www.dublincore.org [7 April 2001].

Electronic Funds Transfer (EFT) Code of Conduct 2002 at http://www.asic. gov.au/asic/pdflib.nsf/LookupByFileNumber/eft_code.pdf/$File/eft_code.pdf [30 June 2004].

Electronic Privacy Information Centre (EPIC) Online Guide to Practical Privacy Tools at http://www.epic.org/privacy/tools.html [31 January 2004].

Eurogeographics at http://www.eurogeographics.org/eng/01_about.asp [16 April 2004].

Federal Communications Commission (FCC) 2004 Enhanced 911 Requirements at http://www.fcc.gov/911/enhanced/ [30 June 2004].

Federal Geographic Data Committee (FGDC) 1998 Policy on Access to Public Information and the Protection of Personal Information Privacy in Federal Geospatial Databases at http://www.fgdc.gov/fgdc/policies/privacypolicy.pdf [30 June 2004].

Free Software and Open Source Foundation for Africa (FOSSFA) 2003 Statement at http://www.admi.net/cgi-bin/wiki?FOSSFA [30 June 2004].

Free Software Foundation (FSF) 1996 'Free software definition' at http://www. fsf.org/philosophy/free-sw.html [30 June 2004].

General Agreement on Trade in Services (GATS) at http://www.dfat.gov.au/ trade/negotiations/services/index.html [30 June 2004].

General Agreement on Trade in Services at http://www.wto.org/english/tratop_e/ serv_e/gatsintr_e.htm [30 June 2004].

GeoOneStop at http://www.geo-one-stop.gov/Intro.html and at http://www. geodata.gov/gos [30 June 2004].

Geoscience Australia data submission guidelines at http://www.ga.gov.au/ oceans/projects/psla_guidelines.jsp [20 April 2004].

GI Gateway at http://www.gigateway.org/ [30 June 2004].

GIS Certification Institute at http://www.gisci.org/why_certification.htm [30 June 2004].

Global Map Data Australia 1M at http://www.ga.gov.au/map/images.jsp [30 June 2004].

GlobeXplorer at www.globexplorer.com [30 June 2004].

GNU General Public License (GPL) at http://www.gnu.org/licenses/gpl.html [14 October 2001].

Information Infrastructure Task Force (IITF) Privacy Working Group at http://www.ibiblio.org/nii/NII_Task_Force.html [30 June 2004].

Internet Engineering Task Force (IETF) at http://www.ietf.org [30 June 2004].

Internet Industry Association (IIA) Code of Conduct Draft at http://www.intiaa.asn.au/codeV2.htm [26 January 2001] and at http://www.iia.net.au/codes.html [26 January 2001].

Inter-university Consortium for Political and Social Research (ICPSR) at http://www.icpsr.umich.edu/org/index.html [30 June 2004].

ISO TC211 Committee at http://www.iso.org/tc211 [30 June 2004].

Legal Information Institute (LII) at http://lii.law.cornell.edu [26 January 2004].

Location Privacy at http://www.allnetdevices.com/wireless/news/2001/07/13/location-privacy.html [22 September 2001] and http://www.techlawjournal.com/cong107/privacy/location/s1164is.asp [22 September 2001].

Location Privacy at http://www.theorator.com/bills108/hr71.html [4 February 2004].

MapBlast! at (http://www.mapblast.com) [30 June 2004].

MapOnUs at (http://www.maponus.com) [30 June 2004].

MapQuest at (http://www.mapquest.com) [30 June 2004].

Multimap at http://www.multimap.com [1 April 2004].

National Office for the Information Economy (NOIE) now the Australian Government Information Management Office (AGIMO) at http://www.agimo.gov.au [30 June 2004].

Naviant website at http://www.naviant.com/ and see http://www.equifax.com/corp/pressroom/pressreleases/ [30 June 2004].

NCGIA Core Curriculum in GIS 1990 Unit 70 (Legal Issues) at http://www.ncgia.ucsb.edu/pubs/core.html [30 June 2004] and at http://www.geog.ubc.ca/courses/klink/gis-notes/ncgia/toc.html [30 June 2004].

NetMap Analytics at http://www.netmapanalytics.com/other.asp?heading=1 [25 May 2004].

New Zealand e-government at http://www.e-government.govt.nz [30 June 2004].

OECD Consumer Protection in Electronic Commerce Guidelines at http://www.treasury.gov.au/ecommerce [30 June 2004].

OECD Privacy Policy Statement Generator at http://cs3-hq.oecd.org/scripts/pwv3/pwhome.htm [2 February 2004].

OpenGIS Consortium (OGC) metadata specifications at http://www.opengis.org/techno/specs.htm [30 June 2004].

OpenGIS Consortium Vision at http://www.opengis.org/about/?page=vision [30 June 2004].

Open Source definition at http://www.opensource.org/osd.html [18 January 2003] and http://www.opensource.org/docs/definition.php [18 January 2003].

Permanent Committee for Geographic Information in Asia and the Pacific (PCGIAP) at http://www.permcom.apgis.gov.au/ [21 April 2004].

Privacy Policy Editors at http://www.p3pwriter.com [4 February 2004] and http://policyeditor.com [4 February 2004].

Public Library of Science (PLOS) at http://www.plos.org [30 June 2004].

Safe Harbour Principles at http://europa.eu.int/comm/internal_market/privacy/index_en.htm [30 June 2004].

Space Imaging at www.spaceimaging.com [30 June 2004].

Spatial Sciences Institute (SSI) Code of Ethics and Mission Statement at http://www.spatialsciences.org.au/Prospectus.htm [30 June 2004].

Steganography software at http://www.petitcolas.net/fabien/steganography/stego_soft.html [30 June 2004] and http://www.stego.com/index.html [30 June 2004]; hiding Jpeg images at http://www.cryptobola.com [30 June 2004]; digital watermarks and photographs at http://www.clickok.co.uk [30 June 2004].

Terraserver at http://www.terraserver.com [30 June 2004].

TRIPS Geographical Indications at http://www.wto.org/english/tratop_e/trips_e/gi_e.htm [29 March 2004].

TRUSTe E.U. Safe Harbor Privacy Program at http://www.truste.org/bus/pub_global.html [30 June 2004].

U.K. Information Commissioner Data Transfer principles at http://www.informationcommissioner.gov.uk/eventual.asp [15 January 2004].

U.S. Copyright Clearance Centre Inc. (CCC) at http://www.copyright.com [30 June 2004].

U.S. Department of Commerce self-certified agencies and companies under Safe Harbour principles at http://www.web.ita.doc.gov/safeharbor/shlist.nsf/webPages/safe+harbor+list [6 February 2004].

U.S. Department of Commerce Safe Harbour site http://www.export.gov/safeharbor [6 February 2004].

U.S. Federal Geographic Data Committee site at http://www.fgdc.gov/clearinghouse/clearinghouse.htm [30 June 2004].

U.S. Patent and Trademark Office (USPTO) at http://www.uspto.gov/patft/index/html [29 March 2004].

URISA-Institute for GIS Professionals and GIS Code of Ethics at http://www.urisa.org/ethics/code_of_ethics.htm [30 June 2004].

W3C Mission at http://www.w3c.org/Consortium/#mission [30 June 2004].

Webopedia at http://www.webopedia.com [30 June 2004].

WIPO Treaties at http://www.wipo.org/about-wipo/en/index.html [30 June 2004].

World Wide Web Consortium (W3C) at http://www.w3.org/P3P/ [30 June 2004].

Index

ABC, Australian Broadcasting Corporation 224
ABGI, Advisory Board for Geographic Information 103
Access and exchange standards 93
  interoperability 93
  SDI spatial data infrastructures 93
Access control of data 32
Acontextual data 214
ACSPRI, Australian Consortium for Social and Political Research Inc. 198
Adhesion contracts 321
ADMA, Australian Direct Marketing Association 234
ADR, alternative dispute resolution 24–5
  remedies 25
    equitable 26
    injunctions 26
Aerial photographs 46, 269
  copyright in 48
  privacy 46
Ageographic 6, 49
AGI, Association for Geographic Information 94
AGPS, assisted GPS 267
Al Gore, Vice President 97, 108
Alabama Gulf Coast boat captains 45
  private reefs 45
Amazon.com 'one click' patent 120, 171
Anonymity services *fn 150* 271
ANSI Z39.50–1995 (ISO 23950–1997) 92–3

ANZLIC metadata core elements 87
  ANZMETA XML Document Type Definition 89
  Guidelines 87
  Spatial Data Transfer AS/NZS 4270 89
  summary *table* 88–9
ANZLIC, Australia New Zealand Land Information Centre 69, 87
AOD, Australia on Disk 261
  AOD Mapper 261
APSDI, Asia–Pacific spatial data infrastructure 104
ArcInfo Export format 87
Arms Export Control Act 1976 379
AS/NZS Joint Technical Committee IT-004 94
ASIBA, Australian Spatial Information Business Association 34, 71, 103, 108
AskGIraffe 94
AT&T Bell Labs 178
ATMs, automatic teller machines 264
Australia
  AAT, Administrative Appeals Tribunal 380
  ACA, Australian Communications Authority 266
  ACIF, Australian Communications Industry Forum code 266
  AGIMO, Australian Government Information Management Office 71, 94
  AMSA, Australian Maritime Safety Authority 357

Geographic Information Science: Mastering the Legal Issues George Cho
© 2005 John Wiley & Sons, Ltd ISBNs: 0-470-85009-4 (HB); 0-470-85010-8 (PB)

Australia (*Continued*)
ASDD, Australian Spatial Data
Directory 32, 33, 71, 89
ASDI, Australian Spatial Data
Infrastructure 16, 103
AUSLIG, Australian Land Information
Group 34
NATMAP 34
CAA, Civil Aviation Authority 354
cadastral databases prices 68
CSIRO, Commonwealth Scientific
Industrial Research Organisation
224
data policy developments 33
ECEG, expert group on e-commerce 333
Global map data Australia 34
SIIAA, Spatial Information Industry
Action Agenda 33, 68, 71,
103, 108
Australian Centre of Excellence in GIS 108
Australian Spatial Information Business
Association, *see* ASIBA, Australian
Spatial Information Business
Association
Australian Standard for Records
Management AS4390 91
AVCC, Australian Vice Chancellors
Committee 194, 313

Bangemann Report 97
Barbra Streisand suit 46
Barnesandnoble.com 171
Beowulf Project 181
Berne Convention and *Universal
Copyright Convention* 1998
122, 128, 166
Bespoke computer software 187, 309
Biometrics 267
British Horseracing Board case 161
Buenos Aires Declaration 97
Business intelligence 50
Business method patents 120, 171,
fn 115 172
EPC, European Patent Convention 172
EPO, European Patent Office 172
international harmonisation 172
prior art 171
USPTO 172

C language 178
Cable Communications Policy Act 1984 242
California anti-paparazzi Act 46
California Coastal Records Project 46
Cambodia war crimes database 40
Canada
CGDI, Canadian Geospatial Data
Infrastructure 100
GeoConnections Discovery Portal
92, 100
GeoInnovations 100
GeoPartners 100
GeoSkills 100
GIAC, Geomatics Industry Association
of Canada 100
Natural Resources Canada 100
Statistics Canada 19, 55
Sustainable Communities 100
The Atlas of Canada 100
Yellow Pages directory 154
Canberra bush fires 16
CANOGGIS, Canadian Oil and Gas
GIS 41
CANRI, NSW Community Access to
Natural Resource Information 92, 107
Case Notes
Caslec Industries 304
Desktop Marketing Systems 151–2
Feist Publications 147–9
Hill v *Gateway* 320–1
ProCD Inc v *Zeidenberg* 321–2
Shroeder Music Publishing 313
Catherine Zeta-Jones litigation 47
Caveat 4
caveat emptor 342, 381
CCTV, closed circuit TV 267
CGIAR, Consultative Group on
International Agricultural
Research 196
*Children's Online Privacy Protection
Act* 1988 242
ChoicePoint database 255
Civil law 23
also Roman law 23
Clearinghouse 32, 91
catalogue services 91
directory services 32
registries 91

CLI, caller line id   261
Clinton, President decree   97
　NSDI, National Spatial Data
　　Infrastructure   97
CLM, collection level metadata   91
CND, calling number display   261
Coffs Harbour, NSW   356
Common law   21
　jurisdictions   4
Communications Act 1934   243
Compilations   146
　Australia low threshold of originality
　　152
　Canada yellow pages directory   154
　computer-produced data   146
　facts not copyrightable   150
　originality   146
　'sweat of the brow'   146, 149
　US position   147
*Computer Matching and Privacy
　Protection Act* 1988   240
Content standards   93
Contract *for* service   12, 307
Contract liability *drafting tips*   368
Contract *of* service   12, 307
Contracts–electronic   335
　draft language   336
　exceptions   338
　formation   335
　forum selection clauses   341
　jurisdiction   337
　machine acceptance   336
　minimum provisions   343
　placement of terms   339
　terms and conditions on Websites   341
　'ticket' cases   336
　valuable consideration   340
　waiver of IPR   341
　writing and digital signatures   340
Contracts–general
　damages   331
　discharged   330
　electronic contracts, *see*
　　Contracts–electronic
　electronic transactions regulations   333
　exclusion clauses   328
　execution   330
　failed   330

machine offers   336
Precedents, list of   348–9
privity of contract   327
　Australian position   327
　Canada   328
　liability implications   327
　NZ   328
　UK   327
summary of legal issues   342
　minimum provisions   343
UK test of reasonableness   329
warranty and liability disclaimer   330
Web-based contracts   332
Contracts–GI   292
　boilerplate contract   294
　'contract *for* service'   12, 307
　contract law   11
　'contract *of* service'   12, 307
　definition   296
　e-mail acceptance   301
　electronic contracts, *see*
　　Contracts–electronic
　electronic transactions regulations   13
　elements of a contract   12, 298
　exclusion clauses   12, 328
　execution   13
　express terms   302
　formalities   294
　GIS product or service   12, 306, 314
　　implied terms   315
　　NSW position   315
　　UK position   315
　　US UCC, Uniform Commercial
　　　Code   315
　implied terms   303
　intention to create legal relations   300
　invitations to treat   297
　lapse of offer   302
　legal duties and responsibilities   12
　licensing   12
　'meeting of the minds'   11
　method of acceptance   300
　necessary and sufficient conditions
　　11, 29, 293
　precedent templates   293
　Precedents, list of   348–9
　relationship building   11, 292
　revocation of offer   301

Contracts–GI (*Continued*)
  Statute of Frauds 299
  statutory requirements 299
  termination of offer 301
  terms implied by law 302, 305
  Web-based 13
  why contract? 296
  in writing 299
Contracts–personnel 308
  Academics as employees? 313
  assignment of copyright 308
  Australia restraint of trade 312
  other features 310
  'poaching' of employees 311
  software development 309
  UK common law rights 311
  US tort remedy 311
copy-right (as in right to copy) 43, 131
Copy right *per se* 8
Copyleft movement 8, 178
  CC Creative Commons 180
  CC 'rights' notice *diagram* 181
  FOSS, free and open source software 178
  FOSSFA initiative for Africa 182
  Founders copyright 180
  FSF, Free Software Foundation 179
  GNU/Linux PC operating system 179
  GPL GNU Public Licence 179
  GPL–OSD open source *definition*
    179, 180
  iCommons project 180
  implications for GI and IPR 181
  OSI, Open Source Initiative 179
  'viral' clause in GPL 179
Copyright traps 42
Corner jumping 45
Cost recovery
  objectives 62
  summary *table* 62
Court system 23
CPNI, Customer Proprietary Network
    Information 267
CRC SI, Australian Cooperative Research
    Centre for Spatial Information 108
CSDGM, Content Standards Digital
    Geospatial Metadata 89
CSIRO, Commonwealth Scientific Industrial
    Research Organisation 224

CSO, community service obligation 67
Cujus est solum ejus est usque ad coelum
    268

Daily Mirror litigation 47
Data access
  international 7
  legal frameworks 7
  policies 7
Data and GI, *see under* GI and data
Data pricing policies
  free data 30
  full price 31
  marginal cost 30
  market price 31
  options 30
  re-balancing price 31
  two-tier price 31
Data protection
  access controls 32
  Australia 31
  competitive neutrality 31
  digital watermarks 32, 43, 141
  policy 31
Data quality improvement *checklist*
    386–7
Data sharing
  CANOGGIS, Canadian Oil and Gas
    GIS 41
  economic imperatives 40
  information infrastructure 41
  private common good 40
Database–aggregators
  ChoicePoint 255
  Equifax Canada 256
  Equifax National Decision Systems 254
  Experian 255
  Gale's Directory 255
  Naviant Technologies 255
  PBL–Acxiom 256
  TransUnion 255
Database–electronic 143
  authorship 144
  copyright protection 144
  low originality requirements 143
  'paternity' tracing 143
  remotely sensed data 144
  UK position on ownership 143

Database–general
  aggregators   39
  Cambodia war crimes   40
  Equifax Canada attack   39
  FBI investigation   38
  intrusions   38
  regulation and use   11, 256
  safeguards   257
  security study   39
  'victims' retaliate   39
Database–rights
  acquired   145
  constructed   145
  copyright issues   164
  derivative works   145
  EU Directive   11, 115, 145, 159, 209
  licensing   165
  maximising protection *notes*   165
  in revisions   145
DBMS, database management systems
  190, 197
DC, Dublin Core Metadata Standard   90
DCW, Digital Chart of the World
  200
DeCSS decryption   177
Deep linking   116, 339
Degraded data distribution   141
Desktop Marketing case   115, 151
DGI, distributed geographic
  information   140
  fair use / fair dealing   141
  fingerprinting   141
  limited permission grants   142
  permissions to copy   142
  protection of IPR   141
  waiver of copyright   142
  watermarks   42, 141
DGPS, differential GPS   267, 383
Digital Agenda   8, 175
  Australia legislation   176
  Australia technology neutral right
    177
  circumvention tools   177
  digital geospatial metadata   89
  DRM, digital rights management
    141, 177
  exceptions to rights   176
  US DMCA   176

Digital Millennium Copyright Act (DMCA)
  1998   176
  court cases   177
  DeCSS decryption   177
  'wilful trafficking'   177
Digital persona   *fn 150* 271
Digital watermarks   32, 43, 141
DMA, Defense Mapping Agency   200, 380
  see also NGA, National
    Geospatial-Intelligence
    Agency   200
DNS, domain name system   265
Docusearch.com   379
Douglas, Wyoming   45
  'corner jumping'   45
Drivers Privacy Protection Act 1974   243
DRM, digital rights management   141, 177
droit de suite   *fn 98* 166
droit moral   129, 166
DTD, Document Type Definition   89
DTM, digital terrain models   357
DXF, document exchange format
  fn 61 87

E-commerce   17
  'three pillars'   17, 305
E-pers   *fn 150* 271
E-tag   264
E.164 number   265
E-911 (US)/ e-112 (Europe) emergency
  number   265
EFT, Electronic Funds Transfer Code of
  Conduct   235
EFTPOS, electronic funds transfer-point
  of sale   264
EGII, European Geographic Information
  Infrastructure *GI2000*   101
Electronic charts for navigation   388
  IHO, International Hydrographic
    Organisation   388
  NAVSSI, Navigational Sensor System
    Interface   388
Electronic Communications Privacy Act
  1986   240
Electronic Transactions Act 1999   333
  principles   334
Ellis Island, NJ   37
ENUM protocol   265

EPC, European Patent Convention   172
EPIC, Electronic Privacy Information
 Centre   271
EPO, European Patent Office   172
Equifax Canada attack   39, 256
Equifax National Decision Systems   254
Equitable remedies   26
escrow   198, *fn 29* 308
ESDC, European Spatial Data
 Committee   103
ESF, European Science Foundation
 fn 168 199
EU Data Protection Directive   277
 Data Controller   278
 definition 'personal data'   277
 limitations on processing   278
 transfers of personal data   277
 UK Data Protection Commissioner   284
 US Fair Information Practice
  Principles   281
EU Database Directive   11, 115, 145,
 159, 209
 Australia adequacy standards   283
 British Horseracing Board case   161
 Canada FIPPs Fair Information
  fn 165 283
 citizens of Member states only   160
 court litigation   160
 duration of right   160
 early case law summary *table*   162–3
 extraction, meaning of   159
 independent of copyright protection   160
 non-substantial parts   160
 re-utilisation   160
 scope   159
 statutory exceptions   160
 substantiality   164
 sui generis database right   159
 two-tier protection   159
 UK *Regulations*   161
 unlicensed use   164
 US Safe Harbour Principle   11, 279
EU Directive on Privacy and Electronic
 Communications   267
EU PSI, public sector information   7
 EGII, European Geographic Information
  Infrastructure   101
 Green Paper on PSI   79, 82

Information Society Technology
 Program   101
Synergy Guidelines   80
user pays   61
Eurogeographics   90
EUROGI, European Umbrella
 Organisation for GI   101, 105
European Commission Joint Research
 Centre   *fn 96* 101
European spatial data infrastructure   102
Exchange standards   93
Experian   255

Fair Credit Reporting Act 1970   241
Fair Debt Collection Practices Act 1978   241
Fair information privacy principles
 reports   247
 core principles   248–9
Fair market value for data   66
*Family Education Rights and Privacy
 Act* 1974   242
Feist case   115, 147
FGDC, Federal Geographic Data
 Committee   105, 239
 Content Standards   89
 CSDGM, Content Standards For Digital
  Geospatial Metadata   89
Fingerprinting maps   141
FIP, fair information practices   210
First Readings GIS Law   34
Fonts   122, 185
 Chinese calligraphic characters   185
 on maps   186
 typography   186
force majeure
 also *vis major*   26
Freedom of Information Act 1996   243
Freeware   *fn 170* 202
FTC, Federal Trade Commission 'Do Not
 Call Register'   287
Fundamental datasets   29, 33, 72

G-commerce (geo-commerce)   2, 17
G7 plus Russia   97, 108
GA, Geoscience Australia   41, 68
 Copyright and Licensing Policy   345–7
 Designated Authority   41
 seismic data   41

Gale's Directory of Databases
    255
GATS, General Agreement on Trade
    in Services   307
GDA94   93
General Education Provisions Act
    2002   242
Geo-codes   357
GeoData Alliance   76
GEODATA TOPO-10M   34
Geodemographics   11, 253
Geographical indications   182
    Australian legislation   183
    Cancún Ministerial Conference 2003
        183
    Doha Mandate 1997   183
    geographical elements in   183
    GI industry   183
    TRIPS Agreement   183
Geography of the cyberworld   50
Geolibrary   91
Geoportal   91
George Orwell 1984   215
    'Big Brother'   215
Geospatial DRM digital rights
    management   143
Geospatial information   1, 86
    digital data standards   86
    frameworks for accessing   86
    info-structure   86
    products   17
Germany
    Geodaten Infrastruktur Deutschland
        GID NRW   92
GI
    as digital databases   115
    Australian economy   18
    Canada   19, 55
    differentiating characteristics   66
    Europe   18
    industry value   18
    maps 'on-the-fly'   115
    models for sharing   63
    national strategies   61
    'public good' feature   60
    UK OS turnover   18
    US   19
    valuable resource   16, 51

GI and data
    as a resource   5, 56
    commodification   57
    consumer product   58
    digitisation   29
    estimates of value   56, 58
    Geomatics Industry Canada survey
        19, 55
    investments in   5
    leakages   58
    market   59
    non-substitutability   58
    personal data intensive   10
    preservation   32
    pricing policy   30
    published works and data   56
    role of law and policy   3, 5, 57
    UK OS, Ordnance Survey value   56
    US NAS, National Academy of
        Science   57
    US NRC, National Research Council   57
    value-adding   58
GI and IPR–Australia   8, 131
    assignment of rights   132
    concept of originality   134
    copy-right   131
    creativity   139
    digitising from maps   136
    duration of protection   132
    'exclusive rights'   132
    'fair dealing'/ 'fair use'   133
    geographical indications, *see under*
        geographical indications
    implications for GIS   139
    Internet maps   140
    'literary work'   133
    low originality requirements   143
    map cases   134
    map copyright notice *example*   138
    map doctrine US court cases   136
    map layers of information *diagram*   135
    maps as graphical representations   135
    maps as literary / artistic works   133
    non-copyrightable elements   135
    originality   139
    'performing right'   137
    'republishing'/reproduction   137
    risks to copyright infringement   137

GI and IPR–Australia (*Continued*)
  royalty   131
  scanning of maps   136
  'translating' works   137
  'works'   132
GI and liability
  duty of care   14
  exposure and risks   14
  ground-truthing   356
  legal standards   13, 352
  liability   13
  liability risk management   14
  liability standards   351
  litigation and defences   13
  map data quality elements   14, 356
  misrepresentation   14, 372
  negligent misstatement   14, 372
  omission to act   14
  strict products liability   14
  tortious   14
GI patents   173
  diagram   174
  effect on mapping   174
  implications of patent   175
  Multimap.com patent   173
  scope   174
  US expansive patent laws   175
  USPTO patent search   173
GI policies   27
  factors influencing   28
  key issues   28
  national   30
  policy making   30
GI policies–Australia
  ANZLIC, Australia New Zealand Land
    Information Council   69, 87
  ASIBA, Australian Spatial Industries
    Business Association   71
  basic access model *diagram*   72
  budget-funded   65
  co-funded/collaborative projects   68
  commissioned/fully-funded   68
  cost recovery arrangements   64
  DOFA, Department of Finance and
    Administration   66
  fair market value   66
  fee-for-service   68
  GA, Geoscience Australia   68

historical/sunk costs   66
information agencies   65
information product categories   66
legal and fiscal framework   64
non-rivalrous/non-excludable   29, 64
pricing classes of users   68
Productivity Commission review   64
PSMA, Public Sector Mapping Agencies
    69, 103
public good character   64
rationale for cost recovery   64
sale price of data   66
spill-over effects   64
true cost   66
whole-of-government   71
GI policies–EU   77
  cost recovery   77
  EU *Synergy Guidelines*   80
  'European model'   77
  fair competition   81
  French Prime Minister Circular   80
  funding model   77
  governance   80
  Green Paper on PSI   79, 82
  information resources a commodity   79
  legislation and policy summary   77
  marginal cost of dissemination   81
  Member States experiences   82
  open access   82
  OS, Ordnance Survey approach   82
  pan-European approach   81
  summary *table*   78
  tradeable information initiative   82
  UK Guidelines Tradeable Information   80
GI policies–New Zealand   64
  least restrictive   67
GI policies–US
  charging for Federal data   74
  cost recovery   74, 75
  guiding principles   74
  OMB Circular A-130   74
  OMB Circular A-25 Revised   75
  OMB Fee Schedule and Guidelines   75
  'open records'   72
  public data   73
  State data distribution   73
  user charge limitations   74
GI quality questions *notes*   385–6

GI2000 101
GIAC, Geomatics Industry Association
    of Canada 100
GICSS, Geographic Information Charter
    Standard Statement 94
GIF, graphic interchange format *fn 44* 141
GIgateway 94
Gigo GI professional 8, 110
Gigo's code 9, 201
GII, global information infrastructure
    7, 97, 108
GINIE, Geographic Information Network
    in Europe 101
    partners *fn 96* 101
GIS Certification Institute *fn 35* 311
GIS, geographic information systems 1
    conflicts in 35
    law nexus 36
    neologism 6
    solving boundary disputes 38
Global Information Society 97
GlobeXplorer 92
GML, geographic mark-up language
    93, 107, 110
GNU / Linux *fn 126* 179
Google for GI 92
GPS, global positioning systems 44, 214,
    234, 383
    EU GALILEO system 44–5
    positional accuracy 44
    Russian GLONASS *fn 40* 44
    'selective availability' 44, 266, 383
    US charter boat captains 45
    US NAVSTAR satellites 44
Gramm–Leach–Bliley Act 1999 241
Gretley Colliery, Newcastle, NSW 353
Grey marketing 188
GSDI, global spatial data infrastructure
    7, 105
GSM, global system for mobile
    communications *fn 127* 265

*Health Insurance Portability and
    Accountability Act* 1996 242
Historical/sunk costs 66
*http://*hypertext transfer protocol 340
*https://*hypertext transfer protocol
    security 340

IBM, anti-trust suit 178
ICCPR, International Covenant on Civil
    and Political Rights 217, 219
ICPSR, Inter-university Consortium for
    Political and Social Research
    fn 168 199
IDEC Catalunya project 92, 107
Identity theft 246
IGGI, Inter-governmental Group on
    Geographic Information 94
IIA, Internet Industry Association
    Code of Conduct 235
IML, Information Mark-Up Language
    fn 2 111
Info-structure 86
Infobahn 97
Information
    as an asset 59
    public trust 59
    stewards and custodians 59
    unique qualities 59
Information brokerage systems 233
Information services 26
Information standard time 24×7
    206
Information superhighway 97
Informational privacy 46, 213
Infrastructure
    common elements 60
    GI as part of 60
    SDI, spatial data infrastructure 60
INSPIRE, Infrastructure for Spatial
    Information in Europe 101–2
Intelligent tags 1
International boundaries 49
International law 27
Internet 'ageographic' 6, 49
    geography of the cyberworld 50
Internet Treaties 128, 130
    WCT WIPO Copyright Treaty 128,
        130, 175
    WPPT WIPO Performances and
        Phonograms Treaty 128,
        130, 175
Interoperability *fn 74* 93
    sectors *diagram* 95
IPPs, information privacy principles
    Australia 220

Index

IPR–Australia 118
  AVCC, Australian Vice Chancellors
      Committee guidelines 194, 313
  basic features *summary table* 123–5
  business method and systems 120,
      fn 115 172
  confidential information 122
  copyright 118
  Copyright Act 1968 118
  copyright in compilations of facts
      119
  designs 120
  fonts 122, 185, 186
  geographical indications, *see under*
      geographical indications
  moral rights 121, 166
  originality threshold low 152
  patents 119
  photographs 48, 122, 184
  trade marks 120
  trade secrets 122
IPR–compilations 146
  Canada Yellow Pages directory 154
  computer-produced data 146
  facts not copyrightable 150
  factual 146
  originality 146
  'sweat of the brow' 146, 149
  US position 147
IPR–employees/contractors 9, 193
  academic researcher 193
  Australia AVCC Guidelines 194, 313
  employer-employee *sample agreement*
      195–6
  independent contractor 193
  inventions 195
  postgraduate students 195
  stock-in-trade 195
  UK academic creations 194
  US academic/teacher exception
      193
  works made for hire 193
IPR–general 111
  boundaries 115
  browsing 117
  bundle of rights *diagram* 114
  copyright in maps 115, 155
  electronic environment 117

  electronic maps 115
  expression, form of 117
  innovation 116
  ownership 111
  printer's monopoly 117
  protection plan 203
  Statute of Anne 1709 117
  sui generis 114
  summary *diagram* 187
IPR–infringements 188
  Australia
    'authorise' infringements 188
    fair dealing/use 188
    grey marketing 188
    importation of works 188
    parallel importation 188
    substantiality requirements 189
    territorial copyright 188
  databases–reproducing facts 190
  defences 191
  Hong Kong
    licence terms 189
    legal position 190
  proprietary rights 43
  remedies 192
  reproduction in material form 190
  UK
    licence terms 189
    substantiality requirements 189
IPR–International 122
  balancing of interests *diagram* 127
  compulsory fee 154
  Conventions, Agreements and
      Treaties 128
  'free-rider' problem 154
  geographical indications, *see under*
      geographical indications
  Paris Revision 122
  quid pro quo 8, 122, 127
  rationales for protection 128
  sui generis rights 154, 168
  TRIPS WTO, agreement on trade-related
      aspects of intellectual property
      rights 128, 154, 183, 1995
  WCT WIPO Copyright Treaty 128,
      130, 154, 175
  WIPO, World Intellectual Property
      Organization 128, 130, 154

WPPT WIPO, Photographs and
Phonograms Treaty   128, 130, 175
WTO, World Trade Organization
128, 307
IPR–International Research   196
access to information   199
CGIAR, Consultative Group on
International Agricultural
Research   196
economic protection   198
jurisdiction   199
ownership: before, during, after
9, 197–8
resolving disputes   199
security   198
UN Millennium Development Goals
196
IPR–maps   155
as factual compilations   155
as mere representations   155
like directories   155
matrix of digits   157
Nordic countries   158
originality/creativity   139, 158
property plan
liability insurance   42
litigation   43
proprietary rights   43
IPR–US
copyright law purpose   126
copyright registration   118
Copyright Term Extension Act
1999   126
President James Madison   126
private works and public domain
126
USPTO, US Patent and Trademark
Office   120
ISDN, international subscriber dialling
number   265
ISO 19100 series Geographic
Information   93–4
ISO 19115 GI/Geomatics Metadata   89
ISO 23950 Information Retrieval
Z39.50   92–3
ISO TC211 International Organization
for Standardization Technical
Committee   93, 211

ITS, intelligent transport systems   247, 271
ITU, International Telecom
Telecommunications Union   97, 265
I-way   97

Japan
'data base works'   158
Guidelines Personal Data   *fn 134* 268
Jhai personal computer   111
JPEG, Joint Photographic Experts Group
fn 44 141
Jurisdiction   21, 49

KPMG Consulting   76
Ku-ring-gai Chase National Park,
NSW   355

Landsat 7 (ENVISAT)   197
Law
common law   21
definition   21
functions   21
jurisdiction   21, 49
legal systems   21
precedent   22
Law making
parliamentary process   23
LBS, location-based services   10, 253, 357
Legal liability theories   359
Australia *Trade Practices Act*
1974   361
breach of standards   359
contracts   360
drafting tips *notes*   368
express and implied terms   362
strict product liability   360, 363
US *Restatement of Torts*   363
US Uniform Commercial Code   360
warranties and conditions for goods   361
Legal traceability   383
Liability–negligence chain *diagram*   360
Liability risks *summary table*   358
Liability–risk management   381
data quality standards   382
disclaimers   381
GI quality questions *notes*   385–6
GI Science standards   391
GPS and map quality standards   383

Liability–risk management (*Continued*)
  legal traceability 383
  minimising liability and damages 393
  quality systems *diagram* 387
  steps in *notes* 390
  strategies 390
Liability–strict product 360, 363
  doctrinal caprice of courts 367
  exclusions and limitations 368
  expression of ideas 367
  geospatial products 364
  information misapplied 365
  maps and charts 365
  navigational charts 365
  professional judgement 366
  US cases 365
Liability–tortious 369
  acts causing purely economic loss 371
    Australian–UK–US positions 371
  common law jurisdictions 369
  concurrent liability with contract
    376
  contributory negligence 374
  criminal liability 379
  'degraded' data 373
  GI context 375
  government immunity 378
  intellectual property rights and 379
  negligent misstatement /
    misrepresentation 372
  NZ 'neighbour' principle 375
  objective standards 369
  on-line settings 374
  privacy and 378
  professional standards 370
  proof of negligence 369
  scope 369
  sovereign immunity 380
  statutory liability 377
  tort 369
  US First Amendment defences 374
  vicarious liability 375
Licensing agreements
  'box-top' 'break seal' licences 317
  checkpoints 325–6
  example of 318–19
  features 319
  non-exclusive non-transferable 317

'parking ticket' licences 318
rationale 316
'shrink wrap' licences 318
WALIS Agreements 316
Limited permission to copy 142
In-line linking / framing 184
Location Privacy Protection Act 2001
  243

Maine, University of 34
Map accuracy standards 387
  NMAS, National Map Accuracy
    Standard 387
  NSSDA, National Standard for
    Spatial Data Accuracy 387
Map data quality elements 384–5
Maps
  as evidence 6
  as legal title 42
  for consumers 42
  copyright traps 42
  electronic environment 6, 115, 117
  liability insurance 42
  protections 156
  special protection 158
  street directories 42
  and the law 42
  'thin copyright' protection 157
    idea in a database 157
    purpose of a database 157
  'watermarks' 32, 42, 141
Marrickville City Council, NSW 354
M-banking 17
M-commerce 17
Metadata 87
  ANZLIC core elements 87
  Australian Standard Records for
    Management AS4390 91
  CLM, collection level metadata 91
  DC, Dublin Core metadata standard 90
  directory services 32
  EuroGeographics 90
  FGDC CSDGM 89
  geospatial metadata record 87
  ISO 19115 GI/Geomatics Metadata 89
  OGC, Open GIS Consortium
    recommendations 91
  standards 17

Michael Douglas litigation   47
Microsoft software   179
M-investing   17
MoLI, Mobile Location Indicator
    266
Monarch of the Seas   355
Moral rights   121, 166
    Australia national treatment   166
    Berne Convention   166
    defences to infringements   168
    droit moral   166
    employees/subcontractors   168
    example of notice   168
    GI/IT industries   167
    in maps   167
    UK non-economic rights   167
    US non-recognition of   167

NAFTA, North America Free Trade
    Agreement   154
Naomi Campbell litigation   47
Nashua, NH   379
NATMAP   34
Naviant Technologies   255
NAVSTAR satellites   44
NCGIA, National Center for Geographic
    Information and Analysis   35
    Core Curriculum   35
Netherlands
    NCGI, National Association for GI
        RAVI   92
NetMap Analytics   50
Netscape   179
Neural networks   289
New Zealand
    District Plan/Scheme   36
    legal copy   36
NGA, National Geospatial-Intelligence
    Agency   200, *fn 72* 380
NIMSA, National Interest Mapping
    Service Agreement   83, 94
NJDEP, New Jersey Department of
    Environmental Protection   37
Non-excludable   29, *fn 13* 64
NOIE, National Office for the
    Information Economy   71, 94
Nordic countries copyright in databases
    158

Non-rivalry   29, *fn 13* 64
Norwegian Maritime Directorate   355
NPPs, National Privacy Principles
    Australia   221, 224–6
    practice notes   222–3
NSDI, national spatial data infrastructure
    7, 97, 106
NSW
    CANRI, Community Access to
        Natural Resource Information
        92, 107
    DMR, Department of Mineral
        Resources   353
    NPWS, National Parks and Wildlife
        Service   355
Nym   *fn 150* 271

ODC, Open Data Consortium Project
    2003   76, 77
ODPM, Office of the Deputy Prime
    Minister   94
OECD, Guidelines for Consumer
    Protection   235
OECD, Privacy Guidelines 1980   219
    Privacy Policy Statement
        Generator   249–50
OGC, Open GIS Consortium   140
    catalogue services specifications
        91–3
    Europe   *fn 96* 101
    recommendations   91
    standards   93
OMB, Office of Management and
    Budget   75, 239
    Circular A-130   75
    Circular A-16   98
    Circular A-25 Revised   75
Open Content Open Publication
    Licence   *fn 131* 181
Open source   *fn 170* 202
OS Ordnance Survey   56, 82
    Executive Agency   83
    NIMSA, National Interest Mapping
        Service Agreement   83, 94
    Plc, public limited company   84
    Quinquennial Review   84
    tradeable information initiative   82
    Trading Fund status   83

Index

Parallel importation 188
PBL–Acxiom 256
PCGIAP, Permanent Committee on GIS Infrastructure for Asia and the Pacific 104
PCIDEA, Permanent Committee for the Americas 105
PDA, personal digital assistant 234
Permissions to copy 142
Personal information *fn 40* 223
Personal privacy 46, 213
PET, privacy-enhancing technologies 270
Photographs
  copyright in 48
  fair use 184
  implications for GI 184
  in-line linking/framing 184
  'thumbnail' pictures 184
PII, personally identifiable information 242
PIT, privacy-invasive technologies 270
PLOS, Public Library of Science *fn 129* 180
Precedent in law 22
Privacy Policy Editor services *fn 151* 272
Privacy policy statement generator 249–51
Privacy Protection Act 1980 240
Privacy–Australia 11, 219
  categories of disclosure 233
  common law 228
    duty of confidence 232
    negligence 228
    negligent reports 229
    nervous shock 230
    passing off 209, 230, 311
    torts 228
  data linkage 220
  data matching 220
  Data-matching Program Act 1991 223
  exceptions 224
  Freedom of Information Act 1982 227
  industry codes of conduct 234
  information privacy 234
  interference with privacy 220
  IPPs, Information Privacy Principles 220

NPPs, National Privacy Principles
  compliance 221, 224–6
Privacy Act 1988 219
Privacy Amendment Act 1990 221
Privacy Amendment Private Sector Act 2000 223
Privacy Commissioner 224
private privacy codes–States 227
protection through law of contract 231
rights to privacy 219
security 224
self-regulation 236
sensitive information protection *fn 40* 223, 224
TFN, tax file number guidelines 220
Privacy–general
Canada 218
common law 217
general remedy 218
ICCPR, International Covenant on Civil and Political Rights 217, 219
informational 46
invasion of privacy tort 218
legal and regulatory framework 217
New Zealand 218
soft-touch self-regulation 272
summary privacy protection *table* 244–5
UDHR, Universal Declaration of Human Rights 217
UK privacy of corporations 218
Privacy–geospatial technologies
acontextual data 214
data aggregation and databases 253
'dataveillance' *definition* 216, 259
enhancing technologies 270
ethical use 214
geodemographics 11, 253
geographical profiling 257
GPS, global positioning systems 44, 214, 234
implications 252
invasive technologies 216, 270
LBS, location based services 10, 253, 357
location *definition* 258
locational data sensitive? 258
regulation and use of databases 256

risk summary *table* 273–6
surveillance 211, 216
sympathetic technologies 270
tracking *definition* 259
transponders 214, 267
Privacy–GI 208
Australian Constitution 208, 217
common law protection 10, 208
data protection 211
definition 208
ethical questions 10, 210, 214
EU Data Protection Directive 209
fair information practices 10, 210
geo-coded data 211
geo-slavery 211
GI not personal data intensive 210
informational picture 211
Judge Louis Brandeis 208
lack of understanding 10, 211
personal information 209
personal privacy 209, 213
philosophical issues 210
privacy of information 211
'right to privacy' 10
threat to personal privacy 210
US Constitution 10, 208
Privacy–tracking technologies 260
aerial photo imagery 46, 269
Bearcat scanner 270
emissions 269
FLIR, forward looking infrared
device 269
home location 260
IETF, Internet Engineering Task Force
ENUM 265
MICR, magnetic ink character
recognition 264
movement of individuals 262
pen register 269
RFID, radio-frequency identity tags
264
Safe-T-Cam NSW RTA 263
telecommunication services 260
Transurban Melbourne City Link
263
trespass to land 268
White Pages Directory 261
wireless communications 267

Privacy–UK
breach of confidence 48
right to privacy vs freedom of
expression 47
Privacy–US 237
California Supreme Court 238
common law 245
invasion of 246
public interest override 246
tort liability 245
Constitutional amendments 237
data matching legislation 240
FIPPs, fair information privacy
principles evolution 246
core principles 248–9
identity theft 246
Privacy Act 1974 239
Privacy Protection Act 1980 240
right to privacy 237
summary of legislation *table* 244–5
zones of privacy 237
Private international law 27
Property plan
liability insurance 42
litigation 43
proprietary rights 43
Proprietary questions 54
Pseudo-identifier *fn 150* 271
PSI, public sector information 7, 54, 61
cost recovery 61
EU 7
EGII, European Geographic
Information Infrastructure 101
Green Paper on PSI 79, 82
Information Society Technology
Program 101
Synergy Guidelines 80
funding and financing 'religions' 61
open records 61
policies 61
summary *table* 78
user pays 61
PSMA, Public Sector Mapping
Agencies 69, 103
pricing access policy 70
revenue generation 69
supply chain positioning *diagram* 70
vision 69

PST, privacy-sympathetic technologies   270
PSTN, Public Switched Telephone
    Network   260
Public Sector Mapping Agencies,
    see PSMA, Public Sector Mapping
    Agencies

Recommendations data purchase
    notes   324
Redrock Communications   356
Reverse White Pages   261
RFID, radio-frequency identity tags
    264, 268
*Rome Convention for the Protection
    of Performers, Producers of
    Phonograms and Broadcasting
    Organizations* 1961   129

SADC, Southern Africa Development
    Community   104
Safe Harbour principles   11, 279
    practice notes   280–81
St Maarten, Netherlands Antilles   355
Sale of Goods Acts   306
Sale price of data   66
SDI *Cookbook*   105
SDI, spatial data infrastructure   60
    dependencies   61
SDTS, spatial data transfer standards   87
Selective availability   44, 266, 383
Semantic interoperability   *fn 74* 93
Sendmail   181
Sensitive information   *fn 40* 223, 224
Service New Brunswick   107
SGML, Standard Generalised Mark-Up
    Language   92
Shareware   *fn 170* 202
Simputer   111
Sioux Falls, South Dakota   92
Smart Card Code of Conduct   235
Smart cards   267
SML, small macro language   110
SOAP, Simple Object Access Protocol   95
Sony Bono Copyright Term Extension Act
    1999   126
South Africa
    National Spatial Information
        Framework   92

South Solitary Island, Queensland   356
Space Imaging   92
Spain
    IDEC Catalunya   92, 107
Spatial data special   323
Spatial Data Warehouse, Alberta,
    Canada   107
Spatialisation   50
Spirograph   50
SPOT, *Systeme Probatoire pour
    l'Observation de la Terra*   197
SSI, Spatial Sciences Institute   103, 108
    Code of ethics   *fn 35* 311
SSL, secure socket layer   *fn 91* 340
Standards
    international and national organisation
        interactions *diagram*   96
    SOAP, simple object access
        protocol   95
    UDDI, Universal Description,
        Discovery and Integration   95
    UML, Unified Modelling Language   95
    WSDL, Web Services Definition
        Language   95
Standards Australia   94
Standards New Zealand   94
Statute law   22
Statute of Liberty   37
Steganography   43
Street directory   42
sui generis rights   154, 160, 168,
    169, 170
    information borne on or near the face
        of   170
    legal hybrid   170
    uncertain protection   169
    US models   168–9
Super-computer cluster   181
SVC, stored valued cards   271
SVG, Scalable Vector Graphics   93
Sweat of the brow   146, 149
Sydney Kingsford Smith Airport   354

Technical interoperability   *fn 74* 93
Telecommunications Act 1996   241
Telecommunications Codes   235
Telstra Corporation   224
Tenants' Union of Queensland   39

Terraserver   92
Territorial conflict
  New York–New Jersey   37
  Supreme Court decision   38
Territorial copyright   188
TICA Default Tenancy Control   39
Tort law   26
TPM, technological protection
  measures   130, 154
Transponders   214, 267
TransUnion   255
TRIPS WIPO, Agreement on Trade
  Related Intellectual Property Rights
  1995   128, 154, 183
True cost of data   66
TRUSTe   279

UDDI, Universal Description, Discovery
  and Integration   95
UDHR, Universal Declaration of Human
  Rights 1948   217, 219
UK
  AGI GIgateway   92, 94
  AGI, Association for Geographic
    Information   94
  askGIraffe   94
  DETR, Department of Environment,
    Transport and the Regions   83
  Freedom of Information Act 2000   85
  GICSS, Geographic Information Charter
    Standard Statement   94
  IGGI, Inter-governmental Group on
    Geographic Information   94
  marginal cost data   84
  NAO, National Audit Office   83
  National Digital Topographic
    Database   83
  NGDF, National Geospatial Data
    Framework   94
UML, Unified Modelling Language   95
UN Guidelines Personal Data 1990   219
UN Millennium Development Goals   196
UNCITRAL Model Law on Electronic
  Commerce   333
UNCITRAL, UN Commission on
  International Trade Law   333
UNECA, UN Economic Commission for
  Africa   105

Universal Copyright Convention   122
UNIX operating system   178
URIs, Uniform Resource Identifiers   93
URISA Institute for GIS Professionals
  fn 35 311
URL, uniform resource locator   5, 93, 340
US
  CAP, Cooperative Agreement
    Program   98
  Coast Guard Office of Investigations and
    Analysis   355
  Constitutional amendments
    Fifth   *fn 70* 237
    First   47, *fn 70* 237
    Fourteenth   *fn 70* 237
    Fourth   *fn 70* 237
    Ninth   *fn 70* 237
  DOT ITS, Dept of Transport Intelligent
    Transportation Systems   262
  IVHS, Intelligent Vehicle Highway
    Systems   *fn 119* 262
  e-government initiative   99
  Executive Order 1994   98
  FCC, Federal Communications
    Commission   266
  FMC, Federal Maritime Commission
    77
  ATFI, Automated Tariff Filing and
    Information System   77
  user fees   77
  Geospatial One-stop portal   99
  HHS, Department of Health and
    Human Services   242
  I-Team initiative   99
  Landsat imagery privatisation   100
  MSC, Mapping and Science
    Committee   98
  NAPA, National Academy of Public
    Administration   98
  NAS, National Academy of Science   57
  *National Information Infrastructure
    Act* 1993   97
  National Spatial Data Council   98
  NGDC, National Geospatial Data
    Clearinghouse   92
  NRC, National Research Council   57
  NSDI, National Spatial Data
    Infrastructure evolution   98

US (*Continued*)
  OMB, Office of Management and
    Budget 75, 239
    Circular A-130 75
    Circular A-16 98
    Circular A-25 revised 75
  open records 7, 61
  Safe Harbour Principles 11, 279
    practice notes 280–1
    self-certification 279
    TRUSTe 279
  top-down approach 99
User agents 'bots' 271
USGS US Geological Survey 76
  cost recovery policies 76
  Eros Data Center 92
  The National Map 99

Video Privacy Protection Act 1988 242
Vienna Convention 306

W3C, World Wide Web Consortium 271
  P3P, Privacy Preferences Project 271
W3C, World Wide Web Consortium 93
  Recommendation *fn 77* 94
  Specification *fn 77* 94
WALIS, Western Australian Land
  Information System 316

Wallaby Track, NSW 355
Watermarks in maps 32, 42, 141
WCT WIPO Copyright Treaty
  128, 130, 154
When question 1
Where question 1
WIPO, World Intellectual Property
  Organization 128
*Wireless Communication and Public
  Safety Act* 1999 267
Wireless Privacy Protection Act 2003 243
WPPT WIPO, Performances and
  Phonograms Treaty 128, 130, 175
WSDL, Web Services Definition
  Language 95
WTO, World Trade Organization
  128, 307
WWW, World Wide Web 179

XML, Document Type Definition 89
XML, eXtensible Markup Language
  fn 63 89, 107

Yale University Cambodian Genocide
  Project 40

Z39.50 protocol 92–3
Zones of privacy 237